Vascular Endothelium

Pharmacologic and Genetic Manipulations

NATO ASI Series

Advanced Science Institutes Series

A series presenting the results of activities sponsored by the NATO Science Committee, which aims at the dissemination of advanced scientific and technological knowledge, with a view to strengthening links between scientific communities.

The series is published by an international board of publishers in conjunction with the NATO Scientific Affairs Division

A	**Life Sciences**	Plenum Publishing Corporation
B	**Physics**	New York and London
C	**Mathematical and Physical Sciences**	Kluwer Academic Publishers Dordrecht, Boston, and London
D	**Behavioral and Social Sciences**	
E	**Applied Sciences**	
F	**Computer and Systems Sciences**	Springer-Verlag
G	**Ecological Sciences**	Berlin, Heidelberg, New York, London,
H	**Cell Biology**	Paris, Tokyo, Hong Kong, and Barcelona
I	**Global Environmental Change**	

PARTNERSHIP SUB-SERIES

1. Disarmament Technologies	Kluwer Academic Publishers
2. Environment	Springer-Verlag
3. High Technology	Kluwer Academic Publishers
4. Science and Technology Policy	Kluwer Academic Publishers
5. Computer Networking	Kluwer Academic Publishers

The Partnership Sub-Series incorporates activities undertaken in collaboration with NATO's Cooperation Partners, the countries of the CIS and Central and Eastern Europe, in Priority Areas of concern to those countries.

Recent Volumes in this Series:

Volume 291 — Recollections of Trauma: Scientific Evidence and Clinical Practice
edited by J. Don Read and D. Stephen Lindsay

Volume 292 — Biosocial Bases of Violence
edited by Adrian Raine, Patricia A. Brennan, David P. Farrington, and Sarnoff A. Mednick

Volume 293 — Vaccine Design: The Role of Cytokine Networks
edited by Gregory Gregoriadis, Brenda McCormack, and Anthony C. Allison

Volume 294 — Vascular Endothelium: Pharmacologic and Genetic Manipulations
edited by John D. Catravas, Allan D. Callow, and Una S. Ryan

Series A: Life Sciences

Vascular Endothelium

Pharmacologic and Genetic Manipulations

Edited by

John D. Catravas

Medical College of Georgia
Augusta, Georgia

Allan D. Callow

Boston University School of Medicine
Boston, Massachusetts

and

Una S. Ryan

T Cell Sciences, Inc.
Needham, Massachusetts

Plenum Press
New York and London
Published in cooperation with NATO Scientific Affairs Division

Proceedings of a NATO Advanced Research Workshop on
Vascular Endothelium: Pharmacologic and Genetic Manipulations,
held June 22–July 1, 1996,
in Crete, Greece

NATO-PCO-DATA BASE

The electronic index to the NATO ASI Series provides full bibliographical references (with keywords and/or abstracts) to about 50,000 contributions from international scientists published in all sections of the NATO ASI Series. Access to the NATO-PCO-DATA BASE is possible via a CD-ROM "NATO Science and Technology Disk" with user-friendly retrieval software in English, French, and German (©WTV GmbH and DATAWARE Technologies, Inc. 1989). The CD-ROM also contains the AGARD Aerospace Database.

The CD-ROM can be ordered through any member of the Board of Publishers or through NATO-PCO, Overijse, Belgium.

ISBN 0-306-45819-5

A Division of Plenum Publishing Corporation
233 Spring Street, New York, N.Y. 10013

http://www.plenum.com

10 9 8 7 6 5 4 3 2 1

Printed in the United States of America

CONTENTS

I. Pulmonary Circulation

II. Gene Manipulation in the Vasculature

III. Thrombosis/Hypertension

ABBREVIATIONS

ACD	allergic contact dermititis
AD	Alzheimer's Disease
AECA	anti-endothelial cell antibodies
ALI	acute lung injury
APP	Amyloid precursor protein
ARDS	adult respiratory distress syndrome
bFGF	basic fibroblast growth factor
CAD	coronary artery disease
cGMP	cyclic guanosine monophosphate
CR1	complement receptor type 1
EC	endothelial cells
FH	familial hypercholesterleia
GMP	guanosine monophosphate
HDL	high density lipoprotein
HEV	high endothelial venules
ICAM	intercellular adhesion molecule
IL-1	interleukin-1
IRE	iron responsive elements
IRP	iron regulatory protein
LAD	leukocyte adhesion deficiency
LDL	low density lipoprotein
LPS	lipopolysaccharide
MAdCAM	mucosal addressin cell adhesion molecule
NIF	neutrophil adhesion inhibitor
NO	nitric oxide
NOS	nitric oxide synthase
PAF	platelet activating factor
PDGF	platelet-derived growth factor
PECAM	platelet endothelial cell adhesion molecule

PHF	paired helical filaments
PKC	protein kinase C
ROS	reactive oxygen species
sCR1	soluble complement receptor type 1
SIV	Simian immunodeficiency virus
SOD	superoxide dismutase
SUV	small unilameller lipid vesicle
TNF	tumor necrosis factor
VCAM	vascular cell adhesion molecule
VE	vascular endothelium
VEGF	vascular endothelial growth factor
VLA	very late antigens

I. PULMONARY CIRCULATION

1

MECHANISMS OF ENDOTHELIAL INJURY AND REPAIR

Alain F. Junod

Clinique de Médecine 1
Department of Medicine
University Hospital
Geneva, Switzerland

Microvascular injury is very commonly involved, in the lung as well as in other organs, to explain the pathogenesis of various disorders, the adult respiratory distress syndrome being a paradigm of such a complex situation where toxic and inflammatory elements interact. The role of cytokines has received prominent attention and their involvement in the expression of adhesion molecules has emphasized the importance of the intricate relationship between endothelial cells and circulating cells and of the potential amplification of inflammatory processes. Endothelial injury, which often initiates a whole set of reactions, can result from exposure to exogenous compounds, toxic (drugs for examples) and of microbial origin, and/or to endogenous products such as the reactive oxygen species (ROS), proteases and the already mentioned cytokines. The effects of oxygen radicals will be discussed more specifically because they represent a well studied model of endothelial injury. They are also associated with other important factors implicated in microvascular damage.

The most common ROS result from the monovalent reduction of oxygen which gives rise to superoxide radical, hydrogen peroxide and hydroxyl radical. The latter can be generated from the interaction between superoxide radical and hydrogen peroxide, the so-called Haber-Weiss reaction, or between hydrogen peroxide and a divalent metal (the Fenton's reaction). Superoxide anion and hydroxyl radical, having one impaired electron, belong to the class of oxygen radicals. This is not the case for hydrogen peroxide, a reactive form of oxygen species, in spite of its paired number of electrons. Other radicals exist, which do not derive from oxygen: nitric oxide radical (NO·) is one of them.

Oxidant injury to endothelium can result from an increase in the endogenous production of reactive ROS, for example during exposure to high O_2 partial pressure (hyperoxia) or to drugs and toxic compounds (paraquat, alkylating agents, bleomycin, etc...). It can also be generated from extra-endothelial sources, neutrophils and monocytes or macrophages, which can release ROS whenever they are activated. Another source of

Vascular Endothelium, edited by Catravas *et al.*
Plenum Press, New York, 1998

extracellularly generated oxygen species is the product of the reaction between hypoxanthine (HX) and the activated form of xanthine dehydrogenase, xanthine oxidase (XO), namely the superoxide radical which dismutates into hydrogen peroxide. The HX-XO reaction is especially involved in the pathogenesis of the ischemia-reperfusion syndrome. The differential effects of intra- and extra-cellularly formed ROS can be accounted for by differences in their rates of production and the balance with antioxidant defenses. Enzymes, such as the superoxide dismutases (SOD), catalase and glutathione peroxidase (GPX), and non-enzymatic compounds, among which reduced glutathione, thioredoxin, ascorbate, alpha-tocopherol, are the two main forms of these defenses (Freeman, 1982; de Bono, 1994; Halliwell, 1994).

Extracellular sources of ROS, whether generated from the HX-XO reaction, the administration of H_2O_2 or activated phagocytes, are the substrate in most experimental conditions for the study of oxidant-related microvascular injury. Marked differences, however, exist between the effects of extra- and intra-cellularly generated ROS, and some of them will be illustrated in this presentation, which will focus on the effects of hyperoxia on one hand, and on those of ROS generated from the HX-XO reaction or H_2O_2 administration on the other hand.

For the sake of clarity, it may be convenient to distinguish various sequences in the effects of ROS on endothelial cells: the **injury**, the **adaptation** and the **activation**, the last two taking place often at the same time and forming the **repair** process (Gerritsen, 1993).

1. THE INJURY

For intracellularly generated ROS, most commonly during hyperoxia (95% O_2), evidence for injury may take hours and even days, although free radical formation in EC can be detected already after 30 min (Sanders, 1993). Decreased DNA synthesis can be measured after 12 hours, in parallel with thymidine kinase inhibition (Clément, 1985). A significant decrease in protein synthesis, at the level of the elongation step, requires three days to be demonstrated (Jornot, 1987). No substantial DNA damage, when assessed from the formation of single-strand breaks, can be shown (Junod, 1989).

The picture is completely different when the toxic effects of H_2O_2 or of the HX-XO reaction are analyzed. It takes only minutes to see DNA damage, with formation of single-strand breaks, activation of poly-ADP-ribose polymerase, cellular depletion of ATP and NAD (Schraufstätter, 1986; Jornot, 1991). Marked inhibition of protein synthesis, at the level of the initiation, is also observed (Jornot, 1989). Some, but not necessarily all, of these effects can be ascribed to a primary effect on increase in cytoplasmic [Ca^{++}], essentially the result of its release from intracellular stores. Thapsigargin, an inhibitor of the Ca^{++}-ATPase of the endoplasmic reticulum, appears to mimic most of the effects of the HX-XO reaction, unlike agonists acting through the IP_3 system, thrombin or histamine, which acts on a different Ca^{++} pool (Dreher, 1995).

2. THE ADAPTATION

Hyperoxia can be considered as the paradigm of the adaptive process in the sense that, following a three day exposure to 95% O_2, there is a significant increase in most of the antioxidant defenses of endothelial cells, whether they are enzymes such as Mn-SOD, CuZn-SOD, GPX, but not catalase, or non-enzymatic components such as reduced glutathione (Jornot, 1992). The antioxidant enzymes CuZn-SOD and GPX exhibit a parallel increase in their mRNA and activities, whereas catalase activity was decreased in spite of an elevated mRNA. Clerch (1994) also finds an increased Mn-SOD mRNA and activity in rat lungs exposed to hyperoxia, with a possible involvement of the G-protein system.

Among the stress proteins, which can also be considered to represent a form of adaptation, no increase in heat shock protein (HSP) 70 or HSP 90 could be elicited following hyperoxic exposure (Jornot, 1991). On the other hand, mRNA for HSP 32 or heme oxygenase was found to be markedly elevated. A similar effect was seen after inhibition of glutathione reductase and the concomitant reduction in total and reduced glutathione (Jornot 1993).

For extracellularly generated ROS, no upregulation for the major antioxidant enzymes takes place when the duration of exposure and the period of recovery are short (Jornot, 1993). In the presence of low H_2O_2 concentrations for up to 24 hrs, however, (Lu, 1993) is able to find evidence for increased synthesis of SOD, catalase and GPX, as well as of a member of the HSP 70 family, a finding also reported by (Jornot, 1993). In another series of experiments, which made an extensive use of a two dimensional electrophoresis analysis, (Dreher, 1995) demonstrates an increased synthesis of certain stress proteins, namely the glucose-regulated protein GRP 78, HSP 27, ubiquitin and cyclophilin. These effects are also generally mimicked by thapsigargin and the concomitant increase in cytoplasmic [Ca^{++}] from the release from intracellular stores. Stress proteins may represent an equivocal form of adaptation which could end up in injury, since autoantibodies against HSP60 have recently been shown to mediate endothelial cytotoxicity (Schett, 1995).

3. THE ACTIVATION

In endothelial cells exposed to hyperoxia, there is an increased mRNA of the constitutive form of NO· synthase as well as in its activity. Conversely, exposure to hypoxia results in decreased mRNA and activity of constitutive NO· synthase. NO· synthase activity appears to be down-regulated under normal conditions since cycloheximide treatment is associated with increased NO· synthase activity (Liao, 1995). Hence the possibility that the hyperoxic effect could be mediated through inhibition of production of certain protein(s). For others, modulation of NO· production by variable pO_2 could be mediated through changes in cytoplasmic calcium (Shaul, 1994). Whichever the mechanism might be, this regulation of NO· production may be of great importance in the control of vasomotor tone. The role of NO· in the control of expression of adhesion molecules under the action of inflammatory cytokines (cf below) should also be considered. Hyperoxia is also known to upregulate the plasminogen activator (Phillips, 1992).

Extracellularly generated ROS have become the subject of a great deal of attention with respect to the problem of endothelial cell activation, especially in view of their relationship to the control of adhesion molecules. Thus, (Bradley, 1993) found that a low concentration of H_2O_2 increased the surface expression of ICAM-1, but not that of E-Selectin. This expression at the surface of the endothelial cells is associated with an increase in mRNA, but only for ICAM-1. H_2O_2 inhibits TNF α-induced gene expression for VCAM-1 and ELAM-1. On the other hand, (Sellak, 1994) reports that the HX-XO reaction products could increase the adherence of neutrophils to endothelial cells, without, however, any upregulation of ICAM-1 synthesis. This phenomenon could be ascribed to the action of superoxide anion, H_2O_2 administration being without effect. (Roebuck, 1995) has greatly clarified the mode of action of hydrogen peroxide on the control of gene expression for ICAM-1. Stimulation of ICAM-1 gene transcription resulting from exposure to low concentrations of H_2O_2 appears to be due to the binding of a distinct region of the promoter by the transcription factors AP-1 and Ets. Interestingly, NF-κB is involved only when the increase in ICAM-1 transcription results from the presence of TNFα. Hydrogen peroxide has no effect on NF-κB binding. Finally, (Bradley, 1995) recently gives evidence for a spectacular rearrangement of certain adhesion molecules at the level of microvascular endothelial cells under the effect of H_2O_2.

(Vischer, 1995) also uses the HX-XO reaction to demonstrate the action of ROS on the exocytosis of von Willebrand factor, an adhesive protein stored in the Weibel-Palade bodies. Calcium mobilization from its stores is thought to be responsible for this exocytosis process.

The interaction between NF-κB and oxidants or reducing agents deserves a few comments. (Schreck, 1991) first finds out that oxidative stress was associated with an increased binding of the transcription factor NF-κB to appropriate sequences of various genes. This observation may be cell-specific, since it has not been confirmed in endothelial cells. On the other hand, the ability of inflammatory cytokines, such as TNF, Il-1, or LPS to promote NF-κB binding in endothelial cells is undisputed. This binding is required for the upregulation of some of the adhesion molecules. That NF-κB binding could be influenced by the redox state of the cell is suggested by the observation that reducing agents, such as pyrrolidine dithiocarbamate or N-acetylcysteine, decrease markedly the NF-κB binding activity (Marui, 1993; Ferran, 1995). NO· itself had exactly the same property (de Caterina, 1995) and this inhibition of the binding activity of NF-κB, which is correlated with the regulation of expression of adhesion molecules by NO, could be related to its reducing effect.

This intriguing interaction introduces the question of the transducing mechanism between ROS and the activation and adaptation of endothelial cells. There is little doubt that increased Ca^{++} concentrations from the mobilization of internal stores play a crucial role. This Ca^{++} effect is not related to the IP_3-dependent release of Ca^{++} from another pool. Whether other transducers, for example kinases, such as protein kinase C, could also play a role as an intermediary is possible. Transcription factors such as NF-κB, AP-1, HSF, and CRE have also been reported to be activated in the presence of oxidants. But the precise sequence of events remains unclear and the precise pathway between exposure to oxidants or reducing agents, membrane signaling, activation of nuclear factors and promotion of transcription has yet to be determined. Future experiments are likely to shed more light on the respective roles of all these potential actors.

REFERENCES

Bradley, J.R., Johnson, D.R., and Pober, J.S. Endothelial activation by hydrogen peroxide. Selective increases of intracellular adhesion molecule-1 and major histocompatibility complex. *Am. J. Pathol.*, 142:1598–1609, 1993.

Bradley, J.R., Thiru, S., and Pober, J.S. Hydrogen peroxide-induced endothelial retraction is accompanied by a loss of the normal spatial organization of endothelial cell adhesion molecules. *Am. J. Pathol.*, 147:627–641, 1995.

De Bono, D.P. Free radicals and antioxidants in vascular biology: the roles of reaction kinetics, environment and sustrate turnover. *Quart. J. Med.*, 87:445–453, 1994.

Clement, A., Huebscher, U., and Junod, A.F. Effects of hyperoxia on DNA synthesis in cultured porcine endothelial cells. *J. Appl. Physiol.*, 59:1110–1116, 1985.

Clerch, L.B., Neithardt, G., Spencer, U., Melendez, G.D., Massaro, G.D., and Massaro, D. Pertussis treatment alters manganese superoxide dismutase activity in lung: Evidence for lung oxygen toxicity in air-breathing rats. *J. Clin. Invest.*, 93:2482–2489, 1994.

De Caterina, R., Libby, P., Peng, H.B., Thannickal, V.J., Rajashisth, T.B., Gimbrone Jr., M.A., Shin, W.S., and Liao, J.K. Nitric oxide decreases cytokine-induced endothelial activation. Nitric oxide selectively reduces endothelial expression of adhesion molecules and proinflammatory cytokines. *J. Clin. Invest.*, 96:60–68, 1995.

Dreher, D., Jornot, L., and Junod, A.F. Effects of hypoxanthine-xanthine oxidase on Ca^{+} stores and protein synthesis in human endothelial cells. *Circ. Res.*, 76:388–395, 1995.

Dreher, D., Vargas, R., Hochstrasser, D., and Junod, A.F. Effects of oxidative stress and Ca^{2+} agonists on molecular chaperones in human umbilical vein endothelial cells. *Electrophoresis*, 16:1205–1214, 1995.

Ferran, C., Millan, M.T., Csizmadia, V., Cooper, J.T., Brostjan, C., Bach, F.H., and Winkler, H. Inhibition of NF-κB by pyrrolidine dithiocarbamate blocks endothelial cell activation. *Bioch. Biophys. Res. Com.*, 214:212–223, 1995.

Freeman, B.A. and Crapo, J.D. Free radicals and tissue injury. *Lab. Invest.*, 47:412–426, 1982.

Gerritsen, M. and Bloor, C.M. Endothelial cell gene response in response to injury. *FASEB J.*, 7:523–532, 1993.

Halliwell, B. Free radicals, antioxidants, and human disease: curiosity, cause or consequence? *Lancet*, 344:721–724, 1994.

Jornot, L., Mirault, M,E., and Junod, A.F. Protein synthesis in hyperoxic endothelial cell: evidence for translational defect. *J. Appl. Physiol.*, 63:457–464, 1987.

Jornot, L. and Junod, A.F. Hypoxanthine-xanthine oxidase related defect in polypeptide chain initiation by endothelium. *J. Appl. Physiol.*, 66:450–457, 1989.

Jornot, L., Petersen, H., and Junod, A.F. Differential protective effects of o-phenanthroline and catalase on H_2O_2-induced DNA damage and inhibition of protein synthesis in endothelial cells. *J. Cell. Physiol.*, 149:408–413, 1991.

Jornot, L., Mirault, M.E., and Junod, A.F. Differential expression of HSP 70 stress proteins in human endothelial cells exposed to heat shock and hydrogen peroxide. *Am. J. Respir. Cell. Mol. Biol.*, 5:265–275, 1991.

Jornot, L. and Junod, A.F. Response of human endothelial cell antioxidant enzymes to hyperoxia. *Am. J. Respir. Cell. Mol. Biol.*, 6:107–115, 1992.

Jornot, L. and Junod, A.F. Variable glutathione levels and expression of antioxidant enzymes in human endothelial cells. *Am. J. Physiol.*, 264:L482–L489, 1993.

Junod, A.F., Jornot, L., and Petersen, H. Differential effects of hyperoxia and hydrogen peroxide on DNA damage, polyadenosine diphosphate-ribose polymerase activity, and nicotinamide adenine nucleotide and adenosine triphosphate contents in cultured endothelial cells and fibroblasts. *J. Cell. Physiol.*, 140:177–185, 1989.

Liao, J.K., Zulueta, J.J., Yu, F.S., Peng, H.B., Cote, C.G., and Hassoun, P.M. Regulation of bovine endothelial constitutive nitric oxide synthase by oxygen. *J. Clin. Invest.*, 96:2661–2666, 1995.

Lu. D., Maulik, N., Moraru, L., Kreutzer, D.L., and Das, D,K. Molecular adaptation of vascular endothelial cells to oxidative stress. *Am. J. Physiol.*, 264:C715–C722, 1993.

Marui, N., Offermann, M.K., Swerlick, R., Kunsch, C., Rosen, C.A., Ahmad, M., Alexander, R.W., and Medford, R.M. Vascular cell adhesion molecule-1 gene transcription and expression are regulated through and antioxidant-sensitive mechanism in human vascular endothelial cells. *J. Clin. Invest.*, 92:1866–1874, 1993.

Natarajan, V. Oxidants and signal transduction in vascular endothelium. *J. Lab. Clin. Med.*, 125:26–37, 1995.

Phillips, P.G., Birnby, L., Di Bernardo, L.A., Ryan, T.J., and Tsan, M.F. Hyperoxia increases plasminogen activator activity in cultured endothelial cells. *Am. J. Physiol.*, 262: L21–L31, 1992.

Roebuk, K.A., Rahman, A., Lakshminarayanan, V., Janakidevi, K., and Malik, A.B. H_2O_2 and tumor necrosis factor-α activate intercellular adhesion molecule 1 (ICAM-1) gene transcription through distinct *cis*-regulatory elements within the ICAM-1 promoter. *J. Biol. Chem.*, 270:18966 18974, 1995.

Sanders, S.P., Zweier, J.L., Kuppusamy, P., Harrison, S.J., Bassett, D.J., Gabrielson, E.W., and Sylvester, J.T. Hyperoxic sheep pulmonary microvascular endothelial cells generate free radicals via mitochondrial electron transport. *J. Clin. Invest.*, 91:46–52, 1993.

Schett, G., Xu, Q., Amberger, A., Van Der Zee, R., Recheis, H., Willeit, J., and Wick, G. Autoantibodies against heat shock protein 60 mediate endothelial cytotoxicity. *J. Clin. Invest.*, 96:2569–2577, 1995.

Schreck, R., Rieber, P., and Bauerle, P.A. Reactive oxygen intermediates as apparently widely used messengers in the the activation of NF-κB transcription factor and HIV-1. *Embo. J.*, 10:2247–2258, 1991.

Sellak, H., Franzini, E., Hakim, J., and Pasquier, C. Reactive oxygen species rapidly increase endothelial ICAM-1 ability to bind neutrophils without detectable upregulation. *Blood*, 83:2669–2677, 1994.

Shaul, P.W. and Wells, L.B. Oxygen modulates nitric oxide production selectively in fetal pulmonary endothelial cells. *Am. J. Respir. Cell. Mol. Biol.*, 11:432–438, 1994.

Vischer, U.M., Jornot, L., Wolheim, C.B., and Theler, J.M. Reactive oxygen intermediates induce regulated secretion of von Willebrand factor from cultured human vascular endothelial cells. *Blood*, 85:3164–3172, 1995.

2

MOLECULAR MECHANISMS OF ENDOTHELIAL LEUKOCYTE ASSOCIATION

The Function of the Selectins

Dietmar Vestweber

Institute of Cell Biology, ZMBE
University of Münster
Münster, Germany

1. INTRODUCTION

Attachment of leukocytes to the blood vessel wall initiates leukocyte extravasation. This enables leukocytes to migrate to and accumulate at sites of tissue injury or infection where they execute host-defense mechanisms. A series of vascular cell adhesion molecules on leukocytes and on endothelial cells mediate and control leukocyte attachment to the endothelium in a stepwise process (Springer, 1994) (Figure 1). Members of the selectin family of adhesion molecules mediate the initial tethering and rolling of leukocytes on the activated endothelium through interactions with carbohydrate-bearing ligands. This allows leukocytes to slow down and to interact with other molecules on the endothelial cell surface which induce activation of the leukocytes. Such activating agents can be the phospholipid platelet activating factor PAF or chemokines which are presented by endothelial proteoglycans. They can bind to various receptors which contain seven membrane spanning regions. Signal cascades which are triggered by these receptors lead to the activation of integrins on the leukocyte cell surface which bind to members of the immunoglobulin (Ig) super family on the endothelial cell surface. Such pairs of cell adhesion molecules are involved in a second adhesion step which is more firm than the selectin mediated initial tethering of leukocytes to the endothelium. Adhesion via activated integrins is a prerequisite for the active migration of leukocytes on the endothelial cell surface and finally through the layer of endothelial cells. Very little is known about this last step of the active transmigration process. It is still a matter of debate, whether leukocytes transmigrate through the junctions between adjacent endothelial cells or directly through a single endothelial cell. This review will mainly focus on the selectin mediated initiation of leukocyte tethering to the endothelium.

Vascular Endothelium, edited by Catravas *et al.*
Plenum Press, New York, 1998

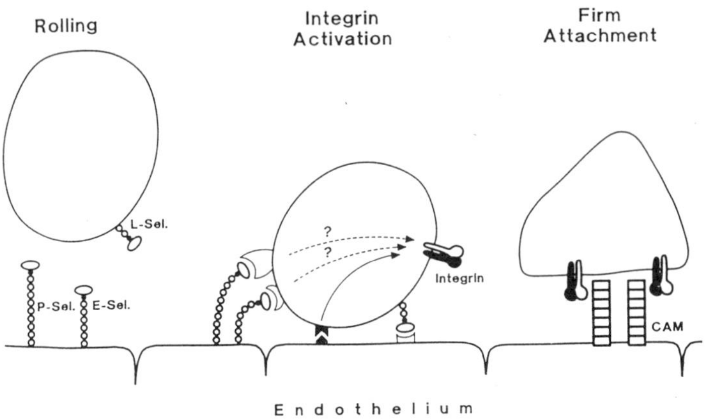

Figure 1. Leukocyte extravasation is controlled by a multistep process that is initiated by the selectins that mediate the rolling of flowing leukocytes on the endothelial surface of the blood vessel wall. L-selectin is found on a large variety of leukocytes while E- and P-selectin are inducible on the endothelial cell surface. Selectin mediated tethering allows other cell surface molecules (see text) to interact with each other and trigger the activation of leukocyte integrins which bind to cytokine inducible members of the Ig-superfamily (CAM). Whether integrin activation can also be induced via the stimulation of selectin ligands is still hypothetical.

2. THE SELECTIN FAMILY

Capturing leukocytes from the rapidly flowing blood stream to the blood vessel wall is a very special example of cell contact formation which differs from most other cell adhesion phenomena since considerable shear forces have to be overcome. The selectins, which mediate the very first steps in this process, seem to be specialized for this task. In contrast to the other gene families of cell adhesion molecules, the selectins are restricted to the leukocyte-vascular system. With only three members, the selectins form the smallest and most recently identified gene family among cell adhesion molecules. In contrast to the vast majority of all other adhesion molecules the selectins mediate cell contact via binding to carbohydrate ligand structures, thus they are lectins. L-selectin is found on most types of leukocytes, E-selectin is specific for endothelial cells and P-selectin is found on endothelium and platelets. The individual members of the group were designated by prefixes, which were chosen according to the cell type where the molecule was first identified.

L-selectin was first found as a “lymphocyte homing” receptor, defined by the monoclonal antibody MEL 14 which blocked the binding of lymphocytes to lymph node high endothelial venules in lymph node tissue (Gallatin et al., 1983). Later, L-selectin was also found on neutrophil granulocytes and monocytes and was shown to be generally involved in leukocyte entry into sites of inflammation. E-selectin was also found by a monoclonal antibody approach when searching for cytokine inducible surface proteins on endothelial cells which would mediate the binding of neutrophils (Bevilacqua et al., 1987). P-selectin was originally found as a membrane protein of unknown function in platelet storage granules which later was also detected in endothelial cells (McEver et al., 1984; Hsu-Lin et al., 1984). Cloning and sequencing of all three selectins happened to occur at the same time and was published in 1989. Only then it became apparent that a new gene family of cell adhesion molecules had been discovered.

3. THE SELECTINS MEDIATE LEUKOCYTE ROLLING

It is an old observation that the first interaction between flowing leukocytes and the endothelium leads to the rolling of the leukocyte along the blood vessel wall. Soon after the identification of the selectins it was shown in vivo, that antibodies against L-selectin could block leukocyte rolling, while antibodies against leukocyte integrins such as LFA-l, only blocked firm attachment ("sticking") (Ley et al., 1991; von Andrian et al., 1991). The fundamental difference between integrin and selectin mediated cell adhesion steps was also very clearly demonstrated in a defined *in vitro* system, where neutrophil adhesion to a glass surface which had been coated with P-selectin or the LFA-l ligand ICAM-l was analyzed under flow conditions. Rolling of cells was only observed on P-selectin but not on ICAM-1, while the combination of both substrates lead to firm adhesion under flow (Lawrence and Springer, 1991). Blocking of leukocyte rolling in blood vessels with selectin antibodies has now been demonstrated for each of the three selectins. It occurs for neutrophils as well as for lymphocytes.

In addition to the rolling of leukocytes on the endothelial cell surface, neutrophils accumulating at the blood vessel wall can also serve as a support on which other neutrophils roll (Bargatze et al., 1994). This rolling seems to be mediated by L-selectin on the newly recruited neutrophils. Carbohydrate presenting L-selectin ligands on neutrophils, which could be involved in this process, might be sialomucin type ligands which are sensitive to an O-sialoglycoprotease. It was shown recently that neutrophil aggregation is mediated by the binding of L-selectin to such type of ligands (Simon et al., 1993; Bennett et al., 1995).

The L-selectin mediated leukocyte tethering and rolling has been analyzed in more detail. It was shown in vitro as well as in vivo that shear above a certain threshold is necessary to maintain rolling interactions through L-selectin (Finger et al., 1996). Interestingly, no such requirement was found for efficient rolling mediated by E- or P-selectin. Another factor which can influence the rate of L-selectin mediated leukocyte rolling, is the down regulation of L-selectin on the cell surface by a surface located protease. It was shown that blocking of this protease by the inhibitor hydroxamic acid reduced neutrophil rolling velocity (Walcheck et al., 1996). Thus, shedding of L-selectin from the leukocyte cell surface may be important to prevent too strong interactions via L-selectin. L-selectin shedding during rolling as well as the necessity of a certain threshold hydrodynamic shear for efficient functioning of L-selectin may both be mechanisms which are important for limiting leukocyte aggregation and accumulation at sites of inflammation.

4. PHYSIOLOGICAL ROLE OF THE SELECTINS

L-selectin is the only member of the selectin family which is involved in the entry of lymphocytes into secondary lymphatic tissue such as lymph nodes. This process requires constitutively expressed ligands which are locally restricted to the specialized endothelium of the high endothelial venules. In contrast to lymphocyte homing, which is a constitutively active process in specialized areas of the vascular system, the migration of leukocytes into sites of inflammation can occur in any tissue and is induced by inflammatory mediators. The regulation of the selectins is of central importance for the initiation of this process. Since L-selectin is constitutively expressed on leukocytes, the involvement of L-selectin in inflammatory processes requires the inducibility of its endothelial ligand(s). One candidate for such a binding partner is E-selectin which seems to be able to bind to carbohydrates present on L-selectin on human neutrophils (but not on lymphocytes)

(Kishimoto et al., 1991; Picker et al., 1991). Additional evidence for this interaction between two selectins is discussed below. Most likely, further endothelial ligands for L-selectin exist which are cytokine inducible (Spertini et al., 1991; Brady et al.. 1992) and are expressed on the cell surface by different kinetics than E-selectin. In addition to these mechanisms, L-selectin may be important in a secondary tethering event as it was described above mediating the rolling of neutrophils on neutrophils which already have accumulated at the blood vessel wall at sites of inflammation.

E-selectin is transcriptionally induced by cytokines such as IL-1β, TNF -α and TNF-β and by lipopolysaccharide (LPS) (Bevilacqua et al.. 1987: Pober et al., 1987). Cell surface expression of the protein on HUVE cells reaches maximal levels 4–6 h after stimulation and rapidly declines again to basal levels after another 12–16 h. Expression kinetics are similar on several mouse endothelioma cell lines, with maximal expression levels at the cell surface 3–4 h after stimulation (Weller et al., 1992, Hahne et al., 1993). It was soon found that NF-κB is one of the intracellular key regulators for transcriptional induction (Montgomery et al., 1991, Whelan et al., 1991).

Upregulation *in vivo* has been analyzed in man (Cotran et al., 1986), baboon (Munro et al., 1989; Redi et al., 1991; Munro et al., 1991), and mouse (Gotsch et al., 1994). In baboon skin, E-selectin was induced within 2 h after local injection of endotoxin, as was assayed by immunostaining of cryostat tissue sections. E-selectin staining was virtually absent again by 9 h (Munro et al., 1991). In another study in baboon, E-selectin was induced within 2 h by intracutaneous injection of TNF-α and was not yet completely absent after 24 h (Munro et al., 1989). Systemic application of live *E. coli* bacteria (injected intravenously) induced widespread expression of E-selectin in baboon endothelium in most tissues (Redl et al., 1991). Similar results were obtained after intravenous injection of TNF-α or LPS in the mouse (Gotsch et al., 1994).

Cell contact between activated T cells and endothelium can also induce E-selectin expression. This was first shown *in vitro* (Damle et al., 1991) and recently also *in vivo* (Sunderkötter et al., 1996), using various inflammation models in normal and in athymic nude mice. In two inflammation models with characteristic involvement of T lymphocytes experimental leishmaniasis and allergic contact dermatitis (ACD)-sustained expression of E-selectin was seen for at least 48 h in normal mice while E-selectin expression was much more transient in T cell deficient nude mice. Substituting these mice with primed T cells reconstituted the ability to maintain E-selectin expression for the same time period as in normal wild type mice. Interestingly this effect was antigen dependent, since T-cells from leishmania-challenged animals were unable to reconstitute normal expression kinetics of E-selectin in the ACD inflammation model (Sunderkötter et al., 1996).

In contrast to E-selectin, P-selectin is stored in the membrane of intracellular storage granules in platelets (α-granules) (Hsu-Lin et al., 1984; McEver et al., 1984) as well as in endothelial cells (Weibel-Paladebodies) (McEver et al., 1989). Fusion of these granules with the plasma membrane is rapidly stimulated by pre-inflammatory mediators such as histamine and thrombin (Stenberg et al., 1985). Thereby P-selectin becomes exposed on the cell surface within minutes after stimulation. It is rapidly downregulated again by internalization. The cytoplasmic domain of P-selectin was found to target newly synthesized P-selectin from the Golgi compartment into storage granules (Disdier et al., 1992; Koedam et al., 1992). This domain was also crucial for the rapid internalization of P-selectin (Greem et al., 1994). In contrast to E-selectin, which is targeted to lysosomes after internalization, P-selectin can recycle from endosomes into storage granules, as was found in transfected AtT 20 cells (Subramaniam et al., 1993). In CHO cells, which lack secretory granules, transfected P-selectin was rapidly internalized and targeted into lysosomes by the cytoplasmic tail domain (Greem et al., 1994). Rapid upregulation of P-selectin on the cell surface was also observed after treatment of endothelial cells with oxygen radicals (Patel et al., 1991). This induction mechanism did not require protein synthesis. Cell surface expression was sustained between 1 and 4 h after stimulation. Thus, oxygen radicals most likely induced transport of stored P-selectin to the plasma membrane and prolonged the half life time of P-selectin at the cell surface.

In addition to the regulation of its transport to the cell surface, P-selectin was found to be inducible by cytokines (TNF-α) and LPS on mouse endothelioma cells. Maximal expression levels of the P-selectin mRNA were reached 2 h after stimulation and maximal protein levels were reached after 3–4 h (Weller et al., 1992, Hahne et al., 1993). Newly synthesized P-selectin reached the cell surface with no need for activating the transport of storage granules to the plasma membrane. In fact, stored P-selectin was still present inside the cell when maximal levels of TNF-α stimulated, newly synthesized P-selectin were present at the cell surface. Additional stimulation of P-selectin secretion at this time point could further increase the amount of P-selectin on the plasma membrane (Hahne et al., 1993).

Cytokine induced upregulation of P-selectin was also found *in vivo* in mouse tissue, on the RNA level (Sanders et al., 1992) as well as on the protein level (Gotsch et al., 1994). Endothelium in mouse brain, in the parenchyme, and in the leptomeninges is devoid of constitutive expression and storage of P-selectin. By immunohistochemistry P-selectin was found to be upregulated upon systemic application of TNF-α or LPS on venules in the leptomeninges and (although much less strongly) in brain parenchyme (Gotsch et al., 1994). Upregulation of both E-selectin and VCAM-l was found in the same blood vessels. However, endothelium of arterioles in the leptomeninges showed only expression of E-selectin and not of P-selectin after stimulation. Thus, although cytokines can induce the synthesis of both endothelial selectins, certain endothelia can respond selectively to such stimuli by upregulation of only one of the two selectins.

The function of the different selectins for extravasation of leukocytes in various pathological and non-pathological processes has been demonstrated *in vivo* in several cases. The role of L-selectin for the homing of lymphocytes into peripheral lymph nodes of the mouse was demonstrated by the blocking effect of the mAb MEL-14 (Gallatin et al., 1983). In this study, no effect with this antibody on the homing of lymphocytes to the gut associated Peyer's Patches had been observed. However using Fab-fragments of MEL-14, homing of lymphocytes into Peyer's Patches was partially blocked (Hamann et al., 1991), suggesting participation of L-selectin in this homing pathway as well.

In addition to lymphocyte homing, L-selectin is essential for the extravasation of neutrophils. Emigration of neutrophils into experimentally induced sites of inflammation in the skin could be blocked *in vivo* with the mAb MEL-14 (Lewinsohn et al., 1987). The same inhibitory effect was seen with this antibody as well as with an L-selectin-immunoglobulin fusion protein on the influx of neutrophils into the inflamed peritoneal cavity of mice (Watson et al., 1991). Such experiments were also performed in rats using the anti-human E-selectin mAb Cl-3 which cross reacts with rat E-selectin (Mulligan et al., 1991). With this antibody the same effect was seen as described above for the anti L-selectin antibody, demonstrating that both L- and E-selectin are necessary for the influx of neutrophils into the peritoneum. The involvement of E-selectin in neutrophil-mediated damage of lung endothelium during acute airway inflammation could be demonstrated with anti E-selectin antibodies in rat (Mulligan et al., 1991) and in monkeys (Gundel et al., 1991). A protective effect of an anti human P-selectin mAb against cobra venom factor induced pulmonary injury in rats could also be demonstrated (Mulligan et al., 1992).

A recently discovered human genetic disease which, beside other defects, causes a markedly reduced ability of neutrophils to adhere to endothelium and causes recurrent episodes of bacterial infection and localized cellulitis without pus formation, is believed to be based on a defect in fucose metabolism (Etzioni et al., 1992). No sialyl Lewis X is found in these patients, which is a fucose containing tetrasaccharide known to bind to all three selectins (see below). Although this tetrasaccharide is not necessarily a physiological ligand for the selectins, it is conceivable that the physiological selectin-ligands might contain fucose as an essential structural element. Indeed, neutrophils of the patients do not bind to E-selectin expressing endothelial cells *in vitro* (Etzioni et al., 1992). This genetic disease has been called leukocyte adhesion deficiency (LAD) II. It demonstrates the importance and the essential role of carbohydrate recognition, probably via the selectins, for host defense mechanisms and inflammation. Another genetic disease which also affects neutrophil extravasation into sites of inflammation, was originally described as LAD and is now called LAD I. This disease is based on the lack of functional integrin β_2-chains (CD18), which disables neutrophils to undergo the second more firm adhesion step and hinders them to enter into sites of inflammation. Such patients suffer from life threatening infections (Kishimoto et al., 1989).

5. GLYCOPROTEIN LIGANDS OF THE SELECTINS

Soluble forms of the selectins have been successfully used as affinity probes to identify glycoprotein ligands. With antibody-like fusion proteins of L-selectin, the two ligands GlyCAM-l (Lasky et al., 1992) and CD34 (Baumhueter et al., 1993) were identified on the endothelium of high endothelial venules of lymph nodes. Later, a third glycoprotein ligand of 2OOkD was found in the same way (Hemmerich et al., 1994). All three ligands can be metabolically labeled with $^{35}SO_4$ and are recognized by the monoclonal antibody MECA 79, which blocks lymphocyte-binding to high endothelial venules (Imai et al., 1991). In addition to these glycoproteins, the anti-carbohydrate mAb MECA 79 recognizes further antigens in lymph node extracts (Berg et al., 1991a) which have not yet been further characterized.

The protein backbones of GlyCAM-l and CD34 are not only expressed in lymph node high endothelial venules. CD34 is also expressed on hematopoetic progenitor cells in the bone marrow, as well as on most endothelial cells throughout the body. GlyCAM-1, which is a secreted protein, was also found in milk (Dowbenko et al., 1993). However, in those sites outside lymph node tissue both ligands lack the correct carbohydrate-modifications which function as selectin-recognition motifs. Sulfation of GlyCAM-l, which was found to be essential for binding to L-selectin (Imai et al., 1993), was not found for GlyCAM-l in milk. Analyzing the carbohydrate capping structures on GlyCAM-1 revealed the presence of a sulfate in position 6 on galactose in sLe^X (Hemmerich and Rosen, 1994). In general, for none of the identified selectin ligands, expression of their protein backbone is restricted to those cells where they serve as ligands (see below). However, only on those cells which are able to bind to a selectin the ligands carry the correct carbohydrate-modifications which define them as selectin ligands. This demonstrates the central importance of the modifying enzymes and their cell type-specific regulation for the selectin-mediated adhesion phenomena.

GlyCAM-l and CD34 are both sialomucins which carry large clusters of sialic acid-rich O-linked carbohydrate side chains. These carbohydrate clusters are essential for the binding to L-selectin. Other ligands are also sialomucins, or at least contain a sialomucin domain, for example the vascular addressin MAdCAM-l. Although MAdCAM-l usually serves as a ligand for the lymphocyte integrin $\alpha_4\beta_7$, a small percentage of these molecules which carry MECA 79 epitopes also carry O-linked carbohydrate modifications which can be recognized by L-selectin (Berg et al., 1993).

The human P-selectin ligand PSGL-1 was identified by expression cloning: COS-7 cells co-transfected with PSGL-1 and fucosyl transferase III were able to bind to human P-selectin–IgG fusion protein coated onto a surface (Sako et al., 1993). The same protein was also found by affinity isolation using purified P-selectin from human platelets as an affinity probe (Moore et al., 1992). Like GlyCAM-1 and CD34, PSGL-1 belongs to the group of sialomucins. Using an adhesion blocking monoclonal antibody against human PSGL-l, it was shown recently that PSGL-l mediates rolling of leukocytes on P-selectin (Moore et al., 1995). A 230/130 kD pair of proteins, which was affinity-purified with mouse P-selectin–IgG from mouse neutrophils (Lenter et al., 1994), is probably the mouse homolog to human PSGL-1.

Human PSGL-1 was also shown to support binding of cells transfected with human E-selectin (Sako et al., 1993). A quantitative comparison of the binding of PSGL-1 transfected cells to immobilized P-or E-selectin, revealed that PSGL-1 binds with higher affinity to P- than to E-selectin (Moore et al., 1994). However, the affinity between PSGL-1 (mouse or human) and mouse E-selectin is sufficient to survive extensive washings in affinity isolation experiments (Lenter et al., 1994).

A more detailed analysis of the binding sites on PSGL-1 has revealed that in addition to the O-linked carbohydrate side chains an additional structural element is necessary for binding and may help to determine the remarkable ligand-specificity (Pouyanis and Seed, 1995; Sako et al., 1995; Wilkins et al., 1995). This element is located in the amino terminal region in front of the 15 consecutive decameric repeats which carry the clustered O-linked carbohydrate side chains. Deletion of amino acids 19 to 57 (with amino acid 1 being the start methionine of the coding sequence) resulted in the complete loss of binding activity of PSGL-l. Adding the first 100 amino acids of PSGL-l to the amino terminus of the mucins CD34 or CD43 made these mucins capable of binding to P-selectin. The region which was responsible for this transfer of binding specificity could be further nailed down to the sequence between amino acids 38 and 57. This sequence contains three consensus sites for tyrosine sulfation and fusion proteins containing this PSGL-l region were shown to become sulfated. Sulfation was abolished by mutating these tyrosine residues. Inhibition of sulfation in PSGL-l transfected COS cells by treating the cells with chlorate blocked P-selectin binding but not binding to E-selectin. Thus in addition to clusters of O-linked carbohydrates, tyrosine sulfation is directly or indirectly necessary for the binding of PSGL-1 to P-selectin. The combination of both structural elements defines PSGL-1 as a ligand for P-selectin.

In contrast to the sialomucin type ligands, the E-selectin ligand ESL-l requires N-linked carbohydrates for binding. ESL-l was found as a 150 kD glycoprotein ligand by affinity isolation with E-selectin-IgG on mouse neutrophil granulocytes (Levinovitz et al., 1993). Like PSGL-l, ESL-l requires sialic acid and fucose for ligand activity. However, ESL-l is not susceptible to degradation with O-sialoglycoprotease (a protease which specifically degrades sialomucins) (Lenter et al., 1994). In contrast to the sialomucin PSGL-l which binds to P-selectin and to E-selectin, ESL-l binds exclusively to E-selectin. This suggests that both ligands, while overlapping in some of their functions, may also serve different functions.

The sequence of ESL-l (Steegmaier et al., 1995) is highly homologous to a novel type of cysteine-rich fibroblast growth factor receptor (CFR) (Burrus et al., 1992). The amino acid sequence of ESL-1 is 94% identical (over 1,078 amino acids) to CFR, except for a unique 70 amino acid amino-terminal domain of mature ESL-1. A splicing variant of ESL-l which lacks 45 amino acids of the unique 70 amino acid domain has recently been identified in brain tissue of 13 day old mouse embryos (Blanks and Vestweber, unpublished). The strong structural homology to the chicken FGF-receptor raises the question whether ESL- 1 could have some signaling function. It is intriguing to speculate that ESL-1 may be involved in triggering signals which lead to integrin activation. Indeed, evidence has been reported that soluble E-selectin can induce activation of the integrin Mac-1 on PMNs (Lo et al., 1991). However, so far no evidence has been presented that ESL-1 or the chicken CFR are able to mediate signal transduction.

Using antibody like fusion proteins of ESL-1 and affinity purified antibodies against ESL-l allowed to demonstrate a role of this ligand in E-selectin mediated adhesion of myeloid cells (Steegmaier et al., 1995). An immunoglobulin fusion protein of ESL-l containing the complete extracellular part of ESL-l and modified by fucose supported binding of E-selectin transfected CHO cells in nonstatic (rotation) adhesion assays. No binding was seen to a fucose-containing L-selectin-Ig fusion protein. Furthermore, affinity-purified polyclonal antibodies against ESL- I partially inhibited the binding of the mouse neutrophilic cell line 32Dcl3 to immobilized E-selectin-IgG. Similarly, the binding of these cells and of mouse PMNs to cytokine-induced mouse endothelioma cells could be partially inhibited by the anti-ESL-1 antibodies.

So far four ligands have been identified and cloned, which can be affinity isolated from cell detergent extracts using selectin-Ig fusion proteins as affinity probes: GlyCAM-1, CD34, PSGL-1 and ESL-l. In addition to these, many ligands have been described for which affinity isolation with a selectin was not shown or which were identified in this way, but have not yet been cloned. Three such ligands were identified by affinity isolation with L-selectin: a 200 kD ligand for mouse L-selectin (Hemmerich et al., 1994), a heparin sulfate containing proteoglycan (Norgard-Sumnicht et al., 1993), and several sulfated proteins in rat lymph nodes (Tamatani et al., 1993). For P-selectin, a 160 kD glycoprotein (reduced 80 kD) was affinity isolated from mouse neutrophil granulocytes with mouse P-selectin–IgG (Lenter et al., 1994), but not yet cloned. Analogously to ESL-1, which does not bind to P-selectin the 160 kD P-selectin–ligand does not bind to E-selectin and requires N-linked carbohydrates for binding. The heat stable antigen (HSA), purified from different mouse leukocytes and immobilized on a surface, supports binding of P-selectin–IgG but not E-selectin–IgG (Sammar et al, 1994). In addition, the binding of neutrophils and monocytic cells to P-selectin expressing platelets or endothelial cells could be blocked with antibodies against HSA (Aigner et al., 1995).

Many ligands were described for E-selectin, of which only one was identified by direct affinity isolation. This is a 250 kD (reduced 280kD) glycoprotein on bovine peripheral γ/δ T-cells (Walcheck et al, 1993). The other E-selectin ligands were defined either by adhesion blocking antibodies or by testing their capacity to support the binding of E-selectin-expressing cells to the purified, immobilized ligand: these proteins are L-selectin from human neutrophils (Picker et al., 1991; Kishimoto et al., 1991), members of the NCA-family (Kuijpers et al., 1992), a subpopulation of sLe^X carrying β_2-integrins (Kotovuori et al., 1993), the sLe^X carrying lamp-1 lysosomal protein on the surface of carcinoma cells (Sawada et al., 1993) and not yet characterized glycoproteins on human T-lymphocytes which carry sLe^a-like carbohydrate structures defined by the mAb HECA 452 (Berg et al., 1991b). Among these indirectly defined ligands for E-selectin, we have recently analyzed the possible binding of L-selectin to E-selectin in more detail. We found that in contrast to L-selectin on mouse neutrophils, L-selectin on human neutrophils could be affinity isolated from detergent extracts of metabolically labeled neutrophils as a major ligand using E-selectin-Ig as affinity matrix. Binding was Ca^{2+}-dependent, required sialic acid on the carbohydrate side chains of L-selectin and was resistant to treatment with endoglycosidase F (Lenter, Zöllner, Borges, Zerwes and Vestweber, manuscript submitted).

In agreement with the biochemical data, attachment of human neutrophils under flow conditions, but not attachment of mouse neutrophils, to the surface of E-selectin coated glass capillaries could be blocked with a mAb against L-selectin.

There is good evidence, that leukocyte tethering to the endothelial surface is not only mediated directly via cell adhesive interactions between endothelium and leukocytes. In a "secondary tethering" event, neutrophils which have already accumulated at the endothelial surface could be involved in the recruitment of further neutrophils which use L-selectin in order to roll on the immobilized neutrophils (Bargatze et al., 1994). Potential L-selectin ligands for this process could be sialomucins, which mediate in vitro aggregation of neutrophils (Bennett et al., 1995). Thus, L-selectin could be involved in neutrophil recruitment to activated endothelium in three ways: first, as a lectin, binding to cytokine inducible, carbohydrate presenting ligands on the endothelial cell surface. Second, as a lectin, binding to carbohydrate based ligands on already immobilized neutrophils. Third, as a carbohydrate presenting ligand for E-selectin. However, the latter function was only observed for human but not for mouse neutrophils.

It is difficult to decide at present, which of the several possible interactions between the selectins and their ligands is of major relevance for leukocyte recruitment into sites of inflammation. Among the many ligands which have been suggested, the best candidates probably are those for which specific binding to a selectin has been directly demonstrated by affinity isolation. Probably the best characterized selectin ligand so far is PSGL- 1, for which its involvement in neutrophil rolling has been demonstrated *in vivo*, analyzing human neutrophils in an animal model (Norman et al., 1995).

We are beginning to understand some of the structural elements of the selectin ligands, which are required for recognition. Again, best characterized in this respect are the cloned, high affinity ligands. They all require sialic acid as well as fucose for binding. For PSGL-l and ESL-l it was directly shown by transfection experiments, that a fucosyl transferase needs to be co-transfected together with the ligand cDNAs in order to generate selectin binding glycoproteins. In addition, PSGL- 1 and GlyCAM- 1 need to be sulfated for ligand activity, a requirement which ESL-l does not share. Both ligands are sulfated by different sulfotransferases, since PSGL-1 carries sulfate on N-terminal tyrosine residues while GlyCAM-1 is sulfated on its carbohydrate side chains. Since the protein scaffolds of the selectin ligands are also expressed on cells where they do not function as ligands, the regulation of modifying enzymes such as the fucosyl transferases and sulfotransferases is the central step which determines the expression of functional ligands. The expression of such ligands seems to be the result of the well controlled expression of certain modifying enzymes and their ability to modify only a few specific protein scaffolds in a cell, which then act as selectin ligands. At present, the mechanisms which determine this specificity are not yet understood.

REFERENCES

Aigner, S., M. Ruppert, M. Hubbe, M. Sammar, Z. Sthoeger, E.C. Butcher, D. Vestweber, and P. Altevogt. Heat-stable antigen (mouse CD24) supports myeloid cell binding to endothelial and platelet P-selectin. *Int. Immunol.*, 7:1557–1565, 1995.

Bargatze, R.F., S. Kurk, E.C. Butcher, and M.A. Jutila. Neutrophils roll on adherent neutrophils bound to cytokine-induced endothelial cells via L-selectin on the rolling cells. *J. Exp. Med.*, 180: 1785–1792, 1994.

Baumheuter, S., M.S. Singer, W. Henzel, S. Hemmerich, M. Renz, S.D. Rosen, and L.A. Lasky. Binding of L-selectin to the vascular sialomucin CD34. *Science.*, 262:436–438, 1993.

Bennett, T.A., C.M.G. Schammel, E.B. Lynam, D.A. Guyer, A. Mellors, B. Edwards, S. Rogeli, and L.A. Sklar. Evidence for a third component in neutrophil aggregation: potential roles of O-linked glycoproteins as L-selectin counter-structures. *J. Leuk. Biol.* 58:510–518, 1995.

Berg, E.L., M.K. Robinson, R.A. Warnock, and E.C. Butcher. The human peripheral lymph node vascular addressin is a ligand for LECAM- 1, the peripheral lymph node homing receptor. *J. Cell Biol.*, 114:343–349, 1991a.

Berg, E.L., T. Yoshino, L.S. Rott, M.K. Robinson, R.A. Warnock, T.K. Kishimoto, L.J. Picker, and E.C. Butcher. The cutaneous lymphocyte antigen is a skin lymphocyte homing receptor for the vascular lectin endothelial cell-leukocyte adhesion molecule 1. *J. Exp. Med.*, 174:1461–1466, 1991b.

Berg, E.L., L.M. McEvoy, C. Berlin, R.F. Bargatze, and E.C. Butcher. L-selectin-mediated lymphocyte rolling on MAdCAM-1. *Nature.*, 366:695–698, 1993.

Bevilacqua, M.P., J.S. Pober, D.L. Mendrick, R.S. Cotran, and M.A. Gimbrone. Identification of an inducible endothelial-leukocyte adhesion molecule. *Proc. Natl. Acad. Sci. USA.*, 84:9238–9242, 1987.

Brady, H.R., O. Spertini, W. Jimene, B.M. Brenner, P.A. Marsden, and T.F. Tedder. Neutrophils, monocytes, and lymphocytes bind to cytokine-activated kidney glomerular endothelial cells through L-selectin (Lam-l) *In vitro. J. Immunol.*, 149:2437–2444, 1992.

Burrus, L.W., M.E. Zuber, B.A. Lueddecke, and B.B. Olwin. Identification of a cysteine-rich receptor for fibroblast growth factors. *Mol. Cell. Biol.*, 12:5600–5609, 1992.

Cotran, R.S., M.A. Gimbrone Jr., M.P. Bevilacqua, D.L. Mendrick, and J.S. Pober. 1986. Induction and detection of a human endothelial activation antigen *in vivo. J. Exp. Med.*, 164:661–666, 1986.

Damle, N.K., C. Eberhardt, and M. Van der Vieren. Direct interaction with primed CD4$^+$ CD45RO$^+$ memory T lymphocytes induces expression of endothelial leukocyte adhesion molecule-l and vascular adhesion molecule-l on the surface of vascular endothelial cells. *Eur. J. Immunol.*, 21:2915–2923, 1991.

Disdier, M., J.H. Morrisse, R.D. Fugate, D.F. Bainton, and R.P. McEver. Cytoplasmic domain of P-selectin (CD-62) contains the signal for sorting into the regulated secretory pathway. *Mol. Biol. of the Cell.*, 3:309–321, 1992.

Dowbenko, D., A. Kikuta, C. Fennie, N. Gillett, and L.A. Lasky. Glycosylation-dependent cell adhesion molecule 1 (GlyCAM-1) mucin is expressed by lactating mammary gland epithelial cells and is present in milk. *J. Clin. Invest.*, 92:952–960, 1993.

Etzioni, A., M. Frydman, S. Pollack, I. Avidor, L. Phillips, J. C. Paulson, and R. Gershoni-Baruch. Recurrent severe infections caused by a novel leukocyte adhesion deficiency. *New Engl. J. Med.*, 327:1789–1792, 1992.

Finger, E.B., K.D. Pun, R. Alon, M.B. Lawrence, U. H. von Andrian, and T. Springer. Adhesion through L-selectin requires a threshold hydrodynamic shear. *Nature.*, 379:266–269, 1996.

Gallatin, W.M., I.L. Weissman, and E.C. Butcher. A cell surface molecule involved in organ-specific homing of lymphocytes. *Nature*, 304:30–34, 1983.

Gotsch, U., U. Jäger, M. Dominis, and D. Vestweber. Expression of P-selectin on endothelial cells is upregulated by LPS and TNF-α *in vivo. Cell Adhesion Commun.*, 2:7–14, 1994.

Greem, S.A., H. Setiadi, R.P. McEver, and R.B. Kelly. The cytoplasmic domain of P-selectin contains a sorting determinant that mediates rapid degradation in lysosomes. *J. Cell Biol.* 124:435–448, 1994.

Gundel, R.H., C.D. Wegner, C.A. Torcellini, C.C. Clarke, N. Haynes, R. Rothlein, C.W. Smith, and L.G. Letts. Endothelial leukocyte adhesion molecule-l mediates antigen-induced acute airway inflammation and late-phase airway obstruction in monkeys. *J. Clin. Invest.*, 88:1407–1411, 1991.

Hahne, M., U. Jäger, S. Isenmann, R. Hallmann, and D. Vestweber. Five TNF-inducible cell adhesion mechanisms on the surface of mouse endothelioma cells mediate the binding of leukocytes. *J. Cell Biol.*, 21:655–664, 1993.

Hamann, A., D. Jablonski-Westrich, P. Jonas, and H.G. Thiele. Homing receptors reexamined: Mouse LECAM-l (MEL-14 antigen) is involved in lymphocyte migration into gut-associated lymphoid tissue. *Eur. J. Immunol.*, 21:2925–2929, 1991.

Hemmerich, S, and S.D. Rosen. 6'-sulfated sialyl Lewis x is a major capping group of GlyCAM- 1. *Biochemistry.*, 33:4830–4835, 1994.

Hemmerich, S, E.C. Butcher, and S. D. Rosen. Sulfation-dependent recognition of high endothelial venules (HEV)-ligands by L-selectin and MECA 79, an adhesion-blocking monoclonal antibody. *J. Exp. Med.*, 180:2219–2226, 1994.

Hsu-Lin, S.C., C.L. Berman, B.C. Furie, D. August, and B. Furie. A platelet membrane protein expressed during platelet activation and secretion. *J. Biol. Chem.*, 259:9121–9126, 1984.

Imai, Y., M.S. Singer, C. Fennie, L.A. Lasky, and S.D. Rosen. Identification of a carbohydrate-based endothelial ligand for a lymphocyte homing receptor. *J. Cell Biol.*, 113:1213–1221, 1993.

Imai, Y., L.A. Lasky, and S.D. Rosen. Sulphation requirement for GlyCAM-1, an endothelial ligand for L-selectin. *Nature*, 361:555–557, 1993.

Kishimoto, T.K., R.S. Larson, A.L. Corbi, M.L. Dustin, D.E. Staunton, and T.A. Springer. The leukocyte integrins. *Adv. Immunol.*, 46:149–181, 1991.

Kishimoto, T. K., R. A. Warnock, M. A. Jutila, E. C. Butcher, C. Lane, D. C. Anderson, and C. W. Smith. Antibodies against human neutrophil LECAM-1 (LAM-1/Leu-8/DREG-56 antigen) and endothelial cell ELAM-1 inhibit a common CDl8-independent adhesion pathway *in vitro. Blood.*, 78:805–811, 1991.

Koedam, J.A., E.M. Cramer, E. Briend, B. Furie, B.C. Furie, and D.D. Wagner. P-selectin, a granule membrane protein of platelets and endothelial cells, follows the regulated secretory pathway in AtT-20 cells. *J. Cell Biol.*, 116:617–625, 1992.

Kotovuori, P., E. Tontti, R. Pigott, M. Shepard, M. Kiso, A. Hasegawa, R. Renkonen, P. Nortano, D.C. Altieri, and C.G. Gahmberg. The vascular E-sectin binds to the leukocyte integrins CD11/CDl8. *Glycobiol.*, 3:131–136, 1993.

Kuijpers, T.W., M. Hoogerwerf, L.C.W. van der Laan, G. Nagel, C.E. van der Schoot, F. Grunert, and D. Roos. CD66 nonspecific cross-reacting antigens are involved in neutrophil adherence to cytokine-activated endothelial cells. *J. Cell Biol.*, 118:457–466, 1992.

Lasky. L.A., M.S. Singer, D. Dowbenko, Y. Imai, W.J. Henzel, C. Grimley, C. Fennie. N. Gillett, S.R. Watson, and S.D. Rosen SD. An endothelial ligand for L-selectin is a novel mucin-like molecule. *Cell.*, 69:927–938, 1992.

Lawrence, M. B., and T. A. Springer. Leukocytes roll on a selectin at physiologic flow rates: distinction from and prerequisite for adhesion through integrins. *Cell.*, 65: 859–873, 1991.

Lenter, M., A. Levinovitz, S. Isenmann, and D. Vestweber. Monospecific and common glycoprotein ligands for E- and P-selectin on myeloid cells. *J. Cell Biol.* 125:471–481, 1994.

Levinovitz, A., J. Mühlhoff, S. Isenmann, and D. Vestweber. Identification of a glycoprotein ligand for E-selectin on mouse myeloid cells. *J. Cell Biol.*, 121:449–459, 1987.

Lewinsohn, D.M., R.F. Bargatze, and E.C. Butcher. Leukocyte-endothelial cell recognition: Evidence of a common molecular mechanism shared by neutrophils, lymphocytes, and other leukocytes. *J. Immunol.*, 138:4313–4321, 1987.

Ley, K., P. Gaethgens, C. Fennie, M.S. Singer, L.A. Lasky, and S.D. Rosen. Lectin-like cell adhesion molecule 1 mediates leukocyte rolling in mesenteric venules *in vivo*. *Blood*, 77: 2553–2555, 1991.

Lo, S.K., S. Lee, R.A. Ramos, R. Lobb, M. Rosa, O. Chi-Rosso, and S.D. Wright. Endothelial-leukocyte adhesion molecule 1 stimulates the adhesive activity of leukocyte integrin CR3 (CD11b/CDl8, Mac-l $\alpha_M\beta_2$) on human neutrophils. *J. Exp. Med.*, 173:1493–1500, 1991.

McEver, R.P. and M.N. Martin. A monoclonal antibody to a membrane glycoprotein binds only to activated platelets. *J. Biol. Chem.*, 259:9799–9804, 1984.

McEver, R.P., J.H. Beckstead, K.L. Moore, L. Marshall-Carlson, and D.F. Bainton. GMP-140, a platelet alpha-granule membrane protein, is also synthesized by vascular endothelial cells and is localized in Weibel-Palade bodies. *J. Clin. Invest.*, 84:92–99, 1989.

Montgomery, K.F., L. Osborn, C. Hession, R. Tizard, D. Goff, C. Vassallo, P.I. Tarr, K. Bomsztyk, R. Lobb, J.M. Harlan, and T.H. Pohlman. Activation of endothelial-leukocyte adhesion molecule 1 (ELAM-1) gene transcription. *Proc. Natl. Acad. Sci. USA.*, 88:6523–6527, 1991.

Moore, K.L., N.L. Stults, S. Diaz, D.F. Smith, R.C. Cummings, A. Varki, and R.P. McEver. Identification of a specific glycoprotein ligand for P-selectin (CD62) on myeloid cells. *J. Cell Biol.*, 118:445–456, 1992.

Moore, K.L., S.F. Eaton, D.E. Lyons, H.S. Lichenstein, R.D. Cummings, and McEver. The P-selectin glycoprotein ligand from human neutrophils displays sialylated, fucosylated, O-linked poly-N-acetyllactosamine. *J. Biol. Chem.*, 269:23318–23327, 1994.

Moore, K.L., K.D. Patel, R.E. Bruehl, L. Fugang, D.A. Johnson, H.S. Lichenstein, R.D. Cummings, D.F. Bainton, and R.P. McEver. P-selectin glycoprotein ligand-1 mediates rolling of human neutrophils on P-selectin. *J. Cell. Biol.*, 128:661–671, 1995.

Mulligan, M.S., J. Varani, M.K. Dame, C.L. Lane, C.W. Smith, D.C. Anderson, and P.A. Ward. Role of endothelial-leukocyte adhesion molecule 1 (ELAM-l) in neutrophil-mediated lung injury in rats. *J. C/in. Invest.*, 88:1396–1406, 1991.

Mulligan, M.S., M.J. Polley, R.J. Bayer, M.F. Nunn, J.C. Paulson, and P.A. Ward. Neutrophil dependent acute lung injury: requirement for P-selectin (GMP-140). *J. Clin. Invest.*, 90:1600–1607, 1992.

Munro, J.M., J.S. Pober, and R.S. Cotran. Tumor necrosis factor and interferon-γ induce distinct patterns of endothelial activation and associated leukocyte accumulation in skin of papio anubis. *Am. J. Pathol.*, 135:121–133, 1989.

Munro, J.M., J.S. Pober, and R.S. Cotran. Recruitment of neutrophils in the local expression response: association with de novo endothelial expression of endothelial leukocyte adhesion molecule-1. *Lab. Invest.*, 64:295–299, 1991.

Norgard-Sumnicht, K.E., N.M. Varki, and A. Varki. Calcium dependent heparin-like ligands for L-selectin in non-lymphoid endothelial cells. *Science.*, 261:480–483, 1993.

Norman, K. E., K. L. Moore, R. P. McEver, and K. Ley. Leukocyte rolling *in vivo* is mediated by P-selectin glycoprotein ligand-1. *Blood.*, 86:4417–4421, 1995.

Patel, K.D., G.A. Zimmerman, S.M. Prescott, R.P. McEver, and T.M. McIntyre. Oxygen radicals induce human endothelial cells to express GMP-140 and bind neutrophils. *J. Cell Biol.*, 112:749–759, 1991.

Picker, L.J, R.A. Warnock, A.R. Burns, C.M. Doerschuk, E.L. Berg, and E.C. Butcher. The neutrophil selectin LECAM-1 presents carbohydrate ligands to the vascular selectins ELAM-1 and GMP-140. *Cell.*, 66:921–933, 1991.

Pober, J.S., L.A. Lapierre, A.H. Stolpen, T.A. Brock, A.T. Springer, W. Fiers,, M.P. Bevilacqua, D.L. Mendrick, and M.A. Gimbrone Jr. Activation of cultured human endothelial cells by recombinant lymphotoxin: comparison with tumor necrosis factor and interleukin 1 species. *J. Immunol.*, 138:3319–3324, 1987.

Pouyani, T. and B. Seed. PSGL-1 recognition of P-selectin is controlled by a tyrosine sulfation consensus at the PSGL-1 amino terminus. *Cell*, 83:333–343, 1995.

Redl, H., H.P. Dinges, W.A. Buurman, C.J. van der Linden, J.S. Pober, RS. Cotran, and G. Schlag. Expression of endothelial leukocyte adhesion molecule-1 in septic but not traumatic/hypovolemic shock in the baboon. *Am. J. Pathol.*, 139:461–466, 1991.

Sako, D., X.J. Chang, K.M. Barone, G. Vachino, H.M. White, G. Shaw, G.M. Veldman, K.M. Bean, T.J. Ahern, B. Furie, D.A. Cummings, and G.R. Larsen. Expression cloning of a functional glycoprotein ligand for P-selectin. *Cell.*, 75:1179–1186, 1993.

Sako, D., K.M. Comess, K.M. Barone, R.T. Camphausen, D.A. Cumming, and G:D. Shaw. A sulfated peptide segment at the amino terminus of PSGL-1 is critical for P-selectin-binding. *Cell.*, 83:323–331, 1995.

Sammar, M., S. Aigner, M. Hubbe, V. Schirrmacher, M. Schachner, D. Vestweber, and P. Altevogt. Heat-stable antigen (CD24) as ligand for mouse P-selectin. *Intern. Immunol.*, 6:1027–1036, 1994.

Sanders, W.E., R.W. Wilson, C.M. Ballantyne, and A.L. Beaudet. Molecular cloning and analysis of *in vivo* expression of murine P-selectin. *Blood.*, 80:795–800, 1992.

Sawada, R., J.B. Lowe, and M. Fukuda. E-selectin-dependent adhesion efficiency of colonic carcinoma cells is increased by genetic manipulation of their cell surface lysosomal membrane glycoprotein-l expression levels. *J. Biol. Chem.*, 268:12675–12681, 1993.

Simon, S. I., Y. P. Rochon, E. B. Lynam, C. W. Smith, D. C. Anderson, and L. A. Sklar. β2-integrin and L-selectin are obligatory receptors in neutrophil aggregation. *Blood.*, 4:1097–1106, 1993.

Spertini, O., F.W. Luscinskas, V.X. Kansas, J.M. Munro, V. Griffin, M.A. Gimbrone Jr., and T.F. Tedder. Leukocyte adhesion molecule-l (LAM-1, L-selectin) interacts with an inducible endothelial cell ligand to support leukocyte adhesion. *J. Immunol.*, 147:2565–2573, 1991.

Springer, T.A. Traffic signals for lymphocyte recirculation and leukocyte emigration: The multistep paradigm. *Cell.*, 76:301–314, 1994.

Steegmaier, M., A. Levinovitz, S. Isenmann, E. Borges, M. Lenter, H. Kocher, B. Kleuser, and D. Vestweber. The E-selectin-ligand ESL-l is a variant of a receptor for fibroblast growth factor. *Nature.*, 3 73:615 -620, 1995.

Stenberg, P.E., R.P. McEver, M.A. Shuman, Y.V. Jacques, and D.F. Bainton. A platelet alpha-granule membrane protein (GMP-140) is expressed on the plasma membrane after activation. *J. Cell Biol.*, 101:880–886, 1985.

Subramaniam, M., J.A. Koedam, and D.D. Wagner. Divergent fates of P- and E-selectins after their expression on the plasma membrane. *Mol. Biol. Cell.*, 4:791–801, 1993.

Sunderkötter, C., K. Stembrink, U. Henseleit, R. Bosse, A. Schwarz, D. Vestweber, and C. Sorg. Activated T cells induce expression of E-selectin *in vitro* and in an antigen dependent manner *in vivo*. *Eur. J. Immunol.*, 26:1571–1579, 1996.

Tamatani, T., K. Kuida, T. Watanabe, S. Koike, and M. Miyasaka. Molecular mechanisms underlying lymphocyte recirculation. *J. Immunol.*, 150:1735–1745, 1993.

Von Andrian, U.N., J.D. Chambers, L.M. McEvoy, R.B. Bargatze, K.E. Arfors, and E.C. Butcher. Two-step model of leukocyte-endothelial cell interaction in inflammation: distinct roles for LECAM-1 and the leukocyte β2 integrins *in vivo*. *Proc. Natl. Acad. Sci. USA.*, 88:7538–7542, 1991.

Walcheck, B., G. Watts, and M.J. Jutila. Bovine γ/δ T cells bind E-selectin via a novel glycoprotein receptor: First characterization of a lymphocyte/E-selectin interaction in an animal model. *J. Exp. Med.*, 178:853–863, 1993.

Walcheck, B., J. Kahn, J.M. Fisher, B.B. Wang, R.S. Fisk, D.G. Payan, C. Freehan, R. Betageri,, K. Darlak, A.F. Spatola, and T.K. Kishimoto. Neutrophil rolling altered by inhibition of L-selectin shedding *in vitro*. *Nature.*, 380:720–723, 1996.

Watson, S.R., C. Fennie, and L.A. Lasky. Neutrophil influx into an inflammatory site inhibited by a soluble homing receptor-IgG chimaera. *Nature.*, 349:164–167, 1991.

Weller, A., S. Isenmann, and D. Vestweber. Cloning of the mouse endothelial selectins: Expression of both E- and P-selectin is inducible by tumor necrosis factor-α. *J. Biol. Chem.*, 267:15176–15183, 1992.

Whelan, J., P. Ghersa, R.H. von Huijsduijene, J. Gray, G. Chandra, F. Talabot, and J.F. DeLamarter. An NFκB-like factor is essential but not sufficient for cytokine induction of endothelial leukocyte adhesion molecule 1 (ELAM-1) gene transcription. *Nucl. Acids Res.*, 19:26452653, 1991.

Wilkins, P.P., K.L. Moore, R.P. McEver, and R.D. Cummings. Tyrosine sulfation of P-selectin glycoprotein ligand-1 is required for high affinity binding to P-selectin. *J. Biol. Chem.*, 270:22677–22680, 1995.

3

COMPLEMENT INHIBITOR THERAPEUTICS AND LUNG INJURY

Una S. Ryan

T Cell Sciences, Inc.
Needham, Massachusetts

1. INTRODUCTION

Lung injury from any cause can have severe clinical outcomes. The Adult Respiratory Distress Syndrome (ARDS) was defined as a clinical entity in 1967.[1] It is generally recognized as acute respiratory failure characterized by relatively normal cardiac function, an increase in vascular permeability leading to pulmonary edema manifested by diffuse pulmonary infiltrates on the chest X-ray and by a major oxygenation defect.[2] Precipitating insults include severe sepsis, diffuse pneumonia, pancreatitis, multiple trauma, aspiration, near-drowning, burns, shock, hypotension, and coagulopathy. ARDS is a major contributor to the morbidity and mortality of patients in intensive care units with 30-day mortality rates varying from 55 to 65%.[1–4] Current estimates are that more than 100,000 patients per year in the U.S. develop ARDS.[5]

Other forms of lung injury, such as chronic emphysema, cancer or cystic fibrosis may only be resolved by lung transplantation. Over the past 20 years, advances in surgical technique, donor organ procurement and preservation, immunosuppressive therapies, and post-surgical care have made lung transplantation a viable therapy for patients with end-stage pulmonary disease. Nonetheless, success is limited by problems such as reperfusion injury to the allograft. Despite the different causes leading to acute or chronic respiratory failure, a common feature is severe injury to the pulmonary endothelium with consequent compromise of endothelial functions. By the time endothelial injury is widespread intervention is problematic, and ARDS leads to multiorgan failure or death. Thus, a therapeutic regimen, whether designed to improve the outcome or sequelae of ARDS, or to counter allograft dysfunction following transplantation would preferably have features which prevent or ameliorate endothelial damage and the consequences thereof.

Vascular Endothelium, edited by Catravas *et al.*
Plenum Press, New York, 1998

2. ENDOTHELIAL FUNCTIONS AND DYSFUNCTIONS IN ARDS

In the uninjured lung pulmonary endothelial cells form a complete barrier both in larger vessels and at the level of the alveolar-capillary unit.[6] In the normal situation circulating blood cells do not interact with the endothelial surface and the constitutive functions of the pulmonary endothelium prevail.[7] Constitutive functions include the metabolic processing of circulating substances by endothelial surface molecules thus determining which biologically active substances flow downstream from the lungs and which are inactivated at the level of the lungs. An example is processing of angiotensin I to angiotensin II by angiotensin converting enzyme on the surface of endothelial cells.[8] In addition, the normal endothelium is capable of releasing into the circulation biologically active substances generated by the endothelial cells in response to signals, for example the ability of the endothelium to release prostacyclin,[9] NO,[10] and endothelin,[11] among others. Further it is now well recognized that the normal endothelium has the possibility of being activated in ways that cause it to interact quite differently with cells and molecules in the circulation. Thus, endothelium has metabolic functions, responsive functions, and signal transducing functions. Cell-to-cell communications in response to activation signals can occur before microscopically-visible damage to the endothelium is apparent and before the barrier function is impaired.

As acute pulmonary inflammation proceeds, the injury to pulmonary vessels results in fibrin deposition, platelet-accumulation, leukocyte activation and degranulation and ultimately complete loss of the endothelium. It is difficult to understand how so many different internal blood enzyme cascades can be activated simultaneously as a result of injury to the endothelium and perhaps it is even more difficult to imagine how, once activated, the downstream effects of these cascades can be reversed either in an experimental or clinical situation.

3. COMPLEMENT ACTIVATION IN ARDS

Looking at the pathogenesis of a very complex lung injury such as ARDS (Figure 1), it is clear that a variety of different insults (trauma, surgery, burns, sepsis, and hypoxia), all trigger conditions that fit the European American Consensus Definition of ARDS.[12] The clinical findings are characterized by acute and severe hypoxemia, diffuse pulmonary infiltrates on the chest X-ray, with no evidence of heart failure. Between acceptance of the multiple causes of ARDS and agreement on the clinical definition there is still disagreement about the precise mechanisms of disease. However, there are several features commonly recognized. One is acute and ongoing inflammation characterized by

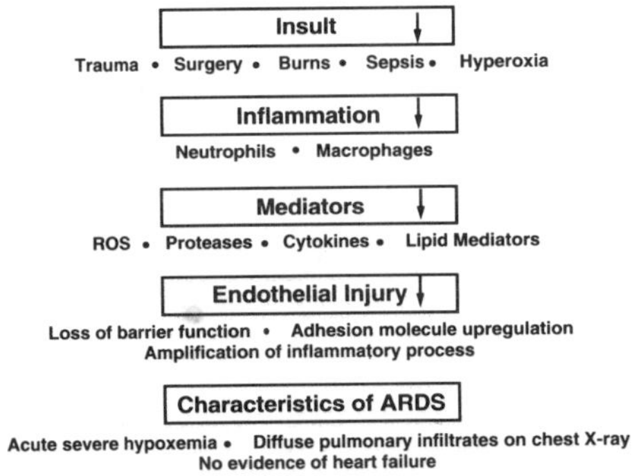

Figure 1. Pathogenesis of ARDS. The steps intervening between the multiple causes and clinical definition of ARDS are complex. Complement activation may be involved in each step in the progression of the clinical characteristics.

large numbers of neutrophils and macrophages in the airways that appear to be activated. In addition, there are a plethora of mediators released including reactive oxygen species, proteases, cytokines and lipid mediators. It is also generally agreed that the leakage of protein-rich fluid into the lungs of ARDS patients is indicative of marked and ongoing endothelial injury. At this point the barrier function of the endothelium is lost, and many of the constitutive or processing functions of the endothelium are lost. When endothelial cell and leukocyte activation occurs, not only is the endothelium the target of the initial injury but it is also the surface on which the inflammatory process progresses.

Despite the multiple causes and the complexity of clinical presentation, a common pathway for the progression of ARDS may revolve around the activation of complement. Complement is involved in the initiation, the progression, or the amplification of each of the steps shown in Figure 1.

Complement activation has been demonstrated to occur both in animal models of ARDS and in patients who either have ARDS or are at high risk of developing ARDS[13]. Complement activation can both initiate an inflammatory response and amplify an ongoing response. Complement activation triggers the activation of other inflammatory cells, such as macrophages, mast cells and neutrophils and they then are responsible for the release of a very large number of inflammatory mediators. All of these attributes are very useful when complement is acting appropriately, i.e. defending the body against invading organisms or foreign substances, but are extremely deleterious when acute inflammation is the problem. A critical step in limiting these deleterious effects on the endothelium is to prevent complement activation.

Figure 2 illustrates the effects of complement components acting on the key cells involved in ARDS. The anaphylatoxins, complement components C3a and C5a, act directly on leukocytes to cause the release of numerous biologically active molecules. These inflammatory mediators have a number of well-characterized effects on the underlying endothelium, including loss of endothelial integrity with increased permeability to molecules, fluids and cells which can now extravasate. The direct effects of complement components on inflammatory and vascular cells as well as indirect effects of complement-induced mediators and growth factors are to induce activation, upregulation of adhesion molecules and cell-to-cell interaction. One might argue that to prevent some of these mediators from being released and some of these adhesion molecules from being upregulated, or to block their effects, might have a benefit on the disease processes. However, there is great redundancy in the system and eliminating one or two culprits still means that there are plenty of others that can effectuate similar pathways. Much of the despondency about being able to intervene in diseases such as ARDS, sepsis and multiorgan failure has arisen as a result of clinical trials that have failed to show benefit to the patient. Perhaps inhibition early in the inflammatory cascade, by prevention of complement activation, would reduce the downstream amplification of pathogenetic mechanisms. A desirable therapeutic candidate would eliminate or reduce all immediate and downstream effects of complement activation (Figure 2).

The effects of the anaphylatoxins on leukocytes include priming and activation resulting in release of granule contents and/or radicals such as nitric oxide and superoxide anion. Activated leukocytes upregulate adhesion molecules and release potent mediators which themselves can damage the underlying endothelium. Like the anaphylatoxins, the terminal complex of complement activation (C5b-9, MAC), can act on the endothelial surface to trigger activation events. The latter appears to be able to trigger release of membrane vesicles from the endothelial surface that lead to activation of the prothrombinase complex[14]. Not only does complement activation lead to these acute responses, but also may lead to chronic effects that involve proliferation of cells underlying the endothelium

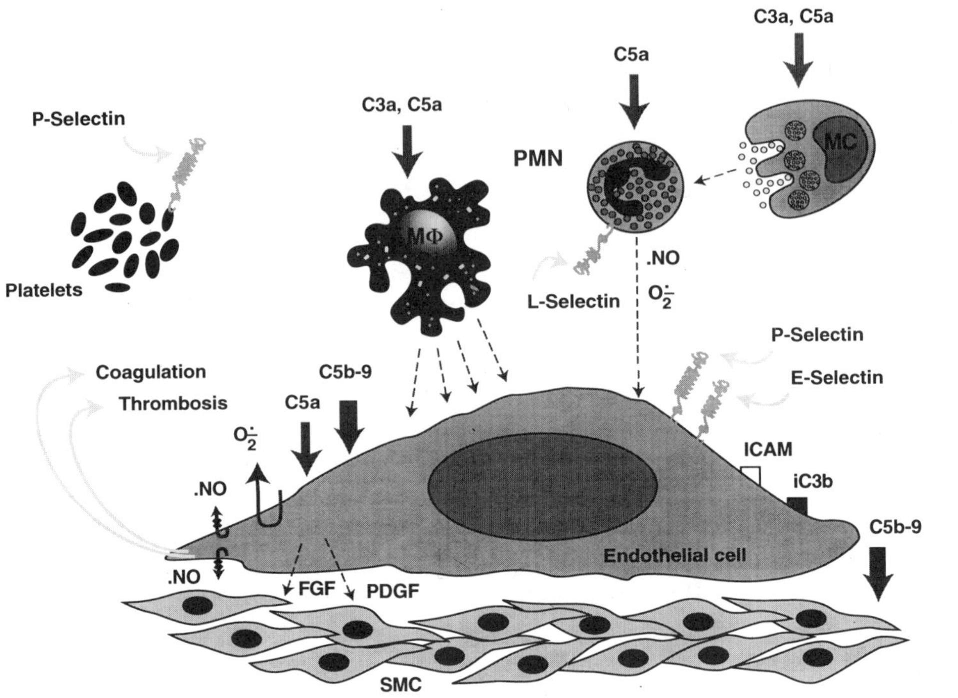

Figure 2. Complement-mediated responses that are inhibited by soluble complement receptor type 1 (sCR1 TP10). Active fragments released during complement activation, such as the anaphylatoxins C3a and C5a, lead to the release of cytokines and other inflammatory mediators (dashed lines) from leukocyte and endothelial cells, as well as upregulation of adhesion molecules. The terminal complex (C5b-9) also has sublytic, agonist-like effects on endothelial cells leading to procoagulant and prothrombotic states and downregulation or destruction of nitric oxide. C5b-9-stimulated release of growth factors such as basic fibroblast growth factor (bFGF) and platelet-derived growth factor (PDGF) might be expected to mediate longer term responses such as smooth muscle and fibroblast hyerplasia. All these effects are ameliorated by treatment with sCR1. New complement inhibitors currently being developed, such as soluble complement receptor type 1 expressing the sialyl Lewis^x tetrasacharide (sCR1sLe^x), bind to selectins both targeting complement inhibition to sites where it is most needed, for example, activated endothelium, and blocking endothelial-neutrophil and neutrophil-platelet adhesive interactions. From ref. 32

such as smooth muscle cells or fibroblasts. C5b-9 acting directly on these smooth muscle cells or fibroblasts is mitogenic either alone or synergistically with platelet-derived growth factor with (PDGF)[15]. Again C5b-9 by acting on endothelial cells causes the release of growth factors, basic fibroblast growth factor (bFGF) and PDGF[16] which stimulate proliferation of smooth muscle cells and fibroblasts. While, in general, acute inflammatory events are recognized to have chronic consequences, there appears to be a clear role for the acute actions of complement on the longer-term remodeling events affecting blood vessels and pulmonary parenchyma.

4. COMPLEMENT INHIBITION BY sCR1

sCR1 (TPl0) is a soluble form of human complement receptor type 1 (CR1), a cell surface glycoprotein that inhibits both the classical and alternative pathways of complement

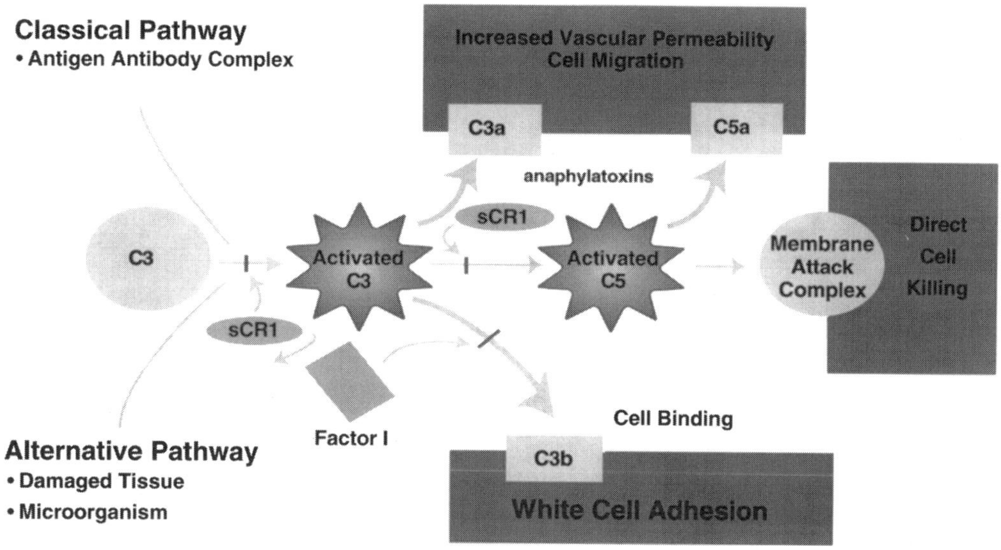

Figure 3. Inhibition of both classical and alternative pathways of complement activation by sCR1 (TP10). This inhibitor is a soluble version of the natural human protein CR1; it prevents downstream consequences of complement activation including assemble of the MAC complex, the release of C3a and C5a, and the deposition of C3b. It is not consumed in the process. From ref. 33.

activation. It acts by preventing C3b and C4b from associating with the C3 and C5 convertases (enzymes that activate C3 and C5 complement proteins) and by accelerating their dissociation. It also acts as a cofactor in the proteolysis of activated C3 and C4 by plasma factor I. The mechanism of action of sCR1 is shown in Figure 3. The complement cascade, activated by either the classical pathway or the alternative pathway, would normally proceed to assemble the membrane attack complex and in so doing the analphylatoxins, C3a and C5a, would be released. Since sCR1 prevents the assembly of the membrane attack complex (C5b-9), it is capable of preventing the direct cell and tissue damage caused by C5b-9. In addition it prevents the release of C3a and C5a both of which are pro-inflammatory and lead to increased vascular permeability and leukocyte recruitment. By acting as a co-factor with Factor I, sCR1 inactivates deposited C3b, a mediator of inflammatory cell function. Very conveniently sCR1 is recycled so it is not consumed in the processes. All in all, sCR1 is an extremely powerful molecule capable of inhibiting complement activation by both pathways and of inhibiting the downstream cell and tissue damaging as well as the pro-inflammatory and procoagulant consequences of complement activation.

sCR1 has been tested in a very large number of animal models including animal models of ischemia-reperfusion, lung injury, autoimmune disease, and transplant rejection.

5. sCR1 IN ANIMAL MODELS OF LUNG INJURY

In pulmonary inflammation models, sCR1 protects against acute immune complex-induced lung injury in rats and lung injury resulting from complement activation by CVF.[17] sCR1 protects rats from lung injury and from pulmonary endothelial deposition of C3 fragments and C5b-9 induced by platelet-activating factor/ lipopolysaccharide (PAF/LPS)[8].

When administered 5 and 15 minutes after injury at total doses of 10 to *25* mg/kg and when administered in a single dose of 25 mg/kg up to 30 minutes after injury, TP10 reduces injury to lung and affected skin of thermally injured rats [17,19].

Thus, in IgG or IgA immune complex injury, cobra factor venom injury to the lungs or remote thermal burns sCR1 will markedly reduce the increased vascular permeability. In the scenario of ARDS, two of the most serious problems are increased edema and influx of neutrophils. Again, in each of the models one to four doses of sCR1 markedly reduced the numbers of infiltrating neutrophils[17,19].

6. sCR1 IN PATIENTS WITH ARDS

Although there is no ideal model for ARDS, data from animal studies provided sufficient evidence that sCR1 might be useful to patients with ARDS or at risk of developing ARDS. T Cell Sciences entered into a Phase I trial, to determine the safety of the drug in these patients, to understand its pharmacokinetics in humans, to examine for immunogenicity, if any, and to examine for its ability to inhibit complement in patients. sCR1 was administered in ascending dosages (0.1 to 10 mg/kg) as an i.v. infusion over 30 minutes.

Analysis of trial data showed that for the first time a complement inhibitor therapeutic, sCR1, was capable of inhibiting the complement cascade in patients with acute lung injury. There were no adverse events directly attributed to sCR1 and there was no apparent immunogenicity. Since sCR1 appears to have a favorable safety and pharmacokinetic profile in this group of patients, the next step was to move to an efficacy trial of sCR1 in patients with ARDS. Currently, a Phase II trial is ongoing at multiple sites in the U.S.

7. sCR1 AND REPERFUSION INJURY IN LUNG TRANSPLANTATION

Complement activation is an integral part of the acute inflammatory response to reperfusion injury. There is mounting evidence that ischemia-reperfusion induces complement activation which, injurious in and of itself, additionally results in neutrophil mediated tissue injury. This type of injury to the lung occurs following lung transplantation. Studies have been undertaken in several species and organs showing the advantage of inhibiting complement activation: sCR1 reduced myocardial injury in rats at 24 hours after coronary artery occlusion[20] and reduced myocardial injury in rats 7 days post-reperfusion after coronary artery occlusion[21]. When administered at a dosage of 0.3 to 3.0 mg/kg before reperfusion, TP10 reduced infarct size by 10 to 46% in pigs in a model of coronary artery occlusion[22]. This last study was significant in the context of clinical treatment of myocardial infarction when TPl0 would be administered at the time of reperfusion rather than prior to ischemia.

T Cell Sciences next conducted a Phase I safety trial in patients who had first-time heart attacks (acute myocardial infarction) using a protocol similar to the one for ARDS patients. These patients were scheduled either for thrombolytic therapy or PTCA (balloon angioplasty) both of which mechanisms of revascularization of the coronary arteries lead to complement activation. In addition to the expected ischemia-reperfusion injury, injury to blood vessels by balloon denudation of the endothelium can activate complement and there is further evidence that thrombolytics activate complement through the massive release of plasmin[23]. The results of the trial showed that complement inhibition with sCR1 in heart attack patients who have had a serious reperfusion injury showed a clear safety and pharmacokinetic profile which was well suited for this group of patients.

It also appears that sCR1 reduces ischemia-reperfusion injury in animal lungs[24]. It has recently been demonstrated in rat models of syngeneic lung transplantation that complement inhibition following sCR1 administration improves lung function caused by reperfusion injury, with a concomitant decrease in neutrophil sequestration. Benefit was also seen in improved arterial oxygenation, and improved survival in rats receiving sCR1 as compared to untreated control animals[24].

In addition to its effects on early graft function, ischemia/reperfusion injury may also have an impact on later events such as graft rejection and long term graft survival. It has been suggested that early allograft dysfunction can influence both acute and chronic rejection. Studies with sCR1 in rats and monkeys has demonstrated prolonged organ survival by inhibiting hyperacute rejection of sensitized allograft[25] and xenograft transplants[26–28]. Renal allograft studies in rats demonstrated reductions in vascular injury and cellular infiltrates in the grafts in the number of activated splenocytes[29] and in the development of allospecific antibodies[30].

In pre-clinical studies, sCR1 protects against injuries that result from organ transplantation, ischemia/reperfusion, or acute lung injury (ALI) induced by immune-complexes, cobra venom factor, IL-2, or LPS/PAF. sCR1 has been shown to prolong organ survival or improve organ function in: isograft lung transplants in rats,[24] allogeneic orthotopic renal transplants in rats,[29] xenogeneic heterotopic cardiac transplants from guinea pigs to rats[26] and from pigs to primates[27,28]. With respect to chronic problems in lung allografts there is now preliminary evidence that sCR1 can significantly suppress the development of experimental obliterative bronchiolitis in a rat tracheal allograft model. This experimental evidence further supports a role for the complement system in this disease process[31].

T Cell Sciences is now conducting a randomized, placebo-controlled, double-blinded, single dose, multi-center, Phase I/II trial of sCR1 in patients with end-stage pulmonary disease undergoing lung transplantation surgery. The trial will determine the ability of sCR1 to reduce reperfusion injury and improve lung function following transplantation. Patients will be followed acutely for 28 days after therapy and then automatically enrolled in a 6-month follow-up study of morbidity and mortality after transplantation.

CONCLUDING COMMENTS

There is a huge amount of work to be done on the etiology, mechanisms, and course of endothelial injury in the lungs by whatever cause, but therapies are now emerging that can counteract some of the symptoms. The remaining challenge will be to see if one can in fact reduce or reverse the causes of injury and it will be interesting to see what role complement inhibitors can play. One message remains, the endothelium is the chief target of injury, and one of the worst culprits in damaging the endothelium is complement activation.

REFERENCES

1. Ashbaugh DG, Bigelow DB, Petty TL, and Levine BE. Acute respiratory distress in adults. Lancet 1967; ii: 319–323.
2. Hyers TM. Prediction of survival and mortality in patients with adult respiratory distress syndrome. New Horizons 1993; 1:466–470.

3. Suchyta MR, Clemmer TP, Elliott CG, Orme JF, Jr., and Weaver LK. The adult respiratory distress syndrome: a report of survival and modifying factors. Chest 1992; 101:74–79.
4. Artigas A, Calet J, Legall JR, et. al. Clinical presentation, prognostic factors and outcome of ARDS in the European Collaborative study (1985–1987). A preliminary report. In: Zapol W, LeMaire F, editors. Adult Respiratory Distress Syndrome, New York: Marcell Dekker, 1991:37–64.
5. Wheeler AP, Carroll FE, Bernard GR. Radiographic issues in adult respiratory distress syndrome. New Horizons 1993; 1:471–477.
6. Ryan US. Structural bases for metabolic activity. Ann Rev Physiol 1982; 44:223–239.
7. Ryan US, Ryan JW. Vital and functional activities of endothelial cells. In: Nossel HL, Vogel HJ, editors. Pathobiology of the Endothelial Cell. New York: Academic Press, 1982:455–469.
8. Ryan US, Ryan JW, Whitaker C, and Chiu A. Localization of angiotensin converting enzyme (kininase II): immunocytochemistry and immunofluorescence. Tissue & Cell 1976; 8:125–146.
9. Crutchley DJ, Ryan JW, Ryan US, and Fisher GH. Bradykinin-induced release of prostacyclin and thromboxanes from bovine pulmonary artery endothelial cells. Studies with lower homologs and calcium antagonists. Biochem Biophys Acta 1983; 751:99–107.
10. Furchgott RF and Zawadzki JV. The obligatory role of endothelial cells in the relaxation of arterial smooth muscle by acetycholine. Nature 1980; 373:299.
11. Yanagisawa M, Kurihara H, Kimura S, et al. A novel potent vasoconstrictor peptide produced by vascular endothelial cells. Nature 1988; 332:411–415.
12. American College of Chest Physicians/Society of Critical Care Medicine Consensus Conference. Definitions for sepsis and organ failure and guidelines for the use of innovative therapies in sepsis. Crit Care Med 1992; 20:864–874.
13. Zilow G, Joka T, Obertacke U, Rother U, and Kirschfink M. Generation of anaphylatoxin C3a in plasma and bronchoalveolar lavage fluid in trauma patients at risk for the adult respiratory distress syndrome. Crit Care Med 1992; 20:468–473.
14. Hamilton KK, Hattori R, Esmon CT, and Sims PJ. Complement proteins C5b-9 induce vesiculation of the endothelial plasma membrane and expose catalytic surface for assembly of the prothrombinase enzyme complex. J Biol Chem 1990; 265:3809–3814.
15. Halperin JA, Taratuska A, Nicholson-Weller A. Terminal complement complex C5b-9 stimulates mitogenesis in 3T3 cells. J Clin Invest 1993; 91:1974–1978.
16. Benzaquen LR, Nicholson-Weller A, and Halperin JA. Terminal complement proteins C5b-9 release basic fibroblast growth factor and platelet-derived growth factor from endothelial cells. J Exp Med 1994; 179:985–992.
17. Mulligan MS, Yeh CG, Rudolph AR, and Ward PA. Protective effects of soluble CR1 in complement-and neutrophil-mediated tissue injury. J Immunol 1992; 148:1479–1485.
18. Rabinovici R., Yeh CG, Hillegass LM, Griswold DC, DiMartino MJ, Vernick J, et al. Role of complement in endotoxin/platelet-activating factor-induced lung injury. J Immunol 1992; 149:1744–1750.
19. DiMartino MJ, Wolfe CE, Slivjak MJ, Minthorn EA, Feuerstein G. Effects of soluble complement receptor (sCR1, BRL55730) on thermal skin injury induced hemoconcentration and lung inflammation in rats. Pharmacol Commun 1993; 3:249–256.
20. Weisman HF, Bartow T, Leppo MK, et. al. Soluble human complement receptor type 1: In vivo inhibitor of complement suppressing post-ischemic myocardial inflammation and necrosis. Science 1990; 249:146–151.
21. Smith EF, III, Griswold DE, Egan JW, Hillegass LM, Smith RAG, Hibbs MJ, et al. Reduction of myocardial reperfusion injury with soluble complement receptor I (BRL 55730). Eur J Pharmacol 1993; 236:477–481.
22. Dupe RJ, Goddard ME, Freeman AM, Hibbs MJ, Lifter J, Mossakowska DE, et al. Utility of complement inhibition during myocardial reperfusion: Pharmacology of soluble complement receptor 1. 13th Congress of the International Society of Thrombosis and Haemostasis. Thrombos Haemostasis 1991; 65:695
23. Schaiff WT and Eisenberg PR. Pharmacologic activation of plasminogen directly induces and enhances complement activation. Suppl. to Circulation 1995; 92:I-342 Abstract.
24. Naka Y, Roy DK, Marsh HC, et. al. Protective effects of complement blockade in an isograft model of lung preservation and transplantation. Am Col of Cardiology 45th Annual Scientific Symposium 1996; Abstract.
25. Pruitt SK and Bollinger RR. The effect of soluble complement receptor type I on hyperacute allograft rejection. J Surg Res 1991; 50:350–355.
26. Pruitt SK, Baldwin WM, Marsh HC, Jr., Lin SS, Yeh CG, Bollinger RR. The effect of soluble complement receptor type I on hyperacute xenograft rejection. Transplantation 1991; 52:868–873.
27. Pruitt SK, Kirk AD, Bollinger RR, Marsh HC, Jr., Collins BH, Levin JL, et al. The effect of soluble complement receptor type I on hyperacute rejection of porcine xenografts. Transplantation 1994; 57:363–370.

28. Davis EA, Pruitt SK, Greene PS, Ibrahim S, Lam TT, Levin JL, et al. Inhibition of complement, evoked antibody, and cellular response prevents rejection of pig-to-primate cardiac xenografts. Transplantation 1996; 62:1018–1023.
29. Pratt JR, Hibbs MJ, Laver AJ, Smith RAG, Sacks SH. Allograft immune response with sCR1 intervention. Transplant Immunol 1996; 4:72–75.
30. Pratt JR, Harmer AW, Smith RAG, and Sacks SH. Influence of complement inhibition with soluble complement receptor (sCR1) on the B cell response in experimental allograft rejection. Transplantation Society 1996; Barcelona, Spain: Abstract.
31. Kallio E, Koskinen P, Krebs R, Ryan U, Hayry P, and Lemstrom K. Soluble recombinant complement receptor type 1 significantly reduces the development of experimental obliterative bronchiolitis in rat tracheal allografts. Int Transplantation Society 1996; Abstract.
32. Ryan US. Complement inhibitory therapeutics and xenotransplantation. Nature Medicine 1995; 1:967–968.
33. Ryan US. Complement inhibition: the sine qua non of xenotransplantation? Xeno 1994; 2:19–22.

II. GENE MANIPULATION IN THE VASCULATURE

4

CONTROL OF iNOS EXPRESSION IN RAT AORTIC SMOOTH MUSCLE CELLS

John D. Catravas[1] and Nandor Marczin[2]

[1]Vascular Biology Center
Medical College of Georgia
Augusta, Georgià 30912
[2]Health Science Center
Imperial College
Harefield Hospital
Harefield, Middlesex, United Kingdom

1. INTRODUCTION

Beyond endothelial NO production through ecNOS (type III NOS), there is increasing evidence for a different mechanism of NO formation via the inducible NO synthase (iNOS or type II NOS) in a variety of cells, including vascular endothelial and smooth muscle cells.[11,3,1,2,6,27] This enzyme also uses L-arginine as substrate and is sensitive to inhibitory L-arginine analogues; furthermore, it does not appear to differ significantly from ecNOS on the basis of cofactor requirements.[17] It can be distinguished, however, by the major characteristics of its activation. In contrast to the endothelial, rapidly responsive constitutive pathway, the activity of the inducible pathway is generally calcium and calmodulin independent, slow in onset, occurs after a delay of several hours, it is transcriptionally regulated and sensitive to dexamethasone.[31]

Since iNOS lacks calcium regulation, it will produce NO continuously for the lifetime of the enzyme once it is induced in the vasculature.[17] This "malignant mechanism" results in overproduction of NO, persistent stimulation of soluble guanyl cyclase (sGC) causing prolonged smooth muscle relaxation. Bacterial lipopolysaccharide, as well as cytokines, IL-1 β and TNF inhibit agonist-induced contractions of isolated vascular preparations in an endothelium-independent manner.[11,3,1,2,6,27] This impairment of contractility is associated with activation of sGC and increased production of cGMP. The clinical finding of increased serum levels of nitrite/nitrate, the stable degradation products of NO, in patients with sepsis, coupled with the ability of NO inhibitors to inhibit endotoxin and cytokine induced hypotension in animal models of sepsis, and in patients with sepsis syndrome, suggest that activation of the iNOS pathway is a major factor responsible for reduced contractile responsiveness of vessels and hypotension associated with septic shock.[16,21,32,23,14,19]

Vascular Endothelium, edited by Catravas *et al.*
Plenum Press, New York, 1998

In an effort to identify regulatory steps unique to the iNOS isozyme, we have studied certain aspects of iNOS induction. We now present data suggesting that protein tyrosine kinase inhibitors and microtubule depolymerizing agents selectively prevent the type II NOS induction, without affecting the function of the other two isozymes.

2. METHODS

Endothelial and Smooth Muscle Cell Culture

Endothelial cells (EC) from bovine pulmonary artery (BPAE) or from bovine aorta (BAE) were harvested non-enzymatically and identified utilizing previously published procedures.[24] Endothelial cells were grown as monolayers in T-75 tissue culture flasks (Corning, NY) in Medium 199 (Mediatech, Washington, DC), supplemented with 10% fetal bovine serum (Hyclone Laboratories, Logan, UT), penicillin and streptomycin. Culture Medium was changed twice weekly. The cells were mechanically subcultured, using a rubber policeman for up to 30 times. Each subculture of endothelial cells was monitored for (a) morphology, using phase-contrast microscopy, (b) size and (c) angiotensin converting enzyme activity, using ^{3}H-benzoyl–Phe–Ala–Pro as the substrate.[37]

Aortic smooth muscle cells (RASM) from Wistar rats (Harlan) were isolated by ezymatic dissociation using standard methods.[13] The cells were positively identified as smooth muscle by indirect immunofluorescent staining for α-actin, using mouse anti-α-actin antibody and anti-mouse IgG FITC conjugate. RASM cells grown in T-75 tissue culture flasks (Corning, Corning, NY) in 50% F12 nutrient medium and 50% Dulbecco's Modified Eagle Medium (GIBCO Laboratories, Grand Island, NY), supplemented with 10% fetal bovine serum, glutamate, penicillin (Sigma, St. Louis, MO, 10,000 U/l) and streptomycin were subcultured into 24-well tissue culture plates (Costar, Cambridge, MA) at 50,000/cm^2 and reached confluence after 5–7 days. All cultures were allowed to grow at 37°C under 5% CO_2, in air. Experiments were performed between passages 5–24 and at 3–6 days post confluence.

cGMP and cAMP Radioimmunoassay

Monoclonal antibody against cGMP was a generous gift from Dr. Ferid Murad (University of Texas at Houston). Antibody for cAMP was provided by Dr. G. Brooker, Georgetown University. The radioligands (^{125}I–succinyl cGMP–tyrosine methyl ester and ^{125}I-succinyl cAMP-tyrosine methyl ester) were prepared in our laboratory. Stock solutions of the succinyl tyrosine methyl esters of cGMP and cAMP (Sigma) were made up in 50 mM sodium acetate buffer, pH 4.75 and iodinated by the method of Hunter and Greenwood[38] using carrier free ^{125}I (Du Pont, NEN, Boston, MA). The iodination reaction products were separated by reverse-phase high-performance liquid chromatography according to Patel and Linden.[39] Radioimmunoassay was performed using the Gammaflow automated RIA system.[33] Standard stock solutions of cGMP and cAMP (20 mM) were prepared in 0.1N HCl, and the absorbance of the solution was routinely monitored spectrophotometrically (Shimadzu, UV 160U). Standard dilutions (0.63–80 nM) were made from the stock solution. The HCl extract containing cGMP was used for RIA directly whereas cAMP samples were first acetylated using triethylamine and acetic anhydride (Aldrich Chemical Comp., Milwaukee, WI) before assay.

Protein Determination

Protein content of the supernatant of the centrifuged (2,000 rpm for 5 min at room temperature) NaOH solubilized samples was measured by the Bradford method.[4] Sample aliquots were combined with the protein binding dye (BIO-RAD, Richmond, CA) and optical density determined at 630 nm using a multiwell plate reader (Dynatech Laboratories Inc.). Bovine albumin, fraction V (Sigma) was used as standard.

Data Analysis

Unless otherwise indicated, data are presented as means ± SEM of four individual cultures from representative experiments of two to three. Data are presented as means ± SEM of the indicated number of individual cultures. Data are expressed either as actual values of CGMP in pmol/mg protein/15 min or as a percent of the control value. Statistical comparisons between groups were performed using the one way ANOVA followed by appropriate post hoc tests. Differences among means were considered significant when $P < 0.05$.

3. PROTEIN TYROSINE KINASE INHIBITION

Results

Protein phosphorylation on tyrosine residues is a widespread signaling mechanism in response to growth factors, oncogenes and cytokines.[7] IL-1 and prooxidant conditions such as exposure to UV light, reagent $H_2 0_2$ and diamide also activate tyrosine kinases.[20,34,35] Coyne and Morrison have demonstrated that genistein blocks IL-1-induced prostaglandin production by mesangial cells.[36] The critical contribution of the tyrosine kinase pathway to iNOS expression has been implicated in the IL-1 induced inhibition of insulin secretion in islet cells. Corbett et al suggested that the tyrosine kinase inhibitor genistein acts by inhibiting NO formation.[28] In fact they demonstrated that genistein inhibited IL-1-induced nitrite accumulation in an insulinoma cell line. On the basis of the above studies, we hypothesized that LPS and cytokines may induce iNOS through a tyrosine kinase mediated pathway in vascular smooth muscle cells. Since Di Salvo et al. found large quantities of tyrosine kinase in smooth muscle preparations and they also showed that both genistein and geldanamycin selectively inhibited tyrosine kinases in vascular smooth muscle cells,[29] we set out to investigate the effects of these inhibitors on LPS and IL-1-induced CGMP accumulation in RASM cells.

A 3 hour exposure of RASM cells to 10 ng/ml LPS resulted in 40 fold increases in CGMP levels in RASM cells. Similarly to LPS, IL-1 (5 U/ml) also elicited about 20 fold increases in intracellular concentrations of cGMP(Figure 1). Although 30 min pretreatment of the cells with increasing concentrations of genistein did not affect baseline CGMP levels, it inhibited IL-1-induced formation of cGMP. Although LPS-induced cGMP was virtually unaffected by lower concentrations of genistein, the agent inhibited cGMP accumulation at 50 and 100 μM (41 ± 4 and 14.4 ± 0.5 % of control cGMP, respectively)(Figure 1,Top). Another chemically distinct tyrosine kinase inhibitor, geldanamycin also inhibited both IL-1 and LPS-elicited CGMP responses, without affecting baseline cGMP levels in the absence of endotoxin and cytokine. Although the inhibition of cGMP accumulation was evident at 50 ng/ml concentration of geldanamycin, complete inhibition was not obtained even at 20 fold higher concentration. (Figure 1, Bottom)

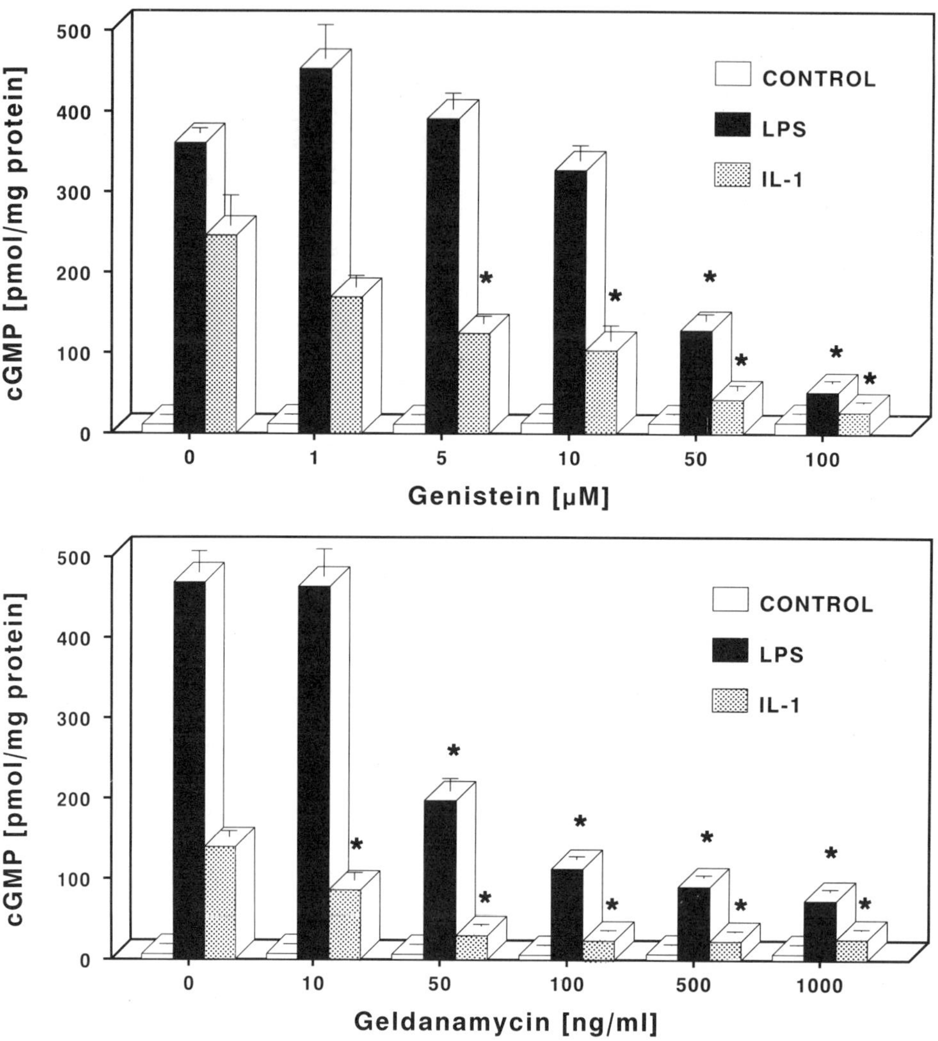

Figure 1. Effects of genistein and geldanamycin on either basal or interleukin-1β (IL-1β) and lipopolysaccharide (LPS)-induced guanosine 3′,5′-cyclic monophosphate (cGMP) formation in rat aortic smooth muscle cells. Cells were pretreated with vehicle (0.1% dimethyl sulfoxide) or with increasing concentrations of genistein (Top) or geldanamycin (Bottom) for 30 min, and then 5 U/ml IL-1β or 10 ng/ml LPS were added and the cells were incubated for an additional 3h in the continuous presence or absence of genistein or geldanamycin. Data are means ± SE of 4 cultures. *$P < 0.05$ from cGMP levels in control cultures in absence of inhibitors.

To investigate the effects of protein tyrosine kinase inhibitors on sGC itself, we investigated the responsiveness of RASM cells to vasodilators either included exogenously or derived endogenously from endothelial cells in response to endothelium-dependent vasodilators. Neither genistein nor geldanamycin affected cGMP accumulation in response to sodium nitroprusside (1 μM) over the same time frame and concentrations that significantly reduced cGMP accumulation in response to endotoxin and IL-1. For instance, sodium nitroprusside produced 675 ± 24 pmol cGMP/mg protein/15 min in the presence of 100 μM genistein, a value not significantly different from cGMP accumulation in the absence of genistein (639 ± 60 pmol cGMP/mg protein/15 min). Similarly,

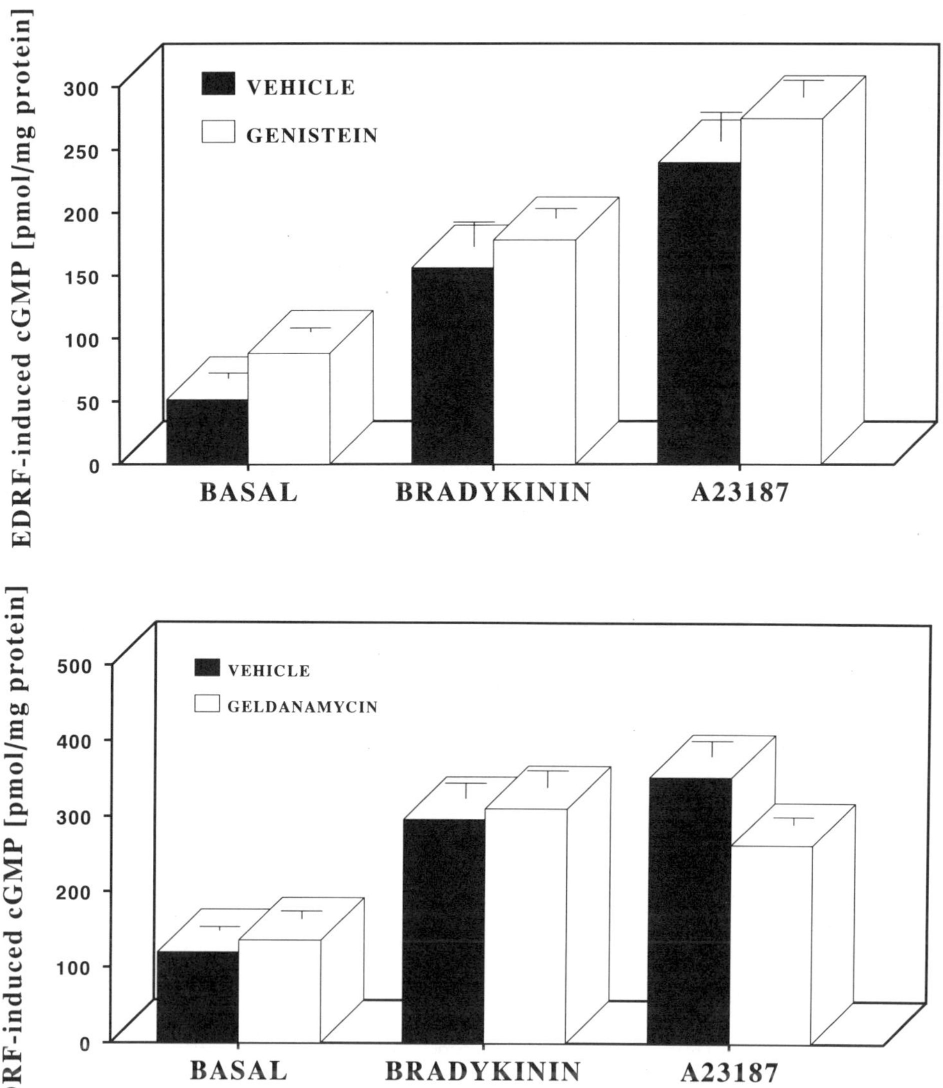

Figure 2. Effects of genistein and geldanamycin on endothelium-derived relaxing factor (EDRF; nitric oxide)-induced cGMP fromation in cocultures of rat aortic smooth muscle cells and bovine aortic endothelial cells. Cells were pretreated with vehicle (0.1% dimethyl sulfoxide) or with either genistein (100 μM, Top) or geldanamycin (1,000 ng/ml, Bottom) for 30 min before determination of cGMP formation either during control conditions or on exposure to bradykinin (10 μM) or calcium ionophore (A23187, 1 μM). nitro-L-arginine methyl ester (l-NAME)-inhibitable cGMP accumulation is plotted with and without tyrosine kinase inhibitors. Data are means ± SE of 4 cultures.

sodium nitroprussside was capable of stimulating guanylate cyclase in the presence of 1000 ng/ml geldanamycin (95.5 ± 4.5 vs. 100 ± 4.6 % of control cGMP).

Responsiveness of RASM cells to endothelium-derived NO was investigated in co-cultures of RASM and BAE cells in the presence or absence of 30 min preincubation with either genistein or geldanamycin (Figure 2). In these co-cultures, high basal levels of cGMP content of 92 ± 4 pmol cGMP/mg protein was observed, suggesting basal release of NO. Exposure of co-cultures to endothellum-dependent vasodilator bradykinin or calcium ionophore further stimulated cGMP accumulation (187 ± 7.9 and 280 ± 13 pmol cGMP/mg protein, respectively) but this stimulation was lost in the presence of LNAME (8.9 ± 1.4 and 5.8 ± 1.4 pmol/mg protein, respectively). Preincubation of the co-cultured cells with 100 μM genistein (Figure 2, Top) or 1000 ng/ml geldanamycin (Figure 2, Bottom) had virtually no effect on cGMP accumulation elicited by endothelium-derived NO released basally or upon stimulation with bradykinin or calcium ionophore.

To elucidate the time course of inhibition by genistein and geldanamycin of LPS-stimulated cGMP accumulation, RASM cells were exposed to LPS for 3 hours and the experimental agents were added at different time points relative to the initiation of LPS stimulation. A similar time course of inhibition was obtained with both genistein and geldanamycin. These agents were capable of inhibiting cGMP accumulation if included during the first 90 minutes of LPS incubation. Their inhibitory potential was lost, however, if they were added after a delay of 90 minutes.[46]

Discussion

Similarly to the results of Corbett et al.[40] with insulinoma cells, genistein also prevented NO formation in response to IL-1 in vascular smooth muscle cells. We extended those results and show that 1) a chemically distinct tyrosine kinase inhibitor geldanamycin also produces the same effect and 2) that these agents are not inhibitors of NO synthases, but specifically interfere with the signaling process to induce iNOS by IL-1. We also suggest that this regulatory mechanism is not restricted to IL-1 alone, but appears to function during endotoxin signaling, as well. The present study thus suggests that these enzymes may be involved in both interleukin and endotoxin-induced signaling mechanisms to induce iNOS in vascular smooth muscle cells.

4. CYTOSKELETAL REGULATION

The cytoskeletal system plays a pivotal role in the regulation of cell shape, intracellular transport, secretion and signal transduction.[30,45] Furthermore, there is evidence for the influence of the cytoskeletal architecture on gene activation. Alterations in the microfilament structure stimulates the expression of certain proteolytic enzymes in fibroblasts and LLC-PKI cells.[41,42] Disassembly of microtubules enhances the mitogenic activity of growth factors[22] and prevents gene expression of positive acute phase proteins in hepatocytes.[8] Moreover, microtubules are involved in certain actions of endotoxin in macrophages, such as in the release of TNF-α and in the inhibition of TNF-α binding.[10,11]

In light of this relationship between the organization of the filamentous apparatus of the cytoplasm and activation of certain genes, we have investigated the role of microfilaments and microtubules, two major components of the cytoskeleton, in the activation of NO pathways in vascular cells. Here we show evidence that microtubules play an essential role in mediating endotoxin- and interleukin- inducible NO-responses in smooth muscle cells but not constitutive release of NO in endothelial cells.

Experimental Design

In the first series of experiments we modulated the assembly of microtubules and microfilaments and monitored the effects of the various treatment protocols on the structural organization of the cytoskeleton. Assembly of cytoskeleton after experimental protocols was assessed by immunofluorescence as well as biochemically by quantifying tubulin content in the detergent resistant cellular compartment. Next, we investigated the effects of cytoskeletal alterations on both the endotoxin and cytokine-inducible NO pathway in RASM cells and on the activation of the constitutive NO pathway in endothelial cells by measuring NOS activity and steady state iNOS mRNA levels.

Methods

Modulation of Microtubule and Micrfilament Assembly. Microtubule polymerization was modulated by either a depolymerization protocol using nocodazole or colchicine, or by promoting microtubule assembly with taxol, a plant-derived drug that has been reported to enhance microtubule assembly and stabilize microtubules against depolymerizing conditions.[12,18,26] To depolymerize microtubules, RASM cells or RASM/BAE co-cultures were chilled to 4°C and incubated with DMSO as vehicle or were exposed to various concentrations of nocodazole or colchicine for 90 min at 4°C according to published procedures[5]. Cells were then rewarmed at 37°C in the continued presence or absence of the agents for 30 min. The experiment of interest was then performed in the continued presence or absence of nocodazole or colchicine. The effects of taxol on microtubule polymerization was investigated by exposing the cells to taxol alone or in combination with nocodazole or colchicine. To depolymerize microfilaments, RASM cells, or RASM-BAE co-cultures were exposed to cytochalasin D or to cytochalasin B for 60 min at 37°C prior to stimulation of NO responses. To visualize microtubules, cells grown and treated on glass coverslips were fixed with 4% formaldehyde, permeabilized with 0.2% Triton X-100 and processed for anti-tubulin indirect immunofluorescence using a mouse monoclonal antibody (TU-27) to b-tubulin and rhodamine labeled polyclonal anti-mouse IgG (DACO corporation). Actin filaments were stained by treating the coverslips with fluorescein-labeled phalloidin (1 μg/ml, Sigma). Images were taken with a NU-200 series cooled CCD camera and relayed to a Macintosh computer. Digital enhancement and gray-scale normalization of 16 bit images were performed identically for each sample. To quantify cytoskeletal tubulin content, cells were exposed to microtubule modulating agents and, after the appropriate incubation, native cells were permeabilized with 0.5% Triton X-100. Detergent resistant material was then solubulized in sample buffer and 25 μl aliquots were loaded for gel electrophoresis and Western blot for b-tubulin.

Results

Cytoskeletal Changes Induced by Nocodazole, Colchicine and Taxol in RASM Cells. In vehicle-treated cells, microtubules were abundant and frequently formed bundles. Cells treated with either nocodazole or colchicine, however, were largely depleted from microtubules, and only a few solitary fragments of microtubules were seen. Taxol alone significantly promoted microtubule assembly in RASM cells, as characteristic dense bundles of microtubules were seen following taxol treatment. Taxol not only promoted microtubule polymerization but also prevented microtubule depolymerization induced either by nocodazole or colchicine. In the presence of both taxol and nocodazole or taxol and colchicine, the microtubule staining was similar to that in control cells in the absence of nocodazole or colchicine. Staining of actin filaments with fluoresceinated phalloidin revealed that microfilament structures remained unaltered after treatment with nocodazole, colchicine, taxol or with the combination of taxol and colchicine when compared to untreated control. In resting RASM cells, about 30% of total tubulin was found in the detergent resistant compartment. Treatment of RASM cells with colchicine reduced cytoskeletal tubulin whereas taxol caused a large increase in tubulin levels in this compartment. Furthermore, in the presence of taxol, colchicine was unable to affect tubulin levels.[47]

Effects of Microtubule Depolymerizing Agents on the Induction of NO Pathway in RASM Cells. Nocodazole treatment, although having no effect on basal cGMP levels, completely prevented LPS responses. Similar treatment with nocodazole, on the other hand, had virtually no effect on the cGMP increase elicited by the NO-releasing sodium nitroprusside. This suggests that the effect of nocodazole was not related to a direct action on guanylate cyclase activity. Similarly to nocodazole, exposure of RASM cells to colchicine also diminished LPS-induced cGMP accumulation, with no effect on that elicited by sodium nitroprusside. We also investigated the role of microtubules in the activation of the NO-pathway in response to interleukin-1 The IL-1 induction of cGMP accumulation (31 ± 3 fold increase) was also inhibited by both nocodazole and colchicine. (Figure 3)

Effects of Taxol on the Induction of NO Pathway in RASM Cells. Although taxol alone had no effect on basal cGMP levels and on those elicited by maximal effective concentrations of either LPS (1 μg/ml) or IL-1 (10 U/ml), it significantly stimulated cGMP accumulation induced by lower concentrations of LPS ($\leq$ 1 ng/ml) in the presence of serum. In addition, taxol also potentiated the relatively low cGMP formation elicited by 10 μM LPS under serum-free conditions (206 ± 23 92 vs. 137 ± 7 pmol/mg protein). These results suggest that promoting microtubule polymerization does not modulate cGMP formation alone, but it might increase the sensitivity of RASM for LPS. Taxol (10 μM) fully protected LPS-induced cGMP production against the action of 1 μM nocodazole and partially protected against 10 μM nocodazole. Moreover, taxol provided full protection against colchicine inhibition of IL-1-induced cGMP formation in RASM cells. (Figure 4)

Effects of Microtubule Modulating Agents on iNOS mRNA Steady State Levels. There was no iNOS mRNA signal detected in resting RASM cells, however stimulation with LPS for 3 hours induced the transcription of the iNOS gene. Microtubule depolymerization by either colchicine or nocodazole appeared to reduce the iNOS mRNA signals. Since the reference "housekeeping" GAPDH signals were even stronger in these cases, nocodazole and colchicine attenuate steady state levels of iNOS message. Although taxol had only a minor effect alone, its administration 30 min prior to the microtubule depolymerization protocol prevented the reduction of iNOS mRNA by both colchicine and nocodazole.

Delayed Effects of Microtubule Depolymerization on the Induction of NO Pathway in RASM Cells. To further characterize the role of microtubules in the transduction of immunmodulatory signals, we identified two phases of LPS-induced cGMP accumulation. RASM cells were allowed to interact with LPS for 15 min only, followed by an additional 3h incubation in the absence of LPS. Even under these conditions, LPS produced a 5–10 fold increase in cGMP levels, suggesting that the continuous presence of LPS is not required to stimulate cGMP generation. Thus two phases of LPS-action can be identified, a short "priming" phase in the presence of LPS and a subsequent "induction" period in the absence of LPS. To investigate the requirement for microtubules during the "induction' phase of LPS action, we depolymerized microtubules at different time points after LPS exposure. Exposure of RASM to either nocodazole or colchicine immediately after the 15 min LPS-treatment, significantly inhibited cGMP accumulation. However, when the mi-

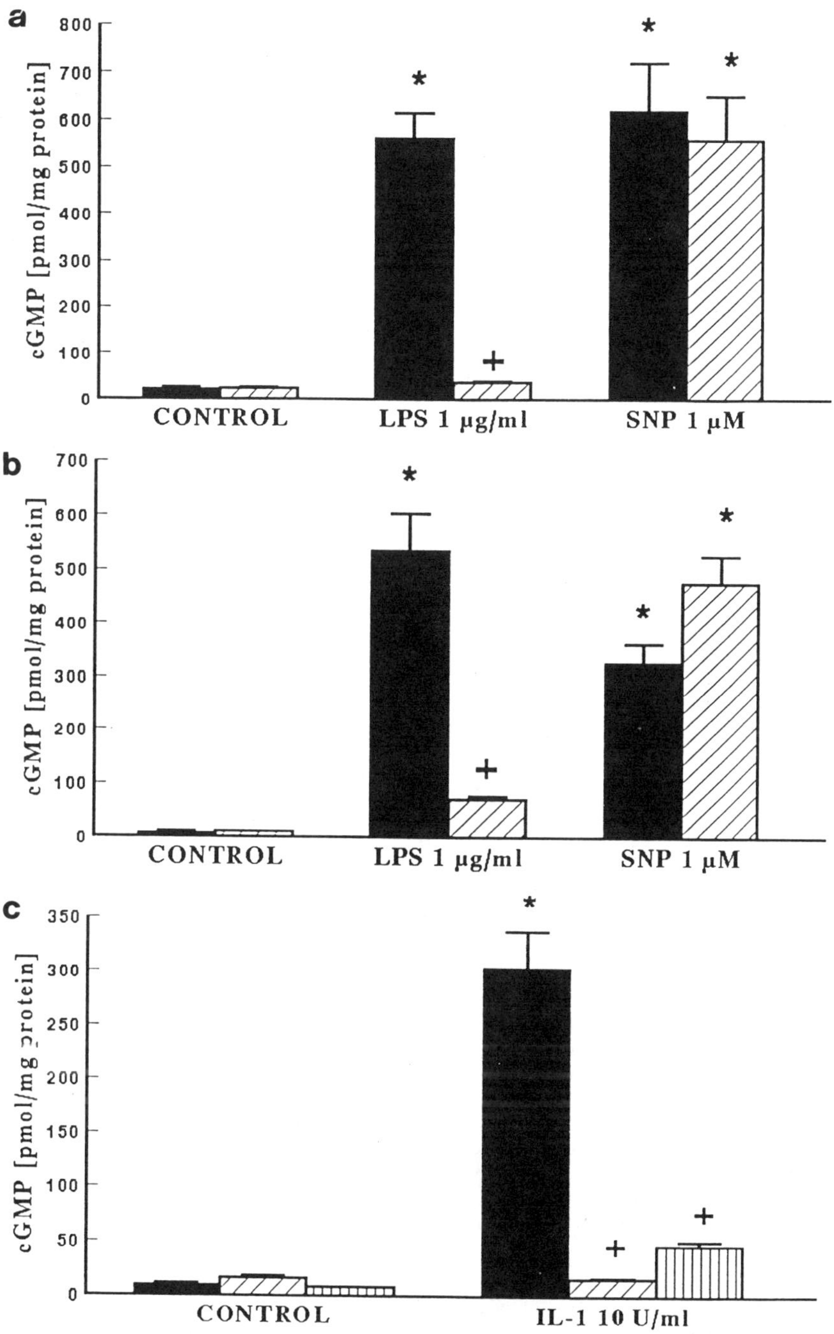

Figure 3. Effects of microtubule disassembly on the induction of cyclic GMP accumulation in cultured rat aortic smooth muscle cells. Effects of nocodazole (33 μM, **a**, hatched bars) or colchicine (5 μM, **b**, hatched bars) on LPS (1 μg/ml) and SNP (1 μM) induced cyclic GMP levels. **(c)** Effects of nocodazole (diagonally hatched bars) or colchicine (vertically hatched bars) on interleukin-1β (IL-1, 10 U/ml) induced cyclic GMP accumulation. All cells were chilled to 4°C and incubated in the absence (VEHICLE, 0.2% DMSO, black bars) or in the presence of nocodazole or colchicine for 90 min. Cells were then warmed to 37°C in the continued absence or presence of nocodazole or colchicine for 30 min, prior to stimulation with or without LPS or IL-1 for 3 hours. *P:<0.05 from cyclic GMP levels in appropriate CONTROL cultures. ⁺P:<0.05 from cyclic GMP levels in appropriate VEHICLE cultures. Data shown are means ± SE of 4–6 cultures.

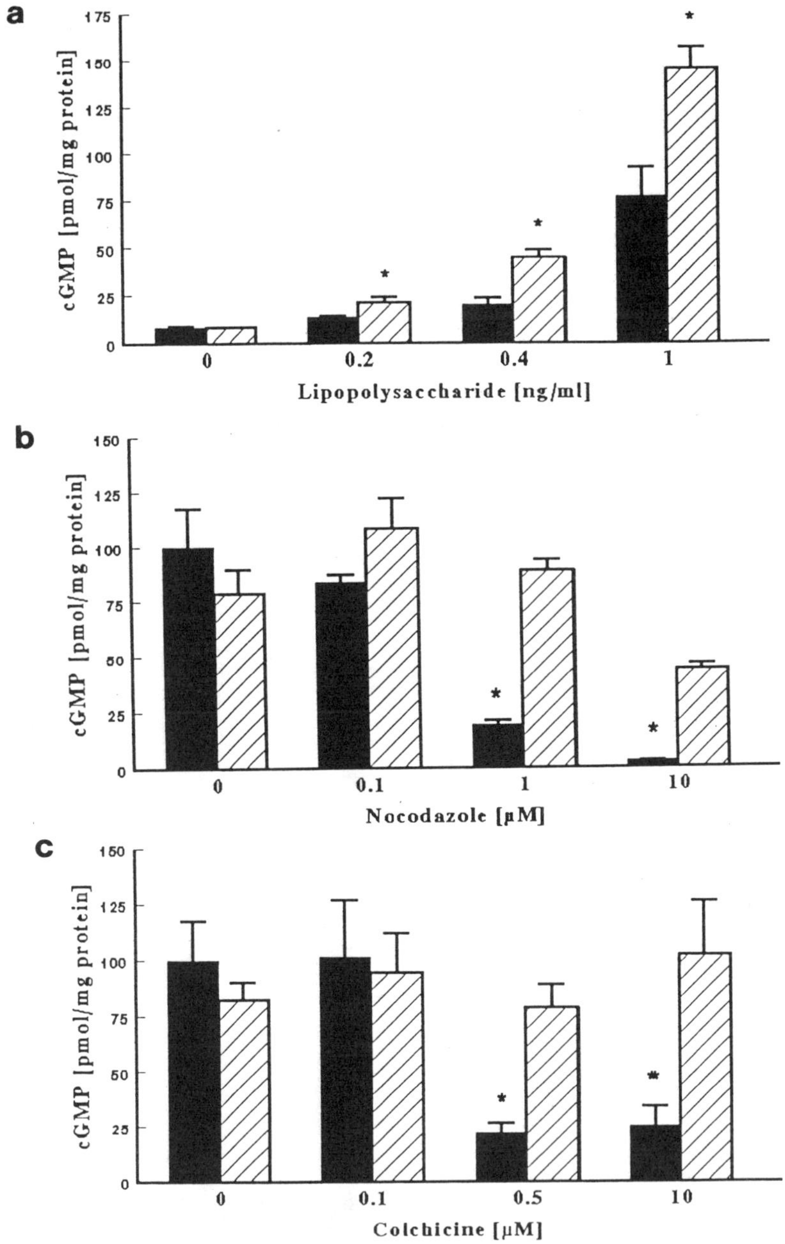

Figure 4. Effects of microtubule stabilization by taxol on the induction of cyclic GMP accumulation in cultured rat aortic smooth muscle cells. (**a**) Effects of taxol (10 µM) on LPS-induced cyclic GMP accumulation. *P<0.05 from cyclic GMP levels in appropriate VEHICLE cultures. (**b**) Effect of taxol on nocodazole-mediated inhibition of LPS-(1 µg/ml) induced cyclic GMP accumulation. *P<0.05 from cyclic GMP levels in VEHICLE cultures in the absence of nocodazole. (**c**) effects of taxol on colchicine-mediated inhibition of IL-1 (10 U/ml) induced cyclic GMP accumulation. *P:<0.05 from cyclic GMP levels in VEHICLE cultures in the absence of colchicine. Cells were pretreated with DMSO as vehicle (black bars) or with taxol (hatched bars) for 30 min prior to the microtubule disassembly protocol using either colchicine or nocdazole prior to LPS challenge for 3 hr. Data shown are means ± SE of 4 individual cultures.

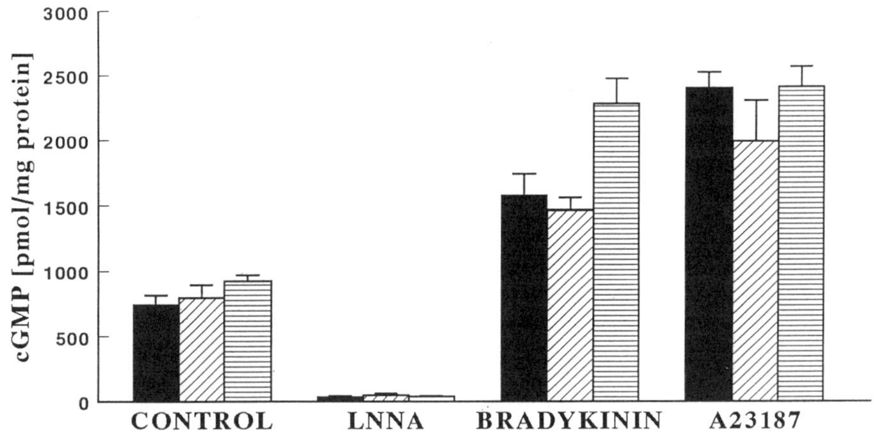

Figure 5. Effects of microtubule modulating agents on basal and stimulated release of NO from endothelial cells. BAE-RASM co-cultures were chilled to 4°C and incubated in the absence (VEHICLE, 0.2% DMSO, black bars) or in the presence of nocodazole (diagonally hatched bars) or taxol (horizontally hatched bars) for 90 min. Cells were then warmed to 37°C in the continued absence or presence of these agents for 30 min, prior to determination of cyclic GMP accumulation in the presence of IBMX and L-arginine (1 mM for 15 min). Accumulation of cyclic GMP was determined either under basal conditions (VEHICLE) or in the presence bradykinin (10 μM, 10 min) or calcium ionophore (A23187, 10 μM, 10 min). LNNA was included 30 min prior to and during IBMX incubations.

crotubule disassembly protocol was initiated 4 hours after LPS exposure, LPS-induced cGMP levels were not affected by nocodazole or colchicine.

Effects of Microtubule Depolymerization on the Activation of the Constitutive NO Pathway in BAE-RASM Cultures. To detect basal or stimulated release of NO from BAE cells, BAE-RASM co-cultures were exposed to vehicle or to the endothelium-dependent vasodilator bradykinin or calcium ionophore, or potent inhibitor of the constitutive NO synthase, N^G-nitro-L-arginine (LNNA). None of these agents affected cGMP accumulation in single cultures of either RASM or BAE. In co-culture, however, the high basal cGMP levels of 747 ± 68 pmol/mg protein were inhibited by 95% by LNMMA, suggesting basal release of NO from BAE. Inclusion of bradykinin or A23187 (both at 10 μM) produced a further 2 and 3 fold increase in cGMP from basal levels suggesting calcium dependence of these responses. Nocodazole or taxol had no effect on either basal or stimulated NO release in these co-cultures. (Figure 5)

Effects of Cytochalasins on the Cytoskeleton and on the Induction of NO Pathway in RASM Cells. Cytochalasins produced dramatic alterations of actin staining, by completely eliminating stress-fiber like structures. The small amounts of remaining polymer actin (i.e. actin that is capable of phalloidin binding) was seen as small patches of fluorescence without any clear orientation. Although cell shape was dramatically altered by cytochalasin treatment, thus microtubule morphology was not without any change, still a large number of bundled microtubules were seen. Cytochalasin B produced similar morphological changes, although to a lesser extent. Exposure of RASM cells to either cytochalasin B or cytochalasin D (both at 10 μM) had no effects on basal cGMP levels. Similarly, they did not modulate cGMP accumulation in response to 1 μM sodium nitroprusside (51.2 ± 4.3, 56.0 ± 5.2 and 57.1 ± 2.7 fold increase for vehicle, cytochalasin B and cytochalasin D, respectively). Both cytochalasins, however, tended to enhance LPS-induced cGMP accumulation.

Discussion

The major novel finding is that the cytoskeleton selectively modulates the activation of NO pathways in cultured vascular cells. This regulation appears to be selective in two ways: (1) the cytoskeleton plays an essential role in bringing about activation of the inducible NO pathway in smooth muscle cells, but not the activation of the constitutive NO pathway in endothelial cells, and (2) there is an absolute requirement for microtubules but not for micro filaments during induction of NO in the smooth muscle cells.

In this study, two chemically distinct agents, nocodazole and colchicine disrupted microtubule assembly and prevented endotoxin-elicited and NO-mediated cGMP accumulation in vascular smooth muscle cells. Their effects on cGMP accumulation appear to correlate well with the polymerization state of microtubules, because the inhibitory action of nocodazole and colchicine was absent, when microtubule depolymerization was prevented by taxol. Since nocodazole and colchicine did not affect microfilament assembly and disruption of microfilaments had no inhibitory effects on LPS responses, we suggest that the activation of the inducible NO pathway is regulated by a microtubule-dependent system in vascular smooth muscle cells. This regulatory mechanism is not restricted to endotoxin alone, but functions during interleukin responses, as well, suggesting a potentially central role in the induction process in these cells.

There are several potential means by which cytoplasmic microtubules might influence the activation of the inducible NO pathway by immunomodulators. Microtubules may play a role in the interaction between the activators and the cell surface, in the internalization process, signal transduction leading to activation of the iNOS gene or transcription of mRNA. Furthermore, microtubules might affect stability and sorting of mRNA and protein synthesis.[43,44] In any instant, this regulatory mechanism is a rather unique phenomenon, since disruption of this component of the cytoskeleton reduces the expression of only certain genes such as positive acute phase proteins in hepatocytes and tissue plasminogen activator in endothelial cells, whereas it does not affect the expression of other genes such as negative acute phase reactants in hepatocytes or actually promotes protein expression such as urokinase in LLC-PK1 cells or plasminogen activator inhibitor in endothelial cells.[42,25] Our initial characterization suggests that signaling mechanism(s) beyond the interaction of LPS with the RASM cell surface, but not the activity of the iNOS once it is induced, or the guanylate cyclase activation process depend(s) upon microtubules.

In conclusion, the present study demonstrates that microtubules mediate the activation of an LPS- and interleukin-lb-inducible pathway to synthesize L-arginine-derived NO and increase cGMP in vascular smooth muscle cells, a proposed pathophysiological mechanism underlying septic shock and side-effects of antitumor therapy with cytokines. This recognition offers an experimental framework for further investigating the participation of the structural organization of cytoskeleton in vascular responses during immunological challenges.

REFERENCES

1. Beasley, D., R. A. Cohen, and N. G. Levinsky. Interleukin-I inhibits contraction of vascular smooth muscle. J.Clin. Invest. 83:331–335, 1989.
2. Beasley, D., J. H. Schwartz, and B. M. Brenner. Interleukin 1 induces prolonged L-arginine-dependent cyclic guanosine monophosphate and nitrite production in rat vascular smooth muscle cells. J. Clin. Invest. 87:602–608, 1991.

3. Beasley, D. Interleukin 1 and endotoxin activate soluble guanylate cyclase in vascular smooth muscle. Am. J. Physiol. 259:R38-R44, 1990.
4. Bradford, M. A rapid and sensitive method for quantitation of microgram quantities of protein utilizing the principle of protein-dye binding. Anal Biochem 1976;72:248–253.
5. Breitfeld, P. P., McKinnon, W. C., and- Mostov, K. E. (1990) J. Cell Biol. 11 1, 2365.
6. Busse, R. and Mulsch, A. Induction of nitric oxide synthase by cytokines in vascular smooth muscle cells. FEBS LeUers 275:87–90, 1990.
7. Cantley, L.C.,Auger, K.R.,Carpenter, C.,Duckworth, B., Graziani, A.,Kapeller, R. and Soltoff, S. Oncogenes an@ signal transduction. Cell 64:281–302, 1991.
8. Carter K.C., Cooper, R., Papaconstantinou, J. and Ritchie, D.G. Microtubule Depolymerization inhibits the regulation of a,-acid glycoprotein MRNA by hepatocyte stimulating factor. J. Biol. Chem. 264:515–519, 1989.
9. Ding, A. H., Porteu, F., Sanchez, E. and Nathan, C. F. Shared actions of endotoxin and taxoi: induction of TNF release and downregulation of TNF receptors. Science 248:370–372,1992.
10. Ding, A.H.,Porteu, F., Sanchez, E. and Nathan, C.F. Down-regulation of tumor necrosis factor receptors on macrophages and endothelial cells by microtubule depolymerizing agents.J.Exp. Med. 171, 715–727, 1990.
11. Fleming, I., G. A. Gray, G. Julou-Schaeffer, J. R. Parratt and J. Stoclet. Incubation with endotoxin activates the L-arginine pathway in vascular tissue. Biochem. Biophys. Res. Commun. 171:562–568, 1990
12. Garland, D. L. (1978) Biochemistry. 17, 4266.
13. Geinsterfer, A. A. T., M. J. Peach, and G. Owens. Angiotensin 11 induces hypertrophy but not hyperplasia of cultured rat aortic smooth muscle cells. Circ. Res. 62: 749–756, 1988.
14. Geroulanos, S., Schilling, J., Cakmakci, M., Jung, H.H.,and Largiader, F. (1992) Inhibition of NO synthesis in septic shock. Lancet. 339, 434–435.
15. Hortelano, S., A. M. Genaro, and L. Bosca. Phorbol esters induce nitric oxide synthase activity in rat hepatocytes. J. Biol. Chem. 267: 24937–24940, 1992.
16. Kilbourne, R.G. and P. Belloni. Endothelial cell production of nitrogen oxides in response to interferon in combination with tumor necrosis factor, interleukin-1, or endotoxin. J. Natt. Cancer Inst. 82:772–776, 1990.
17. Knowles, R.G. and Moncada, S. Nitric oxide as a signal in blood vessels. TIBS 17:399–402, 1992
18. Lee, J. C., F leld, D. J., and Lee, L. L. Y. (1 980) Biochemistry. 19, 6209.
19. Lorente, J.A., Landin, L., De Pablo, R., Renes, E. and Liste D. L-arginiiie pathway in sepsis syndrome. Citical Care Medicine 21 (No 9):1287–1295, 1993.
20. Munoz, E., Zublaga, A., Huang, C-K and Huber, B.T. Interleukin-I induces protein tyrosine phosphorylation in T cells. Eur. J. Immunol. 22:1391–1396, 1992.
21. Ochoa, J.B., Udekwu, A.0., Billar, T.R., et al. 1991. Nitrogen oxide levels in patients after trauma and sepsis. Ann. Surg. 214:621–626.
22. Otto, A.M.,Zumbe, A., Gibson, L., Kubler, A. and de Asua, L.J. Cytoskeleton-disrupting drugs enhance effect of growth factors and hormones on initiation of DNA synthesis. Proc. Natl Acad. Sci. U. S.A. 76:6435–6438, 1979.
23. Petros, A., Bennett, D., and Vallance P. (1991) Effect of nitric oxide synthase inhibitors on hypotension in patients with septic shock. Lancet. 338, 1557–1558.
24. Ryan J.W. & Ryan U.S. Endothelial surface enzymes and the processing of plasma substrates. Int.Rev.Exp.Pathol. 26:1, 1984
25. Santell, L., Marotti, K., Bartfeld, N.S.,Baynham, P. and Levin E.G. Disruption of microtubules inhibits the stimulation of tissue plasminogen adtivator expression and promotes plasminogen activator inhibitor type I expression in human endothelial cells. Exp. Cell Res. 201:358–365, 1992.
26. Schiff, P.B., Fant, J., and Horwitz, S.B. (1979) Nature. 277, 665–667
27. Schini, V. B., D. C. Junquero, T. Scott-Burden, and P. M. Vanhoutte. Interieukin- 1 b inducesthe production of an L-arginine-derived relaxing factor from cultured smooth muscle cells from rat aorta. Biochem. Biophys. Res. Conunun. 176:114–121, 1991.
28. Corbett, J.A., M.A. Sweetland, J.R. Lancaster Jr., and M.L. McDaniel. A 1-hour pulse with IL-1 induces formation of nitric oxide and inhibits insulin secretion by rat islets of L.angerhans: evidence for a tyrosine kinase signaling mechanism. FASEB J. 7:369–374, 1993.
29. Di Salvo, J., A. Steusloff, L. Semenchuk, S. Satoh, K.Kolquist and G. Pfizer. Tyrosine kinase inhibitors supress agonist-induced contraction in smooth muscle. Biochem. Biophys. Res.Commun. 190, 968–974, 1993.
30. Penman, S., D. G. Capco, E. G. Fey, P. Chatterjee, T. Reiter, S. Ermish and K. Wan. The three -dimensional structural networks of cytoplasm and nucleus: Function in cells and tissue. Modern Cell Biol. 2:385–415, 1983.

31. Xie, Q., Cho, H.J., Calaycay, J., Mumford, R.A., Swiderek, K.M., Lee, T.D., Ding, A., Troso, T. and Nathan, C. Cloning and characterization of inducible nitirc oxide synthase from mouse macrophages. Science 256:225–228, 1992.
32. Evans, T., Carpenter, A., and Cohen , J. Inducible nitric-oxide-syntase mRNA is transiently expressed and destroyed by a cycleheximide-sensitive process. Eur. J. Biochem. 219:563–569, 1994.
33. Brooker G., Terasake, W., and Price, M. Gammaflow: A completely automated radioimmunoassay system. Science 194:270–276, 1976.
34. Meyer, M., Schreck R. and Bauerie, P.A. H202 and antioxidants have opposite effects on activation of NF-KB and AP-1 in intact cells: AP-1 as secondary antioxidant-responsive factor. ne EMBO Journal :2005–2015.
35. Devary, Y., Gottlieb, R.A., Smeal, T. and Karin, M. The Mammalian Ultraviolet Response IsTriggered by Activation of Src Tyrosine Kinases. Cell 71:1081–1091, 1992.
36. Coyne, D.W., and A. R. Morrison. Effect of the tyrosine kinase inhibitor-, genistein, on interleukin-I stimulated PGE, production in mesangial cells. *Biochem. Biophys. Res. Commun.* 173: 718–724, 1990.
37. Catravas, J.D. and Watkins, C.A. Plasmalemmal metabolic activities in cultured calf pulmonary arterial endothelial cells. *Res.Commun. Chemic. Path. Pharmacol.* 50:163–179.
38. Hunter, W.M. and Greenwood, F.C. Preparation of iodine-131 labeled juman growth hormone of high specific activity. *Nature,* 194;495–496, 1962.
39. Patel, A. and Linden, J. Purification of ^{125}I-lbeled succinyl cyclic nucleotide tyrosine methyl esters by high-performance liquid chromatography. *Anal Biochem.,* 168:417–420, 1988.
40. Corbett, J.A., Sweetland, M.A., Lancaster, J.R., Jr., and McDaniel, M.L. A 1-hour pulse with IL-1β induces formation of nitric oxide and inhibits insulin secretion by rat isles of Langerhans: evidence for a tyorsine kinase signaling mechanism. *Faseb J.* 7:369–374, 1993.
41. Unemori, E.N. and Werb, Z.J. Reorganization of polymerized actin: a possible trigger for induction of procollagenase in fibroblasts cultured in and on collagen gels. *J. Cell Biol.,* 103;1021–1031, 1986.
42. Botteri, F.M., ballmer-Hofer, K., Rajput, B., and Nagamine, Y. Disruption of cytoskeletal structures results in the induction of the urokinase type plasminogen activator gene expression. *J. Biol. Chem.,* 265:13327–13334, 1990.
43. Walker, P.R. and Whitfield, J.F. Cytoplasmic microtubules are essential for the formatin of membrane bound polyribosomes. *J. Biol. Chem.,* 260:765–770, 1985.
44. Sundell, C.L., and Singer, R.H. Requirement of microfilaments in sorting of actin mRNA. *Science,* 253:1275–1277, 1991.
45. Birchmeier, W. Cytoskeleton structure and function. *Trends Biochem. Sci.,* 9:192–195, 1984.
46. Marczin, N., Papapetropoulos, A., and Catravas, J.D. Tyrosine kinase inhibitors suppress endotoxin- and IL-1β-induced NO synthesis in aortic smooth muscle cells. *Am.Physiol.Soc.* H1014-H1018, 1993.
47. Marczin, N., Jilling, T., Papapetropoulos, A., Go C., and Catravas, J.D. Cytoskeleton-dependent activation of the inducible nitric oxide synthase in cultured aortic smooth muscle cells. *British J. Pharmacol.* 118:1085–1094, 1996.

5

ADHESIVE MOLECULES AT LUMINAL SURFACE AND AT INTERCELLULAR JUNCTIONS OF THE ENDOTHELIUM IN THE REGULATION OF LEUKOCYTE TRANSENDOTHELIAL TRAFFIC

Aldo Del Maschio,[1] Adriana Zanetti,[1] Paraskevi Andriopoulou,[2]
Maria Grazia Lampugnani,[1] and Elisabetta Dejana[1]

[1]Laboratory of Vascular Biology
Instituto di Ricerche Farmacologiche "Mario Negri"
Milano, Italy
[2]University of Patras
Laboratory of Pharmacology
Patras, Greece

1. INTRODUCTION

Circulating leukocytes migrate from the vessels and enter tissues under both normal and pathological situations. Whereas monocytes, lymphocytes and natural killer cells exhibit a significant spontaneous migration through resting endothelium neutrophils and eosinophils require chemotactic stimuli and/or endothelial cell activation. Cell passage across endothelial monolayers involves leukocyte adherence to the endothelium, crawling on the endothelial surface and transmigration, most probably through endothelial clefts.

Endothelial cells (EC) can actively regulate leukocyte infiltration in inflamed tissues through different mechanisms such as vasodilation, release of chemotactic cytokines and expression of adhesive molecules. All these reactions act in concert in localizing leukocytes and in facilitating their passage through the endothelial layer. Inflammatory stimuli are able to activate EC inducing their functional reprogramming toward a "proinflammatory phenotype." Classical mediators of inflammation such as interleukin-l (IL-1) and tumor necrosis factor (TNF) induce production of vasodilatory products such as prostacyclin and nitric oxide as well as the synthesis of a large series of adhesive molecules and chemokines, and cause enhanced permeability of the endothelial layer and alteration of the organization of interendothelial junctions.

Vascular Endothelium, edited by Catravas *et al.*
Plenum Press, New York, 1998

In this review we will focus concisely on the role of endothelial cells in promoting leukocyte infiltration in tissues. In particular we will consider the role of adhesive molecules.

2. CELL SURFACE ADHESIVE MOLECULES: SELECTINS AND THEIR LIGANDS

Selectin family consists of three partially homologous highly glycosilated adhesive molecules found on leukocytes (L-selectin; 100kDa), platelets and EC (P-selectin; 140-kDa) and EC uniquely (E-selectin; 115-kDa), respectively.

L-selectin, originally identified as a lymphocyte homing receptor[1], is expressed by all resting leukocytes and contributes to the transient adhesion to the endothelium termed leukocyte "rolling"[2]. L-selectin is shedded from the cell surface upon leukocyte exposure to EC-derived chemokines and other chemotactic factors[3]. As a result, a soluble form of L-selectin can be detected in plasma[4].

P-selectin (CD62) is stored in platelet α-granules and endothelial Weibel-Palade bodies[5,6] and can be rapidly (within minutes) translocated to the membrane surface upon cell activation by a variety of mediators, including thrombin, histamine, complement fragments, hydrogen peroxide and various neuropeptides[7]. P-selectin can also be regulated transcriptionally upon cell activation with TNF[8]. Once expressed on cell surface, this molecule can be internalized into lysosomes within minutes[9]. P-selectin is implicated in leukocyte accumulation in growing thrombi[10], as well as in the early phases of leukocyte recruitment at sites of inflammation[11].

E-selectin is a selective, inducible molecule of EC surface involved in leukocyte adhesion to inflamed endothelium[12]. Cell surface expression of E-selectin, upregulated by TNF, IL-1, LPS and neuropeptides[8,12,13], is maximal after 4–6 h of stimulation and decreases to basal levels after 24–48 h[7].

Selectin ligands are oligosaccharides present in glycoproteic and glycolipidic structures of leukocyte and EC membranes. They interact with the lectin domain of the selectin, common to all members of the family[7]. The recognized oligosaccharides contain a terminal sialic acid molecule (Neu5Ac) linked a2-3 to galactose (Gal) which is linked in turn to an N-acetylglucosamine (GlcNAc) and a fucose molecule (Fuc) coupled to GlcNAc. Two isomers have been identified: Neu5Ac a2-3 Galb1-4 (Fuc a1-3)-GlcNAc and Neu5Ac a2-3 Galb1-3 (Fuc a1-4)-GlcNAc, respectively known as sialyl-Lewis x (sLe^x) and sialyl-Lewis a (sLe^a)[7]. The three selectins bind sLe^x and sLe^a, but they also recognize other mono and polysaccharidic phosphorylated structures. P-selectin binds also a non sialylated form of sLe^x and sLe^a (Le^x and Le^a).

Several glycoproteic structures which have ligand activity for the three selectins bear sLe^x and sLe^a. Moreover, most of them bind to more than one type of selectin. The glycosylation-dependent cell adhesion molecule-1 (GlyCAM-1) is a specific endothelial adhesion molecule (addressin) which prevalently binds L-selectin but also E-selectin. It is expressed mainly in high endothelial venules (HEV) of lymph nodes were it regulates lymphocyte homing[14].

Another L-selectin ligand is the vascular sialomucin CD34, expressed as a cell surface transmembrane protein on EC and hemopoietic progenitors[15]. This molecule, which may be differently glycosilated or sulfated, mediates lymphocyte homing to lymph nodes and serves as an L-selectin ligand for neutrophils[16]. Since in regions of the skin with massive lymphocyte infiltration, endothelial CD34 expression is low (while E-selectin is elevated), it has been postulated that CD34 is involved in leukocyte circulation under normal, but not under inflammatory conditions[17].

The mucosal addressin cell adhesion molecule-1 (MAdCAM-1) is expressed on HEV of both Peyer's patches and embryonic lymph nodes as well as in intestinal venules[18]. MAdCAM-1 binding to L-selectin and $\alpha 4\beta 7$ integrin[19] mediates lymphocyte homing[46] and adhesion to EC[20].

Two P-selectin ligands have been characterized: P-selectin glycoprotein ligand-1 (PSGL-1) and a 120-kDa ligand. PSGL-1 (110-kDa) is a transmembrane sialomucin expressed on leukocytes[21]. The 120-kDa P-selectin ligand is expressed on myeloid cells[22]. Finally, P-selectin may also be recognized by L-selectin[23].

Three E-selectin ligands have been characterized: the cutaneous leukocyte antigens (CLA), a 250-kDa receptor and the sialyl stage-specific embryonic antigen (SSEA-1). The highly sialylated molecules of the CLA ligand family are found in a population of memory leukocytes in inflammatory lesions of the skin, characterized by elevated expression of E-selectin[24]. A 250-kDa E-selectin ligand has been identified on T lymphocytes and mediates the recruitment of these cells to the skin[25]. SSEA-1 ligand is expressed on concanavalin A-activated natural killer cells and mediates their interaction with the endothelial E-selectin[26].

3. CELL SURFACE ADHESIVE MOLECULES: INTEGRINS AND THEIR LIGANDS

Leukocyte integrins of the beta2 family consist of four heterodimeric glycoproteins sharing a common $\beta 2$ chain (CD18; 95-kDa): lymphocyte function-associated antigen-1 (LFA-1; CD11a/CD18; 180/95-kDa), Mac-1 (CD11b/CD18; 165/95-kDa), p150,95 (CD11c /CD18; 150/95-kDa) and the newly described aTM (CD11d/CD18; 155/95-kDa). These molecules, generally expressed in inactive form, can control stable cell adhesion by at least two distinct mechanisms: alteration of ligand-binding affinity mediated by structural changes and/or their multimerization at the cell surface[27].

LFA-1 is present on the cell surface of mast leukocytes and interacts with the first Ig domain of ICAM-1 and ICAM-2 (see below) of EC[28]. Besides its role in leukocyte-EC interaction, LFA-1 participate in other intercellular relationships, such as T cell activation by antigen-presenting cells or the killing of virus-infected cells by cytotoxic T lymphocytes[29].

Mac-1 is present on granulocytes and macrophages, interacts with ICAM-1, but does not bind to ICAM-2 and ICAM-3. Mac-1 also interacts with the complement fragment C3bi[30] and fibrinogen[31] and it is a receptor for the hookworm-derived neutrophil adhesion inhibitor (NIF)[32].

The p150,95 protein has a similar distribution as Mac-1, although it is also expressed on some activated lymphocytes and is a marker for hairy-cell leukemia[33]. Like Mac-1, p150,95 has a broad role as a general adhesion protein and it possesses some C3bi-binding activity[34].

The fourth member of this family is the recently identified aTM, which was found in canine tissue macrophages and $CD8^+$ T cell subset[35].

Integrins of the beta1 family are heterodimeric glycoproteins sharing a common $\beta 1$ chain (CD29; 130-kDa) with variable alpha subunits ($\alpha 1$ to $\alpha 9$). Members of this family were originally defined as "very late antigens" (VLA), since they appeared on T cell surface after 2–4 weeks of *in vitro* activation[36].

Integrin $\alpha 4\beta 1$ (VLA-4 or (CD49d/CD29; 150/130-kDa) is expressed on monocyte and lymphocyte cell surface and mediates the interaction of these cells with fibronectin and the endothelial ligand VCAM-1[37,38]. VLA-4 on neutrophils mediates migration in arthritic and dermal inflamed areas[39].

Integrin $\alpha 6\beta 1$ (VLA-6 or CD49f/CD29) expressed by EC, promotes homing of T cell progenitors to the thymus[40] and binds laminin[41].

Molecules belonging to the family of $\beta 7$ integrins, such as lymphocyte-Peyer's patch adhesion molecule-1 (LPAM-1 or $\alpha 4\beta 7$ integrin; 200/150-kDa) and $\alpha E\beta 7$ (150/120-kDa), are expressed on a subset of gut-associated lymphocytes, also known as intestinal intraepithelial lymphocytes (IEL). Integrin $\alpha 4\beta 7$ recognizes the mucosal endothelial ligand MAdCAM-1 which is involved in normal lymphocyte recirculation[19], but also VCAM-1, which play an important role in inflammatory reactions[42]. Binding of the $\alpha E\beta 7$ integrin to epithelial E-cadherin mediates adhesive interactions between lymphocytes and epithelial cells[43].

Endothelial ligands for integrins belong to the immunoglobulin superfamily. Five members of this family are involved in leukocyte-EC interaction: intercellular adhesion molecule (ICAM)-1, ICAM-2, ICAM-3, vascular cell adhesion molecule (VCAM)-1 and platelet/endothelial cell adhesion molecule (PECAM)-1.

ICAM-1 (CD54; 95-kDa) is an integral membrane protein weakly expressed in resting endothelium. Exposure of EC to IL-1, TNF or interferon γ (IFN) strongly enhances ICAM-1 expression, with maximal level 12 h after stimulation[44]. ICAM-1 is the endothelial ligand of LFA-1 and Mac-1[45], but it also exerts the function of rhinovirus ligand[46]. A soluble form of ICAM-1 (89-kDa) has been purified from EC supernatants[47]. ICAM-2 (CD102; 46-kDa) is constitutively expressed at a high level on resting EC and its expression is not upregulated by stimulation[48]. ICAM-2 controls both leukocyte adhesion and constitutive basal transmigration binding LFA-1. ICAM-3 (CD50; 125-kDa) is expressed on resting lymphocytes and monocytes and may play a role in the initiation of immune responses[49].

VCAM-1 (CD106; 110-kDa) is the major ligand for VLA-4[38] but it also binds LPAM-1[42]. VCAM-1 is absent on resting EC but its expression is strongly enhanced after exposure to IL-1 or TNF, reaching maximal level 6–12 h after stimulation[58]. A soluble form of VCAM-1 (95–100-kDa) has been purified from EC supernatants[47].

PECAM-1 (CD31; 130-kDa) is a molecule constitutively expressed on platelet, leukocyte and EC surface, which intervene in several physiological events. These include endothelial cell to cell adhesion, as in the endothelial monolayer PECAM-1 is concentrated at intercellular junctions[51]. Cell-to-cell adhesion mediated by PECAM-1 can be both homophilic and heterophilic[51]. A heterophilic recognition is the recently described PECAM-1 interaction with $\alpha v\beta 3$ integrin, which may be considered an additional mechanism of leukocyte adhesion to EC[52]. Moreover, PECAM-1 plays a fundamental role in the regulation of leukocyte transendothelial migration[53].

4. MOLECULES AT INTERENDOTHELIAL JUNCTIONS AND LEUKOCYTE TRANSMIGRATION

Leukocyte traffic to the extravascular space is limited under normal conditions. However, during immunitary and inflammatory reactions a considerable enhancement of leukocyte migration is observed. Leukocytes migrate through the endothelium at sites of reduced blood flow, such as postcapillary venules or inflamed vessels, dilated upon the effect of vasoactive substances[54,55].

Reduced flow allows the circulating leukocytes to remain in contact with the endothelial surface for a longer time. Leukocyte adhesion to EC is initially transient. This process, known as leukocyte "rolling," is essentially mediated by the interaction of L- and P-selectin with their own ligands, as described above.

When the endothelium is strongly activated, as during inflammatory reactions, it generates and releases chemotactic substances able to induce leukocyte activation, integrin upregulation (mainly of the β2 series) which results in enhanced leukocyte recruitment[56]. This effect, which stabilizes leukocyte adhesion, may be considered an amplificatory mechanism mainly operated by chemokines[56], lipid mediators (such as platelet activating factor and leukotriene B4) and colony stimulating factors (such as G-, M-, and GM-CSF). Chemokines, which stabilize leukocyte adhesion upregulating integrin β2, include: interleukin-8 (IL-8) and GRO/melanoma growth stimulatory activity (GRO/MGSA) which are active on neutrophils, and monocyte chemotactic protein-1 (MCP-1) which is active on monocytes[56]. Upregulation of β2 (but also of β1) integrins[27], determines a strong adhesion of leukocytes to EC (or HEV) monolayers through the recognition of the ligands described above. Once firmly adherent, leukocytes can rapidly (within minutes) migrate through the endothelial monolayer[54].

β2 integrins regulate leukocyte transmigration. Indeed, in the presence of antibodies against β2 integrins and/or their ligands, transmigration is strongly reduced[57–59]. Leukocyte transmigration may however also be blocked by antibodies against α4β1 integrin and VCAM-1[53,68,61]. Particularly interesting is the role played by PECAM-1. This molecule distributing at endothelial cell to cell contacts gives rise to an haptotactic apical-basal gradient through the endothelial junctions. This gradient can be sensed by leukocytes, possibly through homophilic interaction with the leukocytic PECAM, and can direct oriented leukocyte migration[53].

Electron microscopy observation suggests that migration might begin by insertion of leukocyte pseudopods into the inner space between two adjacent EC[54,63]. Within minutes, leukocytes penetrate through the endothelium by a process which does not induce damage of the endothelial monolayer under normal conditions[63,64], or, when inflammation occurs without edema[54,65]. Leukocyte-released substances, such as elastase, cathepsin G, collagenase and toxic oxygen metabolites, contribute to severe tissue injury and organ dysfunction in several pathological conditions[66]. However, these substances seem not to play a role in leukocyte traffic[63,64,67,68]. This suggests that leukocytes may regulate their migration through a very specific mechanism of remodeling of interendothelial junctions.

At least four types of junctions coexist in EC: adherens junctions, tight junctions, gap junctions and complexus adherents[69]. However, in postcapillary venules, where the interchanges between blood and tissues need to be dynamic, EC junctions appear rather simple and only adherens junctions are consistently observed[69]. Adherens junctions are formed by transmembrane Ca^{2+}-dependent adhesive proteins called cadherins which are associated inside the cells with a complex network of cytoskeletal molecules. These in turn promote anchorage to actin microfilaments[70,71]. A specific endothelial cadherin has been identified: vascular endothelial (VE)-cadherin[72] which is complexed to cytoplasmic proteins named catenins, such as α-catenin, β-catenin, plakoglobin and $p120^{cas}$[73].

VE-cadherin/catenin complex in adherens junctions play an important role in the control of endothelial permeability. Indeed, VE-cadherin/catenin complex is rapidly disorganized by agents that increase endothelial permeability, such as elastase, thrombin and EGTA[72,74]. A similar effect has been recently observed during leukocyte adhesion to cytokine-treated EC[75]. The VE-cadherin/catenin complex is disorganized in response to neutrophil adhesion to EC and this phenomenon is accompanied by enhanced endothelial permeability in the absence of EC retraction or injury[75]. This data support a model in which leukocytes could regulate the state of endothelial cell-to-cell junctions to promote their migration through the endothelium toward the extravascular compartment.

5. CONCLUDING REMARKS

We begin now to understand that leukocytes and EC are able to communicate and reciprocally modulate their responses. Following inflammatory stimulation EC attract and localize leukocytes through the release of chemokines and expression of adhesive molecules. Leukocytes in turn might transfer signals to the endothelium releasing soluble mediators, such as cytokines, oxidation products and lytic enzymes. Binding of leukocytes to endothelial adhesive molecules might also cause endothelial cell activation facilitating leukocyte passage through the endothelial layer, possibly at interendothelial clefts. Leukocyte extravasation is not always accompanied by endothelial damage and increased permeability. In contrast, it appears that the remodeling of interendothelial junctions is a finely regulated process that is frequently reversible. Future work is required to fully understand how we might modulate the cross-talk between EC and leukocytes. This appears to be a difficult task considering the number of soluble and membrane molecules involved and the complexity of their interplay.

REFERENCES

1. Gallatin, W.M., Weissman I.L., Butcher, E.C. A cell-surface molecule involved in organ-specific homing of lymphocytes. *Nature*, 304:30–34, 1983.
2. Hogg, N. Roll, roll, roll your leukocyte gently down the vein. *Immunol. Today*, 13:113–115, 1992.
3. Smith, C.W., Kishimoto, T.K., Abbass O., et al. Chemotactic factors regulate lectin adhesion molecule 1 (LECAM-1)-dependent neutrophil adhesion to cytokine-stimulated endothelial cells *in vitro*. *J. Clin. Invest.*, 87:609–618, 1991.
4. Schleiffenbaum, B., Spertini, O., and Tedder, T.F.: Soluble L-selectin is present in human plasma at high levels and retains functional activity. *J. Cell. Biol.*, 119:229–238, 1992.
5. Stenberg, P.E., McEver, R.P., Shuman, M.A., et al. A platelet alpha-granule membrane protein (GMP-140) is expressed on the plasma membrane after activation. *J. Cell. Biol.*, 191:880–886, 1985.
6. McEver, R.P., Beckstead, J.H., Moore, K.L., et al. GMP-140, a platelet a-granule membrane protein, is also synthesized by vascular endothelial cells and is located in Weibel-Palade bodies. *J. Clin. Invest.*, 84:92–99, 1989.
7. Bevilacqua, M.P. and Nelson, R.M. Selectins. *J. Clin. Invest.*, 94: 379–387, 1993.
8. Weller, A., Isenmann, S., and Vestweber, D. Cloning of the mouse endothelial selectins: Expression of both E- and P- selectin is inducible by tumor necrosis factor-α. *J. Biol. Chem.*, 267:15176–15183, 1992.
9. Green, S.H., Setiadi, H., McEver, R.P., et al. The cytoplasmic domain of P-selectin contains a sorting determinant that mediates rapid degradation in lysosomes. *J. Cell. Biol.*, 124:435–448, 1994.
10. Palabrica, T.B., Lobb, A., Furie, B.C., et al. Leukocyte accumulation promoting fibrin deposition is mediated in vivo by P-selectin on adherent platelets. Nature, 359:848–851, 1992.
11. Lorant, D.E., Topham, M.K., Whatley, R.E., et al. Inflammatory roles of P-selectin. J. Clin. Invest., 92:559–570, 1993.
12. Bevilacqua, M.P., Pober, J.S., Mendrick, D.L., et al. Identification of an inducible endothelial-leukocyte adhesion molecule. *Proc. Natl. Acad. Sci. USA*, 84:9238–9242, 1987.
13. Smith, C.H., Barker, J.N., Morris, R.W., et al. Neuropeptides induce rapid expression of endothelial cell adhesion molecules and elicit granulocytic infiltration in human skin. *J. Immunol.*, 151:3274–3282, 1993.
14. Lasky, L.E., Singer, M.S., Dowbenko, U., et al. An endothelial ligand for L-selectin is a novel mucin-like molecule. Cell, 69:927–938, 1992.
15. Baumhueter, S., Singer, M.S., Henzel, W., et al. Binding of L-selectin to the vascular sialomucin CD34. *Science*, 262:436–438, 1993.

16. Rosen, S. L-selectin and its biological ligands. *Histochemistry*, 100:185–191, 1993.
17. Norton, J., Sloane, J.P., Delia, D., et al. Reciprocal of CD34 and cell adhesion ELAM-1 on vascular endothelium in acute cutaneous graft-versus-host disease. *J. Pathol.*, 170:173–177, 1993.
18. Streeter, P.R., Berg, E.L., Rouse, B.T.N., et al. A tissue-specific endothelial cell molecule involved in lymphocyte homing. *Nature*, 331:41–46, 1988.
19. Berlin, C., Berg, E.L., Briskin, M.J., et al. $\alpha 4\beta 7$ integrin mediates lymphocyte binding to mucosal vascular addressin MAdCAM-1. Cell, 74: 185–195, 1993.
20. Nakache, M., Berg, E.L., Streeter, P.R., et al. The mucosal vascular addressin is a tissue-specific endothelial cell adhesion molecule for circulating lymphocytes. *Nature*, 337:179–181, 1989.
21. Sako, O., Chang, X.J., Barone, K.M., et al. Expression cloning of a functional glycoprotein ligand for P-selectin. Cell, 75:1179–1186, 1993.
22. Moore, K.L., Stults, N.L., Diaz, S., et al. Identification of a specific glycoprotein ligand for P-selectin (CD62) on myeloid cells. *J. Cell. Biol.*, 118:445–456, 1992.
23. Picker, L.J., Warnock, R.A., Burns, R.R., et al. The neutrophil selectin LECAM-1 presents carbohydrate ligands to the vascular selectins ELAM-1 and GMP-140. Cell, 66:921–933, 1991.
24. Berg, E.L., Yoshino, T., Rott, L.S., et al. The cutaneous lymphocyte antigen is a skin lymphocyte homing receptor for the vascular lectin endothelial cell-leukocyte adhesion molecule 1. *J. Exp. Med.*, 174:1461–1466, 1991.
25. Walchek B., Watts, G., and Jutila, M.A. Bovine g/d T cells bind E-selectin via a novel glycoprotein receptor: Characterization of a lymphocyte/E-selectin interaction in an animal model. *J. Exp. Med.*, 178:853–863, 1993.
26. Ohmori, K., Takada, A., Yaneda, T., et al. Differentiation-dependent expression of sialyl stage-specific embryonic antigen-1 and I-antigens on human lymphoid cells and its implications for carbohydrate-mediated adhesion to vascular endothelium. *Blood*, 81:101–111, 1993.
27. Lub, M., van Kooyk, Y., and Figdor, C.G. Ins and outs of LFA-1. *Immunol. Today*, 16:479–483, 1995.
28. Dustin, M.L. and Springer, T.A. Lymphocyte function-associated antigen-1 (LFA-1) interaction with intercellular adhesion molecule-1 (ICAM-1) is one of at least three mechanisms for lymphocyte adhesion to cultured endothelial cells. *J. Cell. Biol.*, 107:321–331, 1988.
29. Springer, T.A. Adhesion receptors of the immune system. *Nature*, 346:425–434, 1990.
30. Graham, I.L., Gresham H.D., and Brown F.J. An immobile subset of plasma membrane CD11b/CD18 (Mac-1) is involved in phagocytosis of targets recognized by multiple receptors. *J. Immunol.*, 142:2352–2358, 1989.
31. Altieri, D.C., Plescia, J., and Plow, E. The structural motif glycine 190-valine 202 of the fibrinogen g chain interacts with CD11b/CD18 integrin (aMb2, Mac-1) and promote leukocyte adhesion. *J. Biol. Chem.*, 268: 1847–1853, 1993.
32. Rieu, P., Ueda, T., Haruta I., et al. The A-domain of $\beta 2$ integrin CR3 (CD11b/CD18) is a receptor for the hookworm-derived neutrophil adhesion inhibitor NIF. *J. Cell. Biol.*, 127:2081–2091, 1994.
33. Miller, L.J., Schwarting, A., Springer, T.A. Regulated expression of the Mac-1, LFA-1, p150,95 glycoprotein family during leukocyte differentiation. *J. Immunol.*, 157:2891–2900, 1986.
34. Myones, B.L., Dalzell J.G., Hogg, N., et al. Neutrophil and monocyte cell surface p150,95 has a iC3b-receptor (CR4) activity resembling CR3. *J. Clin. Invest.*, 82:640–651, 1988.
35. Cobbold, S., Holmes, M., and Willett, B. The immunology of companion animals: reagents and therapeutic strategies with potential veterinary and human clinical applications. *Immunol. Today*, 15:347–353, 1994.
36. Hemler, M.E. Adhesive protein receptors on hematopoietic cells. *Immunol. Today*, 9:109–113, 1988.
37. Tiisala, S., Hakkarainen, M., Majuri, M.-L., et al. Down-regulation of monocytic VLA-4 leads to a decreased adhesion to VCAM-1. *FEBS*, 332:19–23, 1993.
38. Elices, M.J., Osborn, L., Takada Y., et al. VCAM-1 on activated endothelium interacts with the leukocyte integrin VLA-4 at a site distinct from the VLA-4/fibronectin binding site. *Cell*, 60:577–584, 1996.
39. Issekutz, T.B., Miyasaka M., Issekutz, A.C. Rat blood neutrophils express very late antigen 4 and it mediates migration to arthritic joint and dermal inflammation. *J. Exp. Med.*, 183:2175–2184, 1996.
40. Imhof, B.A., Buiz, P., Hesse, B., et al. EA-1, a novel adhesion molecule involved in the homing of progenitor T lymphocytes to the thymus. *J. Cell. Biol.*, 114:1069–1078, 1991.
41. Sonnenberg, A., Linders, C.J.T., Modderman, P.W., et al. Integrin recognition of different cell-binding fragments of laminin (P1, E3, E8) and evidence that $\alpha 6\beta 1$ but not $\alpha 6\beta 4$ functions as a major receptor for fragment E8. *J. Cell. Biol.*, 110:2145–2155, 1990.
42. Chan, B.M.C., Elices, M.J., Murphy, E., et al. Adhesion to vascular cell adhesion molecule 1 and fibronectin. Comparison of $\alpha 4\beta 1$ (VLA-4) and $\alpha 4\beta 7$ an the human B cell line JY. *J. Biol. Chem.*, 267:8366–8370, 1992.
43. Cepek, K.L., Shaw, S.K., Parker, C.M., et al. Adhesion between epithelial cells and T lymphocytes mediated by E-cadherin and the $\alpha E\beta 7$ integrin. *Nature*, 372:190–193, 1994.

44. Dustin, M.L., Rothlein, A., Bhan, A.K., et al. Induction by IL-1 and interferon-gamma: tissue distribution, biochemistry, and function of a natural adherence molecule (ICAM-I) *J. Immunol.*, 137:245–254, 1986.
45. Diamond, M.S., Staunton, D.E., Marlin, S.D., et al. Binding of the integrin Mac-1 (CD11/CD18) to the third immunoglobulin-like domain of ICAM-1 (CD54) and its regulation by glycosylation. *Cell*, 65:961–971, 1991.
46. Staunton, D.E., Dustin, M.L., Erickson, H.P., et al. The arrangement of the immunoglobulin-like domain of ICAM-1 and the binding sites for LFA-1 and rhinovirus. *Cell*, 61:243–254, 1990.
47. Pigott, R., Dillon, L.P., Hemingway, I.H., et al. Soluble forms of E-selectin, ICAM-1 and VCAM-1 are present in the supernatants of cytokine activated cultured endothelial cells. *Biochem. Biophys. Res. Commun.*, 187:584–589, 1992.
48. Nortamo, P., Li, R., Renkonen R., et al. The expression of human intercellular adhesion molecule-2 is refractory to inflammatory cytokines. *Eur. J. Immunol.*, 21:2629–2632, 1991.
49. de Fougerolles, R.R., and Springer, T.A. Intercellular adhesion molecule 3, a third adhesion counter-receptor for lymphocyte function-associated molecule 1 on resting lymphocytes. *J. Exp. Med.*, 175:185–190, 1992.
50. Osborn, L., Hession, C., Tizard, A., et al. Direct expression cloning of vascular cell adhesion molecule 1, a cytokine-induced endothelial protein that binds to lymphocytes. *Cell*, 59:1203–1211, 1989.
51. DeLisser, H.M., Newman, P.J., and Albelda, S.M. Molecular and functional aspects of PECAM-1/CD31. *Immunol. Today*, 15:490–495, 1994.
52. Piali, L., Hammel, P., Uherek, C., et al. CD31/PECAM-1 is a ligand for $\alpha v \beta 3$ integrin involved in adhesion of leukocytes to endothelium. *J. Cell. Biol.*, 130:451–460, 1995.
53. Muller, W.R., Weigl, S.A., Deng, H., et al. PECAM-1 is required for transendothelial migration of leukocytes. *J. Exp. Med.*, 178: 449–468, 1993.
54. Marchesi, V.T. and Florey, H.W. Electron micro graphic observations on the emigration of leukocytes. *Q. J. Exp. Physiol.*, 45:343–348, 1968.
55. Brenner, B.M., Troy J.L., and Ballermann, B.J. Endothelium dependent vascular responses. Mediators and mechanisms. *J. Clin. Invest.*, 84:1373–1378, 1989.
56. Baggiolini, M., Dewald, B., Moser, B. Interleukin-8 and related chemotactic cytokines-CXC and CC chemokines. In: *Advances in Immunology,* Dixon, F.J., (ed.), vol. 55, pp. 97–148, Academic Press, San Diego, 1994.
57. Smith, C.W., Rothlien, A., Hughes, B.J., et al. Recognition of an endothelial determinant for CD18-dependent human neutrophil adherence and transendothelial migration. *J. Clin. Invest.*, 82:1746–1756, 1989.
58. Hakkert, B.C., Kuijpers, T.W., Leeuwenberg, J.F.M., et al. Neutrophil and monocyte adherence to and migration across monolayers of cytokine-activated endothelial cells: the contribution of CD18, ELAM-1, and VLA-4. *Blood*, 78:2721–2726, 1991.
59. Luscinskas, F.W., Cybulsky, M.I., Kiely J.M., et al. Cytokine activated human endothelial monolayers support enhanced neutrophil transmigration via a mechanism involving both endothelial-leukocyte adhesion molecule-1 and intercellular adhesion molecule-1. *J. Immunol.*, 146:1617–1625, 1991.
60. Chuluyan, H.E. and Issekutz, A.C. VLA-4 integrin can mediate CD11/CDl8-independent transendothelial migration of human monocytes. *J. Clin. Invest.*, 92:2768–2777, 1994.
61. Vaporciyan, A.A., DeLisser, H.M., and Yan, H.C. Involvement of platelet endothelial cell adhesion molecule-1 in neutrophil recruitment *in vivo*. *Science*, 262:1580–1582, 1993.
62. Tanaka, Y., Albelda, S.M., Horgan, K.J., et al. CD31 expressed on distinctive T cell subsets is a preferential amplifier of $\beta 1$ integrin-mediated adhesion. *J. Exp. Med.*, 176:245–253, 1992.
63. Furie, M.B., Naprstek, B.L., and Silverstein, S.C. Migration of neutrophils across monolayers of cultured microvascular endothelial cells. An *in vitro* model of leukocyte extravasation. *J. Cell. Sci.*, 88:161–175, 1987.
64. Moser, R., Schleiffenbaum, B., Groscurth P., et al. Interleukin 1 and tumor necrosis factor stimulate human vascular endothelial cells to promote transendothelial neutrophil passage. *J. Clin. Invest.*, 83:444–455, 1989.
65. Shaw, J.O. Leukocytes in chemotactic-fragment-induced lung inflammation: Vascular emigration and alveolar surface migration. *Am. J. Pathol.*, 101:283–302, 1980.
66. Lehr, H.A., and Arfors, K.E. Mechanisms of tissue damage by leukocytes. *Curr. Opin. Hematol.*, 1:92–99, 1994.
67. Huber, A.R. and Weiss S.J. Disruption of the subendothelial basement membrane during neutrophil diapedesis in an *in vitro* construct of a blood vessel wall. *J. Clin. Invest.*, 83: 1122–1136, 1989.
68. Kuijpers, T.W., Hoogerwerf, M., and Roos, D. Neutrophil migration across monolayers of resting or cytokine-activated endothelial cells. *J. Immunol.*, 148:72–77, 1992.
69. Dejana, E., Corada, M., Lampugnani, M.G. Endothelial cell-to-cell junctions. *FASEB J.*, 9:910–918, 1995.
70. Takeichi, M. Cadherin cell adhesion receptors as a morphogenetic regulator. *Science*, 251:1451–1455, 1991.
71. Kemler, A. From cadherins to catenins: cytoplasmic protein interactions and regulation of cell adhesion. *Trends Genet.*, 9:317–321, 1993.

72. Lampugnani, M.G., Resnati, M., Raiteri, M., et al. A novel endothelial-specific membrane protein is a marker of cell-cell contacts. *J. Cell. Biol.*, 118: 1511–1522, 1992.
73. Lampugnani, M.G., Corada M., Caveda, L., et al. The molecular organization of endothelial cell to cell junctions: differential association of plakoglobin, β-catenin and α-catenin with vascular endothelial cadherin (VE-cadherin*)*. *J. Cell. Biol.*, 129:203–217, 1995.
74. Rabiet, M.J., Plantier, J.L., Rival, V., et al. Thrombin-induced increase in endothelial permeability is associated with changes in cell to cell junction organization. *Arterioscler. Thromb. Vasc. Biol.*, 16:488–496, 1996.
75. Del Maschio, A., Zanetti, A., Corada M., et al. Polymorphonuclear leukocyte adhesion triggers the disorganization of endothelial cell-to-cell adherens junctions. *J. Cell. Biol.*, 135:497–510, 1996.

6

INSERTION/DELETION (I/D) POLYMORPHISM IN THE ANGIOTENSIN CONVERTING ENZYME (ACE) GENE

Kåre Berg

Institute of Medical Genetics
University of Oslo and
Department of Medical Genetics
Ullevaal University Hospital
Oslo, Norway

1. INTRODUCTION

The insertion/deletion (I/D) polymorphism at the locus for the angiotensin I converting enzyme (ACE) has been the focus of considerable interest since the report by Cambien and his co-workers (Cambien et al., 1992) of association between the D allele and coronary heart disease (CHD), particularly in CHD patients without classical risk factors. In the following, this observation by Cambien's group will be compared with observations in our group and with some of the studies in the literature.

A broad spectrum of observations have been reported with respect to this polymorphism, and it is therefore well suited to illustrate problems and phenomena encountered in the study of genetic markers and disease. Accordingly, I will also consider candidate gene approaches, the variability gene concept (Berg, 1994a), gene-gene interaction, "longevity genes" and possible interpretations when conflicting results are obtained in the study of disease associations.

2. CANDIDATE GENE APPROACHES TO THE STUDY OF ATHEROSCLEROTIC/THROMBOTIC DISEASE

With the plethora of genes available for study at the molecular level, present genetic studies of atherosclerotic/thrombotic disease mostly follow a "functional candidate gene" approach. This implies that possible relationship to atherogenic/thrombogenic disease is searched for at any locus whose protein product is or may be involved in processes or

Vascular Endothelium, edited by Catravas *et al.*
Plenum Press, New York, 1998

structures related to atherogenesis/thrombogenesis or thrombolysis/fibrinolysis (Berg, 1994b). Genes that are functioning in atherogenic/thrombogenic lesions or whose protein product may affect the function of other genes in such lesions may also be functional candidate genes. (A different category of candidate genes are genes that we know must reside in a defined, small area of the genome because genetic linkage with a disease or a normal trait has been established with markers in that region; such genes may be termed "positional candidate genes").

The rationale for considering the ACE gene a candidate gene for CHD is that its product is important in the renin-angiotensin system which may be involved in the pathogenesis of hypertension, a well known risk factor for CHD. The renin-angiotensin system is important in the salt-water homeostasis and for vascular tone. Renin and ACE are the two key enzymes in this system. Renin cleaves angiotensinogen to angiotensin I whereas ACE cleaves angiotensin I to angiotensin II and also degrades bradykinin. ACE inhibitors are clinically important antihypertensive drugs. In the myocardium angiotensin II promotes protein synthesis, growth of myocytes and collagen expression in fibroblasts. The tissue renin-angiotensin system is also involved in modulation of local sympathetic activity. Thus, there are ample possibilities for the renin-angiotensin system to contribute to the risk of CHD through one mechanism or the other.

3. GENETICS OF ACE

It has been known for many years that plasma ACE level differ significantly between individuals but is remarkably stabile within a given individual. This suggested a significant influence of genes on plasma ACE level and in 1988 Cambien and his co-workers reported significant correlations of ACE levels between blood relatives (but not between spouses) and genetic analysis suggested that a major gene affected inter-individual variability of plasma ACE, with different co-dominant effects in parents and offspring (Cambien et al., 1988). The authors suggested that the familial resemblance in plasma ACE level could be explained by the segregation of a major gene.

Rigat et al., (1990) reported the I/D polymorphism in the ACE gene and that it accounted for half the variance of serum enzyme levels. The polymorphism is known to reside in intron 16 of the ACE gene (Hubert et al., 1991) and may therefore not affect ACE structure. Its correlation with enzyme level, suggests that it is in linkage disequilibrium with functionally important domains.

Rigat et al., (1990) did not observe any significant effect on blood pressure of this polymorphism, despite the correlation between the polymorphism and enzyme level. Jeunemaitre et al., (1992) failed to detect any correlation between the polymorphism and essential hypertension in humans and neither Higashimori et al., (1993) nor Harrap et al., (1993) found evidence for association between the I/D polymorphism and blood pressure in hypertensive individuals.

Tiret et al., (1992) examined plasma ACE activity and the I/D polymorphism in 98 healthy families. The pattern of familial correlations of ACE level was compatible with no correlation between spouses, and with a 0.24 parent-offspring as well as sibsib correlation. Segregation analysis suggested that this familial resemblance could be explained by transmission of a co-dominant major gene. A combined segregation and linkage analysis provided evidence for a major gene effect at the ACE locus, in strong linkage disequilibrium with the I/D polymorphism. The postulated major gene accounted for 44% of the total variability of ACE level. The authors found no association between the I/D polymorphism and blood pressure but a weak correlation between blood pressure and ACE level in children. No relationship between the I/D polymorphism and high blood pressure could be detected in the Dutch hypertension and offspring study (Schmidt et al., 1993).

McKenzie et al., (1995) found that one quantitative-trait locus (QTL) is located within or close to the ACE locus itself. In a study of African Caribbean families, surprisingly few families showed genetic linkage to the neighboring growth hormone gene. An extended analysis with a two-QTL model where an ACE linked QTL interacts actively with an unlinked QTL, significantly improved the fit of the model and strengthened the evidence for linkage between the ACE linked QTL and polymorphism in the growth hormone gene. The authors concluded that two QTLs jointly influence serum ACE levels in their population. One QTL was located within or close to the ACE locus and explained 27% of the total variability in ACE level. The second QTL was unlinked to the ACE locus and explained 52% of the variability in ACE level.

4. THE I/D POLYMORPHISM AT THE ACE LOCUS AND NORMAL BLOOD PRESSURE LEVEL

My colleague, Knut Erik Berge and I have studied three different series of healthy, unrelated Norwegians, totaling 366 persons in a search for any correlation between the I/D polymorphism and normal blood pressure (Berge and Berg 1994). In the total series, the D allele had a frequency of 0.541 and the I allele a frequency of 0.459. The frequency of the D allele varied from 0.518 in series A with 140 persons to 0.567 in series B with 90 persons. All three series exhibited Hardy-Weinberg equilibrium with respect to this polymorphism (Berge and Berg 1994). There was no significant difference between genotypes with respect to systolic or diastolic blood pressure in any of the three series. We concluded that the I/D polymorphism at the ACE locus does not affect absolute levels of systolic or diastolic blood pressure in healthy people in the population examined.

We have observed heritabilities of 0.64 and 0.51 for systolic and diastolic blood pressure, respectively, in the Norwegian population (Berg, 1993). Genes reflected in the I/D polymorphism at the ACE locus are apparently not among those contributing to the high level of heritability observed for systolic and diastolic blood pressure in this population.

5. THE I/D POLYMORPHISM AT THE ACE LOCUS AND CORONARY HEART DISEASE (CHD)

5.1. Discovery of the Association

Cambien et al., (1992) discovered that homozygosity for the D allele in the I/D polymorphism, which is associated with higher levels of circulating ACE than the two other genotypes, was significantly more frequent in patients with myocardial infarction than in controls ($p = 0.007$). The series examined were large (610 patients with myocardial infarction and 733 controls) and originated from four different centers. The DD genotype was independent of well established risk factors and the authors reasoned that it might be particularly important in persons who previously would have been considered at low risk for myocardial infarction. Accordingly, they divided the series of patients with myocardial infarction into a high risk group and a low risk group. The low risk group included patients not undergoing treatment with lipid-lowering drugs and having body mass index (BMI) as well as apolipoprotein B (apoB) level below the median levels in controls. In this group the association between the D allele and myocardial infarction was highly significant and homogenous across populations (over-all odds ratio, adjusted on population, was 3.2 ($p<0.0001$) with 95% confidence limits 1.7 and 5.9).

Cambien et al. also examined a high risk group, which consisted of all patients that had not been assigned to the low risk group. No association between myocardial infarction and the I/D polymorphism was observed in the high risk group but the strength of the association in the low risk group was such that the association between the DD genotype and myocardial infarction was highly statistically significant also in the total series, after adjustment for age, cigarette consumption, plasma apolipoprotein A-I (apoA-I) level, plasma Lp(a) lipoprotein level, blood pressure and social class. Cambien et al., (1992) inferred that the sensitivity of the D allele as a marker of the postulated "susceptibility" allele was 1 and the specificity 0.77. The authors concluded that the D allele is a very good marker for the susceptibility allele in the Caucasian populations examined.

5.2. Own Studies

We examined 185 male and 49 female survivors of myocardial infarction below 56 and 61 years of age, respectively, and compared them to 366 controls with respect to distribution of genotypes in the I/D polymorphism (Bøhn et al., 1993a). The frequency of the DD genotype was significantly *lower* in patients than controls (22.7% versus 34.9%, $p=0.011$). The association suggested by this difference between patients and controls was in the opposite direction of that observed by Cambien et al., (1992). We constructed a low risk group of people not receiving lipid-lowering drugs and having BMI as well as apoB level below the sex-specific age-adjusted median in healthy people. The prevalence of the DD genotype was not statistically significant between patients and controls in this "low risk group".

In a "high risk" group consisting of those individuals who had not been defined as belonging to the "low risk" group, the prevalence of the DD genotype was 20.9% in male patients and 38.3% in male controls and this difference was highly significant ($p=0.002$). Genotype DD frequency did not differ significantly between female patients and female controls.

We concluded that our results appeared to be at variance with those reported by Cambien et al., (1992) and pointed out the possibility that the different results might be caused by different gene-environment interactions between our population and those examined by Cambien et al., (1992) or that the different results could be a chance event. We emphasized that whereas the patients of Cambien et al. had been examined not later than 9 months after their first myocardial infarction, our patients were examined up to four years after their last myocardial infarction and up to 14 years after the first myocardial infarction, suggesting the possibility that preferential loss of men with the DD genotype could have taken place. However, the Hardy-Weinberg equilibrium that we observed may argue against this notion.

We also emphasized that although the difference did not reach statistical significance, the frequency of the DD genotype was 46.2% in our male "low risk" group versus 32.7% in male controls, a difference in the same direction as that reported by Cambien et al. Clearly, it was still possible that the I/D polymorphism could affect CHD risk even after the seemingly adverse results of our study of people who had suffered premature myocardial infarction.

5.3. Other Studies on the I/D Polymorphism and CHD

Several studies on the apparent association between the I/D polymorphism at the ACE locus and CHD have been reported. In Australian CHD patients a frequency of 0.37 of the DD genotype was observed and this was significantly higher than the frequency of 0.25 in school children (Wang et al., 1996).

In a German study (Gardemann et al., 1995) association between the DD genotype and myocardial infarction was detected in a group of subjects defined to be at low risk according to traditional risk factors whereas ACE activity was not related to disease risk.

In a study from Japan (Nakai et al., 1994) the D allele had a frequency of 0.58 in patients with CHD and of 0.42 in healthy controls and the authors found that this association was particularly pronounced in multiple vessel disease. In another study from Japan (Oike et al., 1995) DD subjects were significantly over-represented in a group of people with coronary artery spasm and the authors hypothesized that the DD genotype may relate to the greater risk of myocardial infarction in patients with coronary artery spasm.

In people with non-insulin dependent diabetes mellitus (NIDDM) (Ruiz et al., 1994), the D allele was associated with early onset CHD, independent of hypertension or hyperlipidemias. A progressively increasing relative risk in individuals heterozygous or homozygous for the D allele was observed (odds ratio 1.41 and 2.35, respectively, $p<0.007$), suggesting a co-dominant effect on the cardiovascular risk. The fraction of CHD attributable to the ACE deletion allele was 24% in the NIDDM population. The authors conclude that identification of NIDDM patients carrying this CHD susceptibility genotype could help to achieve early detection and treatment of CHD (Ruiz et al., 1994).

In a study of cardiomyopathy the DD frequency was higher than in controls in patients with ischemic as well as idiopathic dilated cardiomyopathy (Raynolds et al., 1993). In an Italian population study (Castellano et al., 1995), the DD genotype was associated with significantly increased carotid artery intima thickness.

In the Caerphilly Prospective Heart Disease Study the D allele was marginally more frequent in CHD patients than in controls and the authors concluded that the risk conferred by the DD genotype may be quantitatively small in the general male population (Mattu et al., 1995). Ludwig et al., (1995) found that the frequency of the D allele in 697 angiographically assessed patients (0.54) was virtually identical to the frequency in a sample of 203 representative controls (0.55). There was also no significant difference in frequency of alleles or of genotypes between 362 patients with more than 60% stenosis and 335 patients with less than 10% stenosis.

Different areas of Finland exhibit significant differences in prevalence of CHD and at least part of this variation is believed to be of genetic nature. The frequency of the DD genotype was found to be higher in the high risk area of Eastern Finland than in Western Finland and Perola et al. (1995) suggest that this could be part of the explanation of the differences in CHD frequencies within Finland but also emphasize the point that other risk factors including other genes must contribute to the difference and to the high incidence of myocardial infarction in Finland (Perola et al., 1995). A study of young North Karelian Finns (mean age 41 years) with symptomatic CHD and controls of similar age failed to detect any significant correlation between CHD and the I/D polymorphism (Miettinen et al., 1994).

In a study from Austria (Friedl et al., 1995), 315 CHD patients and 149 controls were examined. No difference in genotype frequencies were observed between patients and controls and no significant association of the DD genotype with coronary heart disease was detectable in sub-groups of patients considered to have a low risk, on the basis of traditional risk factors (Friedl et al., 1995). The authors concluded that the I/D polymorphism at the ACE locus can not be used as a marker for CHD risk.

6. THE I/D POLYMORPHISM AND RESTENOSIS FOLLOWING ANGIOPLASTY

Studies in the literature are controversial with respect to a possible effect of the I/D polymorphism on risk for restenosis following angioplasty. Ohishi et al. (1993) who had previously found a significantly higher frequency of the DD genotype in patients with myocardial infarction than controls (0.45 and 0.16, respectively) studied restenosis in relationship to the I/D polymorphism. They found far more DD individuals (21/32) in patients with restenosis than in patients without restenosis (16/50). At face value, the data strongly suggest an effect of the I/D polymorphism on risk of restenosis. However, van Bockxmeer et al. (1995) found no dear influence of ACE genotype on restenosis and neither Hamon et al. (1995) nor Samani et al. (1995) found any influence of the I/D polymorphism on restenosis.

Although ethnical differences may exist, the available information does not support a major role of the D allele in the development of restenosis.

7. THE I/D POLYMORPHISM AND CEREBROVASCULAR DISEASE

Margaglione et al. (1996) studied 101 persons with stroke and 109 controls and found a higher frequency of people with the DD genotype in stroke patients than in controls (53.5% and 39.4%, respectively, and a risk ratio of 1.76, confidence limits: 1.02–3.05). Markus et al. (1995) also found that the DD genotype was more common in patients with cerebrovascular disease than in control subjects and that the DD genotype conferred a relative risk of any type of cerebrovascular disease of 1.98 (95% confidence interval 1.11–3.51). The association was particularly strong for patients with lacunar stroke. Markus and co-workers found no significant association between genotype in the I/D polymorphism and cerebrovascular disease due to carotid stenosis.

Ueda et al. (1995a) failed to detect association between the DD genotype and ischemic stroke, and Dessi-Fulgheri et al. (1995) failed to detect any correlation between the D allele and carotid arteriosclerosis but emphasized the need for more extensive studies.

Although there is some evidence favoring a role of the I/D polymorphism in cerebrovascular disease, the situation is unresolved.

8. SOME PROBLEMS IN ASSOCIATION STUDIES

8.1. The Meaning of Associations

There are three main causes of association between genetic markers and diseases:

1. There is an effect of a gene, closely linked to the marker gene, on a risk factor or on the pathological process itself.

2. The population examined contains a sub-population characterized by both high frequency of a given genetic marker and of the disease under study (ethnic heterogeneity).
3. There is an effect of the marker gene itself on a risk factor or on the disease process itself.

8.2. Inconsistencies between Association Studies

The inconsistencies between studies on a possible association between the I/D polymorphism and CHD is by no means a new situation. Several possible explanations need to be considered each time inconsistencies are encountered and some of them are well known from studies on other associations.

True differences between populations in disease or risk factor associations could be due to different gene-environment interactions between populations. The product of a given gene may contribute to a disease process only if certain environmental factors are present, or its effect could be obliterated by environmental factors. One example is the polymorphism in the gene for cholesteryl ester transfer protein (CETP) that affects HDL level as well as apoA-I level but this effect is practically obliterated in cigarette smokers (Berg, 1993).

The candidate gene under study could carry mutations in one population that are not present in an ethnically different population and these mutations could render the gene more or less harmful.

A far more trivial reason for differences between association studies is the possibility that the study that led to the first report of an association reflected a random finding in a study where perhaps a large number of comparisons were made. The more analyses are made, the greater the risk of spurious associations. Although a "false" association becomes less likely with increasing level of significance, it remains good advice to always try to confirm an apparent association in an independent second series. "Positive findings" are more likely to find their way into the literature than tests for associations that have come out negatively. Thus, there may always be a risk that many more studies than those actually published have come out negatively. One remedy could be to establish a system whereby any journal which has accepted a paper reporting a particular association would be obliged to also publish very brief reports, according to a standard format, of any studies that have failed to confirm the apparent association. It would be most helpful to the scientific community if information on failed attempts to confirm reported associations were to be easily available.

Another confounding factor could be that comparisons between the examined patient population and the control population are not valid. This could be the case if gene frequencies for the normal genetic marker under study are different in the chosen control population from those in healthy members of the population that the patients were drawn from. For example, there are considerable differences in apoB gene frequencies between Northern and Southern Europe and the inadvertent comparison in an ethically heterogeneous population between patients mainly of Northern European stock and controls mostly of Southern European stock could lead to wrong conclusions. In areas where several ethnic groups are living together, this is far more than a hypothetical worry, since it is not given that a hospital population that patients are drawn from are recruited from the same ethnic group(s) that a control series of people who are not hospitalized, are drawn from. One remedy is to add a control series of hospitalized patients.

Differences between studies, in ascertainment of patients or controls could be one reason for differences between association studies. If people who have recently had their first myocardial infarction are ascertained in one study and people whose first myocardial infarction happened many years ago are ascertained in a second, differences between studies could reflect non-random loss among patients in the series where the first myocardial infarction took place many years ago, or the latter series could have a more benign variant of the disease (since there has been survival for many years).

In rare cases, a presumably healthy control sample does not exhibit Hardy-Weinberg equilibrium, indicating that it may not be representative of the healthy part of the population that the patients were drawn from. The reason is likely to be some unidentified selection bias, and appropriate checks for Hardy-Weinberg equilibrium should always be done with control series.

Differences between studies may also be caused by technical problems in the laboratory. In some cases this possibility may be substantiated from comparing the performance of a commercially available test kit with highly standardized methods of measurements in research laboratories. Confounding factors could be of a biological nature, for example where a heterozygous genotype is more difficult to score that a homozygous, in studies at the protein level.

With respect to the I/D polymorphism at the ACE locus, there is a need to be aware of ethnic differences between populations with respect to gene frequencies (Bloem et al., 1996). With well-standardized PCR and other techniques it should be possible to eliminate technical problems as a cause for inconsistencies between studies.

Each of the above, and maybe other, possible explanations discrepancies between association studies need to be considered in attempts to interpret differences between ACE studies, but no easy single explanation offers itself. It is necessary to remember that in all published series, the "low risk" groups where the association with the D allele may be particularly important were small. The difference, for example between the study between Cambien et al. (1992) and our study (Bøhn et al., 1993a) may well reflect a chance event in a small study.

9. GENETIC MARKERS AND PARENTAL HISTORY OF MYOCARDIAL INFARCTION

9.1. Usefulness of Study of Parental Disease History

The study of parental disease history, such as parental history of myocardial infarction, offers a possibility to circumvent potential problems relating to preferential deaths of people with certain genotypes, in retrospective studies, provided that definite and complete medical information on parents can be obtained. However, ascertainment biases in offspring studies may be difficult to assess. Furthermore, a genetic marker in a child may have come from either the mother or the father so it can not be taken for granted that a given marker originates from the affected parent in the (vast majority of) cases where only one of the parents has CHD.

If proper attention is paid to confounding matters, studies on parental disease could add useful information. An example of important confirmatory evidence resulting from offspring studies, is an early study (Berg et al., 1979) on CHD experiences of parents to individuals with high Lp(a) lipoprotein level that appeared shortly after the association between Lp(a) lipoprotein and CHD had been established (Berg et al., 1974).

9.2. I/D Polymorphism and Parental History of Myocardial Infarction

Tiret et al. (1993) found association between the DD genotype and parental history of fatal myocardial infarction and there was also an excess of people with the ID genotype among those having a father or mother with fatal myocardial infarction. The odds ratio was 2.6 (p=0.02) for the DD genotype and 1.9 (p=0.08) for the ID genotype.

This was part of the same multicenter study that the case-control study of Cambien et al. (1992) originated from. One may have expected that the odds ratios should be lower for parental myocardial infarction because of the "diluting" effect of offspring studies (since a given marker can come from either one of the two parents) but the confidence limits of the odds ratio estimates were sufficiently wide to encompass true values that are somewhat different from those arrived at in this study. The authors argue that although part of the same material was used in both studies, the new study can be considered an independent replication of the previous findings because comparison was based on different groups. They argue that the finding that affected parents transmitted the D allele more frequently than did unaffected parents confirm that genetic variation at the ACE locus could be involved in the risk of myocardial infarction. This is plausible but the question of how representative these studies are with respect to myocardial infarction in general remains open, since any ascertainment bias could apply to both the case-control study of Cambien et al. (1992) and the study by Tiret et al. (1993).

We have examined the I/D polymorphism with respect to parental history of myocardial infarction in Norwegians (Bøhn et al., 1993b). Our study comprised 181 male and 48 female survivors of myocardial infarction as well as 172 male and 194 female controls. Our information on premature myocardial infarction in close relatives originated from the survivors and the controls themselves. A 78% agreement between self-reported history of myocardial infarction in first degree relatives and the diagnosis from doctors' records, hospital records and death certificates has been reported for this population (Førde and Thelle, 1977).

In our total series a larger proportion of DD individuals than of II individuals reported premature parental myocardial infarction (odds ratio =3.1, p=0.009) and a larger proportion of DD individuals than of II individuals reported premature paternal infarction (odds ratio =2.8, p=0.049). The trend across all sub-groups was in harmony with the result in our total series. The frequency of people reporting premature parental myocardial infarction was 14% in DD homozygotes, 10.6% in ID heterozygotes and 6.1% in II homozygotes. We concluded that the ACE polymorphism may be an important genetic marker of myocardial infarction risk and contribute to clustering of cases of premature myocardial infarction in families.

Badenhop et al. (1995) correlated ID genotypes to number of grandparents with a vascular event. Among children with one or no grandparent with a vascular event, 22.6% had the DD genotype. Among children with two or more affected grandparents, 38.4% had the DD genotype. There was an excess of both the DD and the ID genotypes in children with two or more grandparents who had had coronary events, and the association between the number of grandparents who had had coronary events and ACE genotype was significant at the 1% level. Thus, this study is in agreement with the two previous studies showing a relationship between parental history of myocardial infarction in near family and the D allele in the I/D polymorphism at the ACE locus.

The three studies form strong evidence that the D allele in this polymorphism is a marker of familial occurrence of premature CHD. The agreement across three different studies indicate that the finding of association between the D allele and parental CHD is representative for premature CHD in the populations examined.

10. THE I/D POLYMORPHISM AND OTHER PATHOLOGICAL CONDITIONS

Missouris et al. (1996) studying 56 Caucasian subjects with atherosclerotic renal artery stenosis and 74 controls found a higher frequency of the D allele in patients than controls (0.66 and 0.51, respectively), and suggested that the D allele may be associated with increased risk of vascular disease in sites other than the coronary circulation. Other workers have reported evidence that the I/D polymorphism may play a role in genetic predisposition to diabetic nephropathy (Marre et al., 1994; Doria et al., 1994). Tarnow et al., (1995) have suggested that the polymorphism may influence the frequency of life-threatening cardiac complications in patients with insulin dependent diabetes mellitus who have developed diabetic nephropathy, a condition characterized by increased plasma ACE concentration. A possible role of the I/D polymorphism in insulin resistance and glucose intolerance has been suggested (Panahloo et al., 1995; Katsuya et al., 1995). The D allele appears to be associated with progression of IgA nephropathy (Hunley et al., 1996).

Thus, evidence for contribution of the I/D polymorphism to atherosclerotic/thrombotic and related processes originate from several fields of research.

Highly significant associations between the DD genotype and left ventricular hypertrophy have been reported (Iwai et al., 1994; Schunkert et al., 1994; Gharavi et al., 1996), indicating that the I/D polymorphism may form part of a genetic predisposition to develop left ventricular hypertrophy. At face value, this finding further strengthens the notion that this genetic polymorphism may be of considerable importance for the development of cardiovascular disease. However, no association was found between the DD genotype and left ventricular mass in a very recent study by Lindpainter et al., (1996). The reason for the discrepancy is unknown.

11. GENE-GENE INTERACTION

11.1. Gene-Gene Interaction and CHD Risk Factors

We (Pedersen and Berg, 1990) reported the first case of gene-gene interaction affecting CHD risk factors. It had been well established that an allele in the apolipoprotein E (apoE) polymorphism, the apoE4 allele, is associated with significantly elevated cholesterol level and this was the case also in the population examined by us. The population had also been examined with respect to other variables relevant to lipid metabolism, including a normal DNA polymorphism at the locus for the low density lipoprotein (LDL) receptor (LDLR). It turned out that the cholesterol-increasing effect of the apoE4 allele was totally obliterated in people having one specific, normal allele in an LDLR polymorphism. The finding was confirmed in a second series of people and it is reasonable to expect that additional examples of gene-gene interaction affecting CHD risk will be detected.

11.2. Are I/D Genes Involved in Gene-Gene Interaction?

Several traditional CHD risk factors are influenced by genes. Already the discoverers of the association between myocardial infarction and the D allele carefully searched for any association between genotype in the polymorphism and traditional CHD risk factors (including Lp(a) lipoprotein) (Cambien et al., 1992). No differences between genotypic groups were detected and the authors concluded that the D allele is an independent risk factor for CHD.

We have conducted comparisons between genotypic groups in 3 different series of unrelated people with respect to total cholesterol, high density lipoprotein (HDL) cholesterol, LDL cholesterol, triglycerides, apoA-I, Lp(a) lipoprotein and BMI. No significant difference between genotypic groups was observed with respect to these risk factors (Berge and Berg, in preparation) or with respect to blood pressure (as reported previously; Berge and Berg 1993).

The aggregated information from 3 series of unrelated people examined in our laboratory and series from 4 different centers examined by Cambien et al. (1992) makes it highly unlikely that a significant association exists between the I/D polymorphism and any of the risk factors examined. Thus, the contribution of the I/D polymorphism to the development of atherosclerotic/thrombotic disease is not affected through traditional risk factors.

Nevertheless, some claims of correlation between the I/D polymorphism and other risk factors have appeared in the literature. Thus, Badenhop et al. (1995) have reported association between ACE genotype and Lp(a) lipoprotein concentration, the highest Lp(a) lipoprotein level being observed in the DD genotype. However, the difference between genotypes did not reach formal statistical significance. Several comparisons were made, which increases the risk of "false positive" correlations and with the above studies that did not uncover any correlation between the I/D polymorphism and Lp(a) lipoprotein concentration, it seems plausible that the borderline association reported by Badenhop et al. was a chance event. The study by Mattu et al. (1995) also failed to uncover any association between the I/D polymorphism and Lp(a) lipoprotein level.

A suggestion of gene-gene interaction involving the I/D polymorphism was reported by van Bockxmeer (1994). He found that there was no effect of ACE genotype with respect to restenosis in people who did not possess the E4 allele in the apoE polymorphism, whereas there was a significant effect of the D allele in people possessing the apoE4 allele. Further studies are needed to confirm or refute this interesting observation.

12. VARIABILITY GENES

12.1. The Variability Gene Concept

The study of marker genes in relation to diseases or disease risk factors is no longer limited to attempts to detect association between genetic markers and absolute risk factor levels. In addition to contributing to absolute risk factor level, genes are presumably important also for determining the framework within which environmental factors can cause variation in a given risk factor. The latter category of genes may be referred to as "variability genes" as opposed to "level genes" that affect absolute risk factor level (Berg, 1994a). We have devised a way to detect variability genes based on the fact that monozygotic (MZ) twin pairs have identical genes so that any difference in a quantitative parameter between the two members of an MZ pair must be caused by environmental factors. A restrictive variability gene would be detected by a low within-pair difference in a given parameter in MZ pairs possessing as opposed to those lacking the gene, whereas a permissive variability gene would be detected by its opposite effect.

Early studies on random genetic markers as well as studies of candidate genes with respect to atherosclerotic disease have confirmed the validity of the variability gene concept (for review, see Berg, 1994a). Thus, normal genetic variants in DNA at the apoB locus appear to have variability gene effects with respect to apoB concentration as well as BMI. LPA genes (for review, see Berg, 1994c) may have level gene as well as variability gene effects on fibrinogen (Berg, 1996).

The rationale behind the development of the variability gene concept was the fact that high responder and low responder phenotypes with respect to response to dietary cholesterol intake that are almost certainly genetically determined exist in animals and are likely to exist in man. Furthermore, there was information in the literature to suggest that the amount of variability over time in a risk factor such as cholesterol may be as important as absolute risk factor concentration, with respect to CHD risk (for review, see Berg, 1994a).

Finally, we hypothesized that a given person's total genetic risk for CHD may be the result of his combination of level genes and variability genes. If this were to be proven, it may over time have consequences for patient management, for example with respect to how long one would want to try to achieve risk factor reduction by non-pharmacological means, in a given patient. It seems plausible that a person with high risk factor level and restrictive variability genes should not waste too long time striving to achieve a perhaps moderate risk factor reduction by non-pharmacological means.

12.2. Applying the Variability Gene Concept to the I/D Polymorphism and Blood Pressure Variability

Berge and Berg (1994) analyzed two independent series of MZ twins with respect to any variability gene effect of alleles in the I/D polymorphism on blood pressure. There were no differences between genotypic groups in within-pair variation in systolic or diastolic blood pressure to suggest variability gene effects of this polymorphism, on normal blood pressure.

12.3. Are There Other Suggestions of Variability Gene Effects of the I/D Polymorphism?

To our knowledge, search for variability gene effects of alleles in this polymorphism have not been conducted with respect to ACE activity. Such a study might be of interest and could help clarify the absence of a direct effect of the I/D polymorphism on blood pressure despite the effect on ACE activity.

The response of patients to anti-hypertensive therapy is variable. Dudley et al. (1996) attempted to correlate differences in responses to drugs, to genotypes in the ACE polymorphism but failed to find any significant difference between genotypic groups. The DD homozygotes exhibited the largest fall in systolic as well as diastolic blood pressure on atenolol treatment but the difference from the other genotypic groups did not reach statistical significance. Accordingly, there is as yet no evidence that the I/D polymorphism may exert variability gene effects with respect to response to pharmacologic treatment.

Ueda et al. (1995b) examined pressor responses in 10 DD and 10 II homozygotes. Responses were expressed as PD_{20} which reflects the angiotensin dose required to raise mean blood pressure by 20 mm Hg. The PD_{20} for angiotensin I in subjects with DD genotype was significantly lower than that in II subjects (8.8 versus 14.8 ng/kg per minute;

p=0.0091). If confirmed, it would seem that the I/D polymorphism may exert variability gene effect with respect to pressor responses.

Tiret et al. (1994) reported synergistic effect of genotype in the ACE polymorphism and genotype in an angiotensin-II type 1 receptor (AGT_1R) gene polymorphism with respect to myocardial infarction and suggested that the previously reported association between the DD genotype in the ACE polymorphism and myocardial infarction may be restricted to individuals who also carry the C allele in the AGT_1R polymorphism. If this observation is confirmed, it would be yet another example of gene-gene interaction of importance for risk of cardiovascular disease.

13. LONGEVITY GENES

13.1. Genetic Predisposition to CHD and Longevity

Cardiovascular disease is the major cause of death in Western societies and genes conferring predisposition to or protection against such disease may therefore well affect longevity.

The first example of an effect on longevity of genes that influence CHD risk was published by Berg and Rø (1991). The Lp(a) lipoprotein (Berg, 1963) is under strict genetic control by the LPA locus near the end of the long arm of chromosome 6 and high level of Lp(a) lipoprotein is a risk factor for CHD (for review, see Berg, 1994c). Among people who had reached the age of 83 or more and were still in good health and managing on their own, there was a shortage of persons with Lp(a) lipoprotein values corresponding to the top quartile of the distribution of levels in healthy middle-aged people. Presumably, this shortage was caused by selective loss of persons with high Lp(a) lipoprotein level in the sample of old people. Genes specifying lower levels of Lp(a) lipoprotein therefore appeared to be "longevity genes". Interestingly, the corresponding excess of old people with lower Lp(a) lipoprotein levels were not found in the bottom quartile but in the second to lowest quartile, perhaps suggesting that having a moderate amount of Lp(a) lipoprotein rather than a minimal amount, increases the chance of reaching old age in good health.

There are recent data in the literature to indicate that genes in a different polymorphism that influences CHD risk, the apoE polymorphism, affect longevity (Davignon et al., 1988; Kervinen et al., 1994; Schächter et al., 1994). The paper by Schächter et al. (1994) is particularly interesting, showing both a shortage of people possessing the apoE4 allele and an excess of people possessing the apoE2 allele in 325 centenarians compared to 161 controls. The shortage of people possessing an apoE4 allele presumably reflects selective loss of people with that allele whereas the disproportional excess of people possessing the apoE2 allele points to a protective effect of the apoE2 gene. Hence, the apoE2 gene appears to be a "longevity gene".

13.2. Longevity Genes in the I/D Polymorphism at the ACE Locus?

Schächter et al. (1994) reported genotypes in the I/D polymorphism at the ACE locus in 310 centenarians and 164 controls. They made the most surprising observation that the frequency of the DD genotype was markedly *higher* in centenarians than in controls (0.396 *versus* 0.256). The authors calculated that 18.6% of the centenarians could be attributable to the DD genotype. There was a trend towards a negative correlation between the apoE2 allele in the apoE polymorphism and the D allele in the I/D polymorphism at the ACE locus, but the trend did not reach statistical significance. The authors speculate that the unexpected association between the D allele and longevity may indicate a late reversal of the presumed negative survival value of the D allele.

CONCLUDING REMARKS

The keen interest in the I/D polymorphism and its possible relationship to disease in man has led to a plethora of papers and new information is being reported continuously. This flow of information has, however, not settled all matters. Although there is support to the finding by Cambien et al. (1992) of an association between the D allele and myocardial infarction, there are also papers reporting failure to confirm the association. There are conflicting papers with respect to the possible effect of the I/D polymorphism on restenosis following angioplasty and also with respect to the possible role of the polymorphism in the development of left ventricular hypertrophy. The sum of information indicates that any effect of the I/D polymorphism on cardiovascular disease is not mediated through traditional risk factors and the claim of an association between the D allele and high level of Lp(a) lipoprotein is contradicted in extensive studies on several different series of people in at least three laboratories.

The most consistent evidence connecting the I/D polymorphism with CHD risk is from the studies of healthy offspring or grandchildren of people who have suffered myocardial infarction. These studies form a bridge between CHD and the I/D polymorphism that can not be disregarded. Since the I/D polymorphism resides within an intron in the ACE gene the plausible interpretation is that any effect on CHD risk is caused by close proximity to functionally important domains in or near the ACE gene. Further studies are needed to identify these domains, to resolve the above issues and to understand underlying mechanisms at the molecular level. Much work remains to be done also to solve questions relating to gene-gene interaction, possible variability gene effects or a role as "longevity genes" of genes in this polymorphism. There is little doubt that the I/D polymorphism at the ACE locus will continue to pose considerable research challenges in the time to come.

ACKNOWLEDGMENTS

Work in the author's laboratory was supported by the Norwegian Council on Cardiovascular Disease, Anders Jahres Foundation for the Promotion of Science and the Norwegian Research Council for Science and the Humanities.

REFERENCES

Badenhop, R.F., Wang, X.L., and Wilcken D.E.L. Angiotensin-converting enzyme genotype in children and coronary events in their grandparents. *Circulation,* 91:1655–1658, 1995.

Berg, K. A new serum type system in man - the Lp system. *Acta. Path. Microbiol. Scand.,* 59:369–382, 1963.

Berg, K. Genetic and environmental factors in the development of cardiovascular disease. In: *Biologie prospective. Comptes rendus du 8e Collogue de Pont-a-Mousson*, Galteau, M.M., Siest, G., and Henry, J., editors. John Libbey Eurotext, Paris pp. 471–480, 1993.

Berg, K. Gene-environment interaction: variability gene concept. In: *Genetic Factors in Coronary Heart Disease*, Goldbourt, U., de Faire, U., Berg, K., editors. Kluwer Academic Publishers, Dordrecht/Boston/London, pp. 373–383, 1994a.

Berg, K. From random genetic markers to candidate genes in association and linkage studies of coronary heart disease and its risk factors. In: *Genetic Factors in Coronary Heart Disease*, Goldbourt, U., de Faire, U., Berg, K., editors. Kluwer Academic Publishers, Dordrecht/Boston/London, pp. 301–308, 1994b.

Berg, K. Lp(a) lipoprotein: a monogenic risk factor for cardiovascular disease. In: *Genetic Factors in Coronary Heart Disease*, Goldbourt U, de Faire U, Berg K, editors. Kluwer Academic Publishers, Dordrecht/Boston/London, pp. 275–287, 1994c.

Berg, K. Lp(a) lipoprotein and risk of atherosclerotic/thrombotic disease. Third International Fibrinogen Symposium, *Ulm Fibrinolysis*, 1996, In Press.

Berg, K, Dahlén G., and Børresen, A.L. Lp(a) phenotypes, other lipoprotein parameters, and a family history of coronary heart disease in middle-aged males. *Clin. Genet.*, 16:347–352, 1979.

Berg, K., Dahlén, G., and Frick, M.H. Lp(a) lipoprotein and pre-β_1-lipoprotein in patients with coronary heart disease. *Clin. Genet.*, 6:230–235, 1974.

Berg, K., and Rø, O.C. Lp(a) lipoprotein level and longevity. *Annales de Génétique*, 34:264–269, 1991.

Berge, K.E. and Berg K. No effect of insertion/deletion polymorphism at the ACE locus on normal blood pressure level or variability. *Clin. Genet.*, 45:169–174, 1994.

Bloem, L.J., Manatunga, A.K., and Pratt, JH. Racial difference in the relationship of an angiotensin I-converting enzyme gene polymorphism to serum angiotensin I-converting enzyme activity. *Hypertension*, 27:62–66, 1996.

van Bockxmeer, F.M. ApoE and ACE genes: impact on human longevity. *Nature Genet.*, 6:4–5, 1994.

van Bockxmeer, F.M., Mamotte, C.D.S., Gibbons, F.A., Burke, V., and Taylor, R.R. Angiotensin-converting enzyme and apolipoprotein E genotypes and restenosis after coronary angioplasty. *Circulation*, 92:2066–2071, 1995.

Bøhn, M., Berge, K.E., Bakken, A., Erikssen, J., and Berg, K. Insertion/deletion (I/D) polymorphism at the locus for angiotensin I-converting enzyme and myocardial infarction. *Clin. Genet.*, 44: 292–297, 1993a.

Bøhn, M., Berge, K.E., Bakken, A., Erikssen, J., and Berg K. Insertion/deletion (I/D) polymorphism at the locus for angiotensin I-converting enzyme and parental history of myocardial infarction. *Clin. Genet.*, 44: 298–301, 1993b.

Cambien, E., Alhenc-Gelas, F., Herbeth, B., Andre, J.L., Rakotovao, R., Gonzales, M.F., Allegrini, J., and Bloch, C. Familial resemblance of plasma angiotensinconverting enzyme level: The Nancy Study. *Am. J. Hum. Genet.*, 43:774–780, 1988.

Cambien, E., Poirier, O., Lecerf, L., Evans, A., Cambou, J.P., Arveiler, D., Luc, G., Bard, J.P., Bara, L., Ricard, S., Tiret, L., Amouyel, P., Alhenc-Gelas, F., and Soubrier, F. Deletion polymorphism in the gene for angiotensin-converting enzyme is a potent risk factor for myocardial infarction. *Nature*, 359:641–644, 1992.

Castellano, M., Muiesan, M.L., Rizzoni, D., Beschi, M., Pasini, G., Cinelli, A., Salvetti, M., Porteri, L.E., Bettoni, G., Kreutz, R., Lindpainter, K., and Rosei, E.A. Angiotensin-converting enzyme I/D polymorphism and arterial wall thickness in a general population. *Circulation*, 91:2721–2724, 1995.

Davignon, J., Gregg, R.E., and Sing, CF. Apolipoprotein E polymorphism and atherosclerosis. *Arteriosclerosis*, 8:1–21, 1988.

Dessi-Fulgheri, P., Catalini, R., Sarzani, R., Sturbini, S., Siragusa, N., Guazzarotti, F., Offidani, M., Tamburrini, P., Zingaretti, O., and Rappelli, A. Angiotensin converting enzyme gene polymorphism and carotid atherosclerosis in a low-risk population. *J. Hypertension*, 13:1593–1596, 1995.

Doria, A., Warram, J.H., and Krolewski, A.S. Genetic predisposition to diabetic nephropathy. Evidence for a role of the angiotensin I-converting enzyme gene. *Diabetes*, 43:690–695, 1994.

Dudley, C., Keavney, B., Casadei, B., Conway, J., Bird, R., and Ratcliffe, P. Prediction of patient responses to antihypertensive drugs using genetic polymorphisms: investigation of renin-angiotensin system genes. *J. Hypertension*, 14:259–262, 1996.

Friedl, W., Krempler, F., Paulweber, B., Pichler, M., and Sandhofer, F. A deletion polymorphism in the angiotensin converting enzyme gene is not associated with coronary heart disease in an Austrian population. *Atherosclerosis*, 112:137–143, 1995.

Førde, O.H. and Thelle, D.S. The Tromsø Heart Study: risk factors for coronary heart disease related to the occurrence of myocardial infarction in first-degree relatives. *Am. J. Epidemiol*, 105:192–199, 1977.

Gardemann, A., Weiss, T., Schwartz, O., Eberbach, A., Katz, N., Hehrlein, F.W., Tillmanns, H., Waas, W., and Haberbosch, W. Gene polymorphism but not catalytic activity of angiotensin 1-converting enzyme is associated with coronary artery disease and myocardial infarction in low-risk patients. *Circulation*, 92:2796–2799, 1995.

Gharavi, A.G., Lipkowitz, M.S., Diamond, J.A., Jhang, J.S., and Phillips, R.A. Deletion polymorphism of the angiotensin-converting enzyme gene is independently associated with left ventricular mass and geometric remodeling in systemic hypertension. *Am. J. Cardiol.*, 77:1315–1319, 1996.

Hamon, M., Bauters, C., Amant, C., McFadden, E.P., Helbecque, N., Lablanche, J.M., Bertrand, M.E., and Amouyel, P. Relation between the deletion polymorphism of the angiotensin-converting enzyme gene and late luminal narrowing after coronary angioplasty. *Circulation,* 92:296–299, 1995.

Harrap, S.B., Davidson, H.R., Conner, J.M., Soubrier, F., Corvol, P., Fraser, R., Foy, C.J.W., and Watt, G.C.M. The angiotensin I converting enzyme gene and predisposition to high blood pressure. *Hypertension,* 21:455–460, 1993.

Higashimori, K., Zhao, Y., Higaki, J., Kamitani, A., Katsuya, T., Nakura, J., Miki, T., Mikami, H., and Ogihara, T. Association analysis of a polymorphism of the angiotensin converting enzyme gene with essential hypertension in the Japanese population. *Biochem. Biophys. Res. Commun.,* 191:399–404, 1993.

Hubert, C., Houot, A.M., Corvol, P., and Soubrier, F. Structure of the angiotensin I-converting enzyme gene. *J. Biol. Chem.,* 266:15377–15383, 1991.

Hunley, T.E., Julian, B.A., Phillips, J.A., Summar, M.L., Yoshida, H., Horn, R.G., Brown, N.J., Fogo, A., Ichikawa, I., and Kon, V. Angiotensin converting enzyme gene polymorphism: potential silencer motif and impact on progression in IgA nephropathy. *Kindey International,* 49:571–577, 1996.

Iwai, N., Ohmichi, N. Nakamura, Y., and Kinoshita, M. DD genotype of the angiotensin converting enzyme gene is a risk factor for left ventricular hypertrophy. *Circulation,* 90:2622–2628, 1994.

Jeunemaitre, X., Lifton, R.P., Hunt, S.C., Williams, R.R., and Lalouel, J.M. Absence of linkage between the angiotensin converting enzyme locus and human essential hypertension. *Nature Genetics,* 1:72–75, 1992.

Katsuya, T., Horiuchi, M., Chen, Y-D.I., Koike, G., Pratt, R.E., Dzau, V.J., and Reaven, G.M. Relations between deletion polymorphism of the angiotensin converting enzyme gene and insulin resistance, glucose intolerance, hyperinsulinemia, and dyslipidemia. *Arterioscler. Thromb. Vasc. Biol.,* 15:779–782, 1995.

Kervinen, K., Savolainen, M.J., Salokannel, J., Hynninen, A., Heikkinen, J., Ehnholm, C., Koistinen, M.J., Kesäniemi, Y.A. Apolipoprotein E and B polymorphisms-longevity factors assessed in nonagenarias. *Atherosclerosis,* 105:89–95, 1994.

Lindpainter, K., Lee, M., Larson, M.G., Rao, V.S., Pfeffer, M.A., Ordovas, J.M., Schaefer, E.J., Wilson, A.F., Wilson, P.W.F., Vasan, R.S., Myers, R.H., and Levy, D. Absence of association or genetic linkage between the angiotensin-converting-enzyme gene and left ventricular mass. *N. Engl. J. Med.,* 334:1023–1028, 1996.

Ludwig, E., Corneli, P.S., Anderson, J.L., Marshall, H.W., Lalouel, J.M., and Ward, R.H. Angiotensin-converting enzyme gene polymorphism is associated with myocardial infarction but not with development of coronary stenosis. *Circulation,* 91:2120–2124, 1995.

Margaglione, M., Celentano, E., Grandone, E., Vecchione, G., Cappucci, G., Giuliani, N., Colaizzo, D., Panico, S., Mancini, F.P., and Di Minno, G. Deletion polymorphism in the angiotensin-converting enzyme gene in patients with a history of ischemic stroke. *Arterioscler. Thromb. Vasc. Biol.,* 16:304–309, 1996.

Markus, H.S., Barley, J., Lunt, R., Bland, J.M., Jeffery, S., Carter, N.D., and Brown, M.M. Angiotensin-converting enzyme gene deletion polymorphism. *Stroke,* 26:1329–1333, 1995.

Marre, M., Bernadet, P., Gallois, Y., Savagner, F., Guyene, T.T., Hallab, M., Cambien, F., Passa, P., and Alhenc-Gelas, F. Relationships between angiotensin I converting enzyme gene polymorphism, plasma levels, and diabetic retinal and renal complications. *Diabetes,* 43:384–388, 1994.

Mattu, R.K., Needham, E.W.A., Galton, D.J., Frangos, E., Clark, A.J.L., and Caulfied, N. A DNA variant at the angiotensin-converting enzyme gene locus associates with coronary artery disease in the Caerphilly Heart Study. *Circulation,* 91:270–274, 1995.

McKenzie, C.A., Julier, C., Forrester, T., McFarlane-Anderson, N., Keavney, B., Lathrop, G.M., Ratcliffe, P.J., and Farrall, M. Segregation and linkage analysis of serum angiotensin I-converting enzyme levels: evidence for two quantitative trait loci. *Am. J. Hum. Genet.,* 57:1426–1435, 1995.

Miettinen, H.E., Korpela, K., Hämäläinen, L., and Kontula, K. Polymorphisms of the apolipoprotein and angiotensin converting enzyme genes in young North Karelian patients with coronary heart disease. *Hum. Genet.,* 94:189–192, 1994.

Missouris, C.G., Barley, J., Jeffery, S., Carter, N.D., Singer, D.R.J., and MacGregor, G.A. Genetic risk for renal artery stenosis: association with deletion polymorphism in angiotensin 1-converting enzyme gene. *Kidney International,* 49:534–537, 1996.

Nakai, K., Itoh, C., Miura, Y., Hotta, K., Musha, T., Itoh, T., Miyakawa, T., Iwasaki, R. and Hiramori, K. Deletion polymorphism of the angiotensin I-converting enzyme gene is associated with serum ACE concentration and increased risk for CAD in the Japanese. *Circulation,* 90:2199–2202, 1995.

Ohishi, M., Fujii, K., Minamino, T., Higaki, J., Kamitani, A., Rakugi, H., Zhao, Y., Mikami, H., Miki, T., and Ogihara, T. A potent genetic risk factor for restenosis. *Nature Genet.,* 5:324–325,1993.

Oike, Y., Hata, A., Ogata, Y., Numata, Y., Shido, K., and Kondo, K. Angiotensin converting enzyme as a genetic risk factor for coronary artery spasm. *J. Clin. Invest.,* 96:2975–2979, 1995.

Panahloo, A., Andres, C., Mohamed-Ali, V., Gould, M.M., Talmud, P., Humphries, S.E., and Yudkin, J.S. The insertion allele of the ACE gene I/D polymorphism. *Circulation,* 92:3390–3393, 1995.

Pedersen, J.C. and Berg, K. Gene-gene interaction between the low density lipoprotein receptor and apolipoprotein E loci affects lipid levels. *Clin. Genet.*, 38:287–294, 1990.

Perola, N., Sajantila, A., Sarti, C., Stengård, J., Tamminen, M., Puska, P., Juttunen, J., Tuomilehto, J., and Peltonen, L. Angiotensin-converting enzyme genotypes in the high- and low-risk area for coronary heart disease in Finland. *Genet. Epidemiol.*, 12:391–399, 1995.

Raynolds, M.V., Bristow, M.R., Bush, E.W., Abraham, W.T., Lowes, B.D. Zisman, L.S., Taft, C.S., and Perryman, M.B. Angiotensin-converting enzyme DD genotype in patients with ischemic or idiopathic dilated cardiomyopathy. *Lancet*, 342:1073–1075, 1993.

Rigat, B., Hubert, C., Alhenc-Gelas, F., Cambien, F., Corvol, P., and Soubrier, F. An insertion/deletion polymorphism in the angiotensin I-converting enzyme gene accounting for half the variance of serum enzyme levels. *J. Clin. Invest.*, 86:1343–1346, 1990.

Ruiz, J., Blanche, H., Cohen, N., Velho, G., Cambien, F., Cohen, D., Passa, P., and Froguel, P. Insertion/deletion polymorphism of the angiotensin-converting enzyme gene is strongly associated with coronary heart disease in non-insulin-dependent diabetes mellitus. *PNAS*, 91:3662–3665, 1994.

Samani, N.J., Martin, D.S., Brack, M., Cullen, J., Chauhan, *A.*, Lodwick, D., Harley, A., Swales, J.D., de Bono, D.P., and Gershlick, AH. Insertion/deletion polymorphism in the angiotensin-converting enzyme gene and risk of restenosis after coronary angioplasty. *Lancet*, 345:1013–1016, 1995.

Schmidt, S., van Hooft, I.M.S., Grobbee, D.E., Ganten, D., and Ritz, E. Polymorphism of the angiotensin I converting enzyme gene is apparently not related to high blood pressure: Dutch hypertension and offspring study. *J. Hypertension*, 11:345–348, 1993.

Schunkert, H., Hense, H.W., Holmer, S.R., Stender, M., Perz, S., Keil, U., Lorell, B.H., and Riegger, G.A.J. Association between a deletion polymorphism of the angiotensin converting enzyme gene and left ventricular hypertrophy. *N. Engl. J. Med.*, 330:1634–1638, 1994.

Schächter, F., Faure-Delanef, L., Cuenot, F., Rouger, H., Froguel, P., Lesueur-Ginot, L., and Cohen, D. Genetic associations with human longevity at the APOE and ACE loci. *Nature Genetics*, 6:29–32, 1994.

Tarnow, L., Cambien, F., Rossing, P., Nielsen, F.S., Hansen, B.V., Lecerf, L., Poirier, O., Danilov, S., Boelskifte, S., Borch-Johnsen, K., Parving, H.H. Insertion/deletion polymorphism in the angiotensin I-converting enzyme gene is associated with coronary heart disease in IDDM patients with diabetic nephropathy. *Diabetologia*, 38:798–803, 1995.

Tiret, L., Bonnardeaux, A., Poirier, 0., Ricard, S., Margues-Vidal, P., Evans, A., Arveiler, D., Luc, S., Kee, F., Ducimetiere, P., Soubrier, F., and Cambien, F. Synergistic effects of angiotensin-converting enzyme and angiotensin-II type 1 receptor gene polymorphisms on risk of myocardial infarction. *Lancet*, 344:910–913, 1994.

Tiret, L., Kee, F., Poirier, 0., Nicaud, V., Lecerf, L., Evans, A., Cambou, J.P., Arveiler, D., Luc, S., Amouyel, P., and Cambien, F. Deletion polymorphism in angiotensin converting enzyme gene associated with parental history of myocardial infarction. *Lancet*, 341:991–992,1993.

Tiret, L., Rigat, B., Visvikis, S., Breda, C., Corvol, P., Cambien, F., and Soubrier, F. Evidence, from combined segregation and linkage analysis, that a variant of the angiotensin 1-converting enzyme (ACE) gene controls plasma ACE levels. *Am. J. Hum. Genet.*, 51:197–205, 1992.

Ueda, S., Weir, C.J., Inglis, S.C., Murray, G.D., Muir, K.W., and Lees, K.R. Lack of association between angiotensin converting enzyme gene insertion/deletion polymorphism and stroke. *J. Hypertension*, 13:1597–1601, 1995a.

Ueda, S., Elliott, H.L., Morton, J.J., and Connell, J.M.C. Enhanced pressor response to angiotensin I in normotensive men with the deletion genotype (DD) for angiotensin-converting enzyme. *Hypertension*, 25:1266–1269, 1995b.

Wang, X.L., McCredie, R.M., and Wilcken, D.E.L. Genotype distribution of angiotensin-converting enzyme polymorphism in Australian healthy and coronary populations and relevance to myocardial infarction and coronary artery disease. *Arterioscler. Thromb. Vasc. Biol.*, 16:115–119, 1996.

III. THROMBOSIS/HYPERTENSION

7

ROLE OF ENDOTHELIAL PLASMINOGEN ACTIVATORS IN FIBRINOLYSIS AND REPAIR-ASSOCIATED ANGIOGENESIS

Two Sides of a Coin

Victor W. M. van Hinsbergh, Pieter Koolwijk, Erik Ponfoort, Roeland Hanemaaijer, Jef. J. Emeis, Teake Kooistra, and Paul H. A. Quax

Gaubius Laboratory TNO-PG
Leiden, The Netherlands

1. INTRODUCTION

Fibrin is a temporary matrix, which is formed after wounding a blood vessel and when plasma leaks from blood vessels forming a fibrous exudate, often seen in areas of inflammation and in tumors (Dvorak et al., 1992). The fibrin matrix not only acts as a barrier preventing further blood loss, but also provides a structure in which new microvessels can infiltrate during wound healing. Proper timing of the outgrowth of microvessels as well as the subsequent (partial) disappearance of these vessels is essential to ensure adequate wound healing and to prevent the formation of scar tissue. It is generally believed that plasminogen activators play an important role in the migration and invasion of leukocytes and endothelial cells, and in the dissolution of the fibrin matrix (Kwaan, 1966; Pepper et al., 1990; Koolwijk et al., 1996). Plasminogen activators are serine proteases, which enzymatically convert the zymogen plasminogen into the active protease plasmin, the prime protease that degrades fibrin. The production of plasminogen activators by endothelial cells not only contributes to the proteolytic events related to the formation of microvessels in a wound, but also plays a crucial role in the prevention of thrombosis. If fibrin becomes deposited within the lumen of a blood vessel, cessation of the blood flow may occur accompanied by ischemia and eventually death of the distal tissues. The endothelium contributes considerably to the maintenance of blood fluidity by exposing anticoagulant molecules, by providing factors that interfere with platelet aggregation, and by its ability to stimulate fibrinolysis. Lysis of intravascularly generated fibrin must occur rapidly. However, it should be limited to a local area, because a general elevation of fibrinolysis upon wounding would result in recurrent bleeding. Hence, endothelial cells apply

Vascular Endothelium, edited by Catravas *et al.*
Plenum Press, New York, 1998

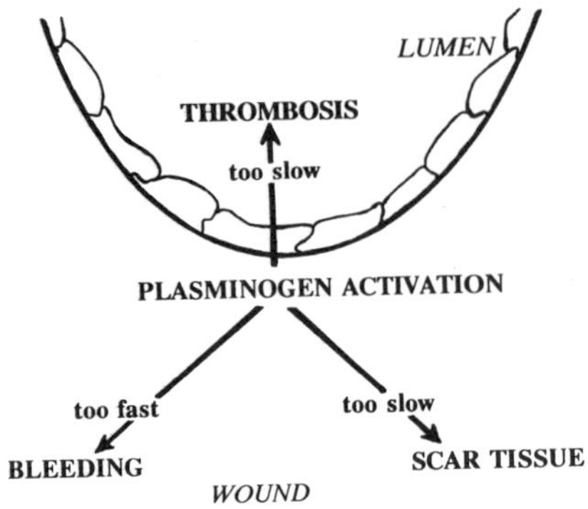

Figure 1. Plasminogen activation by the endothelium is controlled in time and space to meet various functions: prevention of thrombosis without bleeding and angiogenesis, which is necessary for adequate wound repair. (From: van Hinsbergh, Ann. N.Y. Acad. Sci. 667:151, 1992; with permission).

plasminogen activators for initial events in angiogenesis, for the degradation of the temporary fibrin matrix during wound healing, and for the immediate dissolution of a fibrinous thrombus originating within a blood vessel (Van Hinsbergh, 1992) (Figure 1). These cells are able to execute all these processes adequately by orchestrating in time and space the synthesis and action of two types of plasminogen activators as well as that of specific inhibitors and cellular receptors for plasminogen activators and plasminogen.

2. COMPONENTS OF THE PLASMIN/PLASMINOGEN ACTIVATOR SYSTEM

Figure 2 summarizes the proteases and the inhibitors involved in fibrinolysis. Fibrin degradation and probably also activation of several matrix metalloproteinases is accomplished by the serine protease plasmin, which is formed from its zymogen plasminogen by

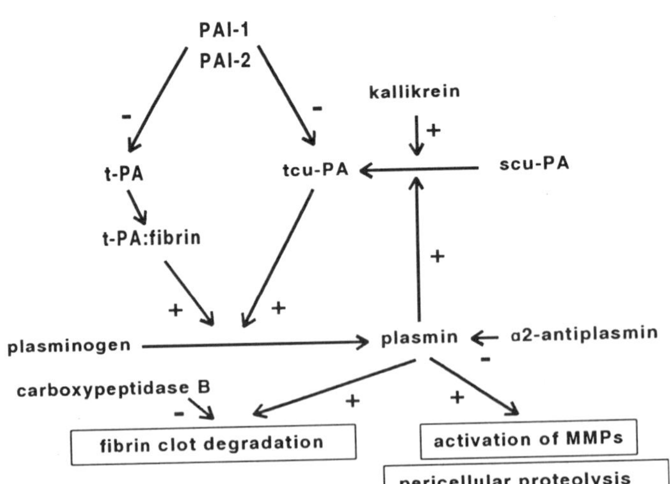

Figure 2. Schematic representation of the plasminogen activation system. +: activation; –: inhibition. Inhibitors are indicated in italics. PA: plasminogen activator; t-PA: tissue-type PA; u-PA: urokinase-type PA; sc-u-PA; single chain u-PA; PAI: PA inhibitor.

plasminogen activators (PAs). The actual activities of plasmin and the PAs are regulated not only by their concentration and activation, but also by their interaction with inhibitors, cellular receptors, and matrix proteins.

2.1. Proteases

Plasmin is formed from its zymogen plasminogen by proteolytic activation by PAs. Two types of mammalian PAs are presently known: tissue-type plasminogen activator (t-PA) and urokinase-type plasminogen activator (u-PA) (Bachmann, 1987; Wallén, 1987). The three serine proteases, plasminogen, t-PA and u-PA, are synthesized as single polypeptide chains, and each of them is converted by specific proteolytic cleavage to a molecule with two polypeptide chains connected by a disulphide bond. The carboxy-terminal part of the molecule (the so-called B-chain) contains the proteolytically active site, whereas the amino-terminal part of the molecule (the A-chain) is built up of domains that determine the interaction of the proteases with matrix proteins and cellular receptors. The proteolytic cleavage of plasminogen and single-chain u-PA to their respective two-chain forms is necessary to disclose the proteolytically active site and to activate the molecule. In contrast, t-PA activity markedly increases by interaction of t-PA with fibrin. Once bound to this substrate, both the single-chain and two-chain form of t-PA are active. The interaction of plasminogen with fibrin or the cell surface occurs predominantly via binding sites in the kringle structures, which recognize lysine residues of proteins, in particular carboxy-terminal lysines. Because B-type carboxypeptidases remove carboxy-terminal lysine residues from potential binding sites for plasminogen in fibrin or on the cell surface, they can act as negative regulators of the fibrinolytic system (Redlitz et al., 1995).

2.2. Inhibitors

The activities of the proteases of the fibrinolytic system are controlled by potent inhibitors, which are members of the serine protease inhibitor (serpin) superfamily. Plasmin, if not bound to fibrin, is instantaneously inhibited by α2-antiplasmin (Holmes et al., 1987). Because this interaction is facilitated by the lysine binding domain of plasmin, it is attenuated when plasmin is bound to fibrin. The predominant regulators of t-PA and u-PA activities are PAI-1 and PAI-2 (Fearns et al., 1996; Bachman, 1995). PAI-1 is a 50 kD glycoprotein present in blood platelets and synthesized by endothelial cells, smooth muscle cells and many other cell types in culture (Sprengers and Kluft, 1987; Loskutoff, 1991). PAI activity in human plasma is normally exclusively PAI-1. PAI-1 binds to vitronectin, which stabilizes its inhibitory activity. PAI-1 is the main if not the sole inhibitor of PAs synthesized by endothelial cells, vascular smooth muscle cells and hepatocytes. PAI-2 is produced by monocytes/macrophages, placental trophoblasts and certain tumor cell lines (Bachman, 1995). It can be found as a glycosylated secreted molecule and as an non-glycosylated molecule intracellularly (Wohlwend et al., 1987).

2.3. Receptors

Regulation of fibrinolytic activity also occurs by cellular receptors. These receptors direct the action of PAs and plasmin to focal areas on the cell surface, or are involved in the clearance of the PAs (Figure 3). High affinity binding sites for plasminogen (Miles et al., 1988; Plow et al., 1991; Nachman, 1992, Hajjar, 1995), t-PA (Hajjar, 1991, 1995) and

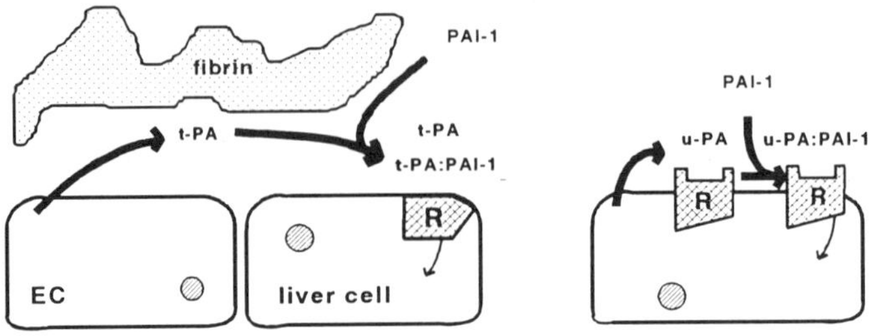

Figure 3. Schematic presentation of the synthesis and cellular uptake of plasminogen activators (PAs). Left: fibrinolysis; right: pericellular proteolysis. After secretion the PA binds to a matrix or cellular receptor (R) where it becomes active and performs its proteolytic activity until it is inhibited by its inhibitor PAI-1. The PA:PAI-1 complex is subsequently internalized by the same or another cell. PAs that do not interact with a suitable matrix or receptor are cleared by the liver. EC: endothelial cell; uPAR: u-PA receptor.

u-PA (Vasalli, 1994; Blasi et al., 1994; Danø et al., 1994) are found on various types of cells including endothelial cells.

Endothelial cells in vitro bind plasminogen with a moderate affinity (120 to 340 nM depending on whether the Lys- or Glu-form of plasminogen is used) but with a high capacity (3.9 to 14 × 10^5 molecules per cell) (Hajjar and Nachman, 1988; Plow et al., 1991). This binding, which is also observed with many other cell types, is mediated by the lysine binding sites of kringles 1–3 of the plasmin(ogen) molecule. Because these lysine binding sites are also involved in the interaction of plasmin with α_2-antiplasmin, occupation of lysine binding sites protects plasmin from instantaneous inhibition by α2-antiplasmin not only when plasmin is bound to fibrin (see above) but also when it is bound to the cellular receptors. The nature of the plasminogen receptors is not fully resolved. In addition to gangliosides, which directly or indirectly contribute to the plasminogen binding (Miles et al., 1989b), at least eight proteins have been reported to be involved in plasminogen binding. Among them are members of the low density lipoprotein receptor family, such as gp330 and LRP; annexin II; a not yet identified 45 kD protein; GbIIb/IIIa and α-enolase (see Hajjar, 1995 for review). In neural cells plasminogen binding to amphoterin was found. Lipoprotein Lp(a), which has strong structural homology with a large part of the plasminogen molecule, can compete for plasminogen binding to endothelial cells (Miles et al., 1989a; Nachman, 1992). This competition also involves lysine binding sites.

Specific binding of t-PA to human endothelial cells has been reported (Hajjar et al., 1987; Beebe, 1987; Barnathan et al., 1988). t-PA binds via its growth factor domain with high affinity to annexin II on human endothelial cells in culture (Hajjar et al., 1994). Different epitopes of the annexin II molecule are involved in plasminogen and t-PA binding. It is conceivable that annexin II, like fibrin, forms a ternary complex with t-PA and plasminogen on the endothelial cell surface (Hajjar, 1995). In addition, t-PA interacts with matrix-bound PAI-1 (Barnathan et al., 1988).

Binding of u-PA with the cell surface limits plasminogen activation to focal areas such as the focal attachment sites and cellular protrusions involved in cell migration and invasion. Furthermore, u-PA interaction with the cell evokes signal transduction and phosphorylation of several proteins (Dumler et al., 1993; Rao et al., 1995). A specific u-PA receptor has been identified and cloned. It is present on many cell types, including endothelial cells (Barnathan, 1992). It is a glycosyl phosphatidyl inositol- (GPI-) anchored glycoprotein (Danø et al., 1994), which binds both single-chain u-PA and two-chain u-PA via their growth factor domain (Apella et al., 1987). The u-PA receptor is heavily glycosylated. It belongs to the cysteine-rich cell surface proteins. After synthesis it is proteolytically processed at its carboxyl-terminus and subsequently anchored in the plasma membrane by a GPI-group (Ploug et al., 1991). It comprises three domains, which are structurally homologous to snake venom α-toxins (Danø et al., 1994). The u-PA receptor has been found in focal attachment sites, where integrin-matrix interactions occur, and in cell-cell contact areas (Pöllänen et al., 1988; Conforti et al., 1994). Human endothelial cells in vitro contain about 140,000 u-PA receptors per cell (Haddock et al., 1991). The expression of the u-PA receptor is enhanced by angiogenic growth factors including basic and acidic fibroblast growth factor (bFGF, aFGF) and vascular endothelial growth factor (VEGF) (Mignatti et al., 1991; Mandriota et al., 1995; Koolwijk et al., 1996). The number of u-PA receptors on human and bovine endothelial cells is also enhanced by the activation of protein kinase C and by the elevation of the cellular cAMP concentration (Langer et al., 1993; van Hinsbergh, 1992).

The u-PA receptor both acts as a site for focal pericellular proteolysis by u-PA and is involved in the clearance of the u-PA:PAI-1 complex. Upon secretion, single-chain u-PA binds to the u-PA receptor and is subsequently converted to the proteolytically active two-chain u-PA. Since the endothelial cell also contains plasmin(ogen) receptors, an interplay between receptor-bound u-PA and receptor-bound plasmin(ogen), and plasmin formation is likely to happen. The generated plasmin can degrade a number of matrix proteins. In addition, a direct plasmin-independent proteolytic action of u-PA on matrix proteins may also occur (Quigley et al., 1987). Like free u-PA activity, receptor-bound two-chain u-PA is subject to inhibition by PAI-1. As a consequence u-PA is only active over a short period of time. In contrast to receptor-bound single-chain or non-inhibited two-chain u-PA, the u-PA:PAI-1 complex is rapidly internalized together with the u-PA receptor (Olson et al., 1992), followed by the degradation of the u-PA:PAI-1 complex and the return of the empty u-PA receptor to the plasma membrane. Internalization of the GPI-linked u-PA receptor occurs probably after interaction with another receptor, such as the α2-macroglobulin/low density lipoprotein receptor-related protein (LRP) (Nykjær et al., 1992) or the VLDL-receptor (Heegaard et al., 1995). VLDL-receptors were demonstrated on capillary and arteriolar endothelial cells in vivo (Wyne et al., 1996). Recently it has been found that the u-PA receptor may have an additional role. The u-PA receptor occupied by u-PA interacts avidly with vitronectin (Wei et al., 1994). Hence, cell adhesion may represent an additional function of the u-PA receptor.

The LRP and LRP-like proteins on liver hepatocytes are involved in the clearance of plasmin-α2-antiplasmin, PA:PAI-1 complexes (Orth et al., 1992; Bu et al., 1992, 1994) and probably free PAs from the circulation. In addition, t-PA can also be cleared by mannose receptors present on macrophages and liver endothelial cells (Otter et al., 1991) and by α-fucose receptors on hepatocytes (Hajjar and Reynolds, 1994).

3. ROLE OF t-PA IN THE PREVENTION OF INTRAVASCULAR FIBRIN DEPOSITION

The fibrinolytic activity in blood is largely determined by the concentration of t-PA, which is synthesized by the endothelium (Rijken et al., 1980; Wun and Capuano, 1985). After release into the blood stream, t-PA resides only for a short period in the blood, unless it encounters a specific binding site, in particular fibrin. This is due to the short half life time of t-PA in the circulation, which is 5 to 10 minutes in man. This rapid clearance, which occurs in the liver, and the ability of endothelial cells to release a relatively large amount of t-PA immediately after exposure to vasoactive agents together can cause the plasma concentration of t-PA to change rapidly (Emeis, 1992).

Endothelial cells contain a storage pool of t-PA, which can be rapidly released after exposure of the cells to vasoactive substances, such as bradykinin, platelet activating factor and thrombin (Emeis, 1992; van den Eijnden-Schrauwen et al., 1995). Intracellularly stored t-PA has been demonstrated to be present in small vesicles (Emeis et al., 1996), which are different from the Weibel-Palade bodies, the storage organelles for von Willebrand factor and P-selectin (Bonfanti et al., 1989). The acute t-PA release mechanism enables an enhanced t-PA concentration, exclusively at those sites of the vascular system where fibrin deposition is pending or occurs. Hence, it contributes to the local protection against an emerging unwanted thrombus. If a generalized stimulation of the endothelium occurs, for example by catecholamines or after dDAVP infusion, the acute release mechanism causes a rapid temporary increase in the systemic blood t-PA concentration. Recent studies have demonstrated that under the proper experimental conditions, acute t-PA release can also be demonstrated to exist in cultured endothelial cells (Figure 4) (Schrauwen, 1994; van den Eijnden-Schrauwen et al., 1995).

The size of the intracellular t-PA pool depends on the rate of t-PA synthesis (van den Eijnden-Schrauwen et al., 1995). Hence, influencing t-PA synthesis can influence both constitutive t-PA production and the amounts of acutely released t-PA. The t-PA synthesis rate is different in various types of blood vessels, e.g. veins produce more t-PA than arteries. Furthermore, t-PA synthesis can be enhanced pharmacologically or physiologically by various mediators (Kooistra et al., 1994). Insight has been gained regarding the regulation of the synthesis of t-PA in endothelial cells in vitro. Activation of protein kinase C has been implicated in the regulation of transcription of the t-PA gene and t-PA synthesis in human endothelial cells. The stimulation of t-PA synthesis by histamine and thrombin is caused by this process. The induction of t-PA by protein kinase C activation is potentiated by a simultaneous increase of the cellular cAMP concentration (Levin et al., 1989; Kooistra et al., 1991a). It has been suggested that the proto-oncogenes c-fos and c-jun, which can form a homo- (jun-jun) or heterodimers (jun-fos) called AP1, are involved in the regulation of the t-PA gene in endothelial cells by interacting with one or more AP1-binding site(s) of the t-PA promoter (Kooistra et al., 1991a). In favor of the

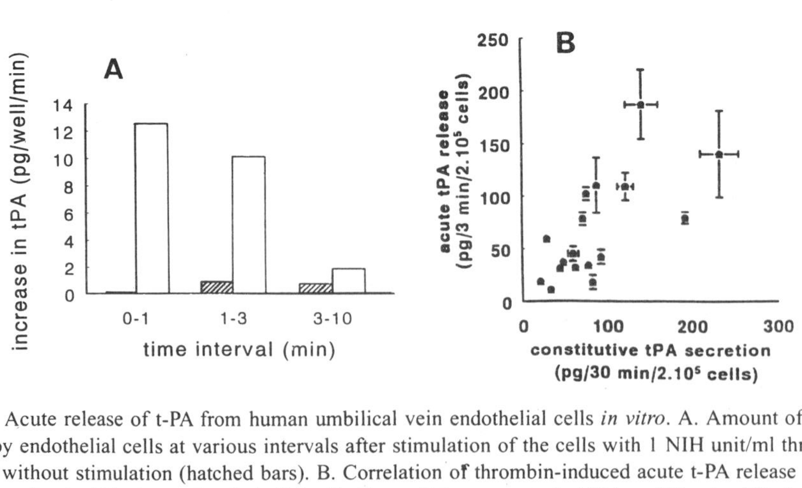

Figure 4. Acute release of t-PA from human umbilical vein endothelial cells *in vitro*. A. Amount of t-PA antigen secreted by endothelial cells at various intervals after stimulation of the cells with 1 NIH unit/ml thrombin (open bars) and without stimulation (hatched bars). B. Correlation of thrombin-induced acute t-PA release and constitutive t-PA section in 17 isolates of endothelial cells (From: Schrauwen et al., 1994, and Van den Eijnden-Schrauwen et al., 1995, with permission).

involvement of AP-1 is an experiment of nature, in which a single mutation in an AP-1 binding site of the t-PA promoter (–TGACATCA–) alters this PMA-responsive element in a cAMP-responsive element (–TGACGTCA–) (Feng et al., 1990). In contrast to human and mouse cells, rat endothelial cells enhance their t-PA synthesis markedly upon stimulation of the cAMP generation alone, whereas rat t-PA synthesis does not respond to the sole addition of PMA (Emeis and Kooistra, 1993).

Other—pharmacologically interesting—agents that can enhance t-PA synthesis are retinoids (Kooistra et al., 1991b; Thompson et al., 1991) and certain benzodiazepines (Kooistra et al., 1993). Retinoic acid and vitamin A have been demonstrated also to enhance t-PA synthesis in the rat, an elevation which is still present after a feeding period of six weeks (van Bennekum et al., 1993). As these components have little effect on the production of PAI-1, vitamin A derivatives are interesting candidates for pharmacological enhancement of t-PA synthesis. They act via specific nuclear receptors in the endothelial cell (Kooistra et al., 1995; Lansink and Kooistra, 1996). Furthermore, exposure of endothelial cells to high shear forces causes an increase in the production of t-PA by these cells (Diamond et al., 1989; Malek et al., 1994). A short promoter sequence has been identified, which is present in a number of genes that respond to shear forces in endothelial cells (Resnick and Gimbrone, 1995).

4. EFFECT OF INFLAMMATORY MEDIATORS ON THE REGULATION OF PLASMINOGEN ACTIVATION AND PERICELLULAR PROTEOLYSIS

The primary cytokines interleukin-1 (IL-1) and tumor necrosis factor-α (TNF α) exert many effects on the vascular endothelium. Their most prominent feature is the induction or increase of the transcription of many genes. These genes include amongst others the leukocyte adhesion molecules E-selectin, VCAM-1 and ICAM-1, cyclooxygenase-2, and a number of proteases and protease inhibitors. It was early recognized that TNFα and IL-1, as well as bacterial lipopolysaccharide (LPS), markedly increase the production of PAI-1 in endothelial cells in vitro (Colucci et al., 1985; Emeis and Kooistra, 1986; Schleef et al., 1988; van Hinsbergh et al., 1988). This induction was also demonstrated at the transcriptional level, and was largely inhibited by the isoflavone compound genistein (van Hinsbergh et al., 1994). *In vivo*, administration of TNFα, IL-1 or LPS causes an increase in PAI-1 concentration in the circulation. After infusion of LPS in animals, PAI-1 mRNA increased in vascularized tissues and PAI-1 mRNA was elevated in the endothelium of various organs (Quax et al., 1990; Keeton et al., 1993). Administration of LPS or TNFα to tumor patients or healthy volunteers caused after about 2 hours a large increase in blood PAI-1 levels, which was preceded by a rapid and sustained increase in t-PA concentration (Suffredini et al., 1989; van Hinsbergh et al., 1990b; van Deventer et al., 1990). An increase in t-PA gene transcription after TNFα exposure was demonstrated in human umbilical vein endothelial cells (van Hinsbergh et al., 1994), but an increase in t-PA synthesis was only found in microvascular endothelial cells (van Hinsbergh et al., 1988; Koolwijk et al., 1996), but not in those of the umbilical vein (Schleef et al., 1988; van Hinsbergh et al., 1988; Fearns et al., 1996). The mechanism underlying the stimulation of t-PA synthesis in vivo by LPS or TNFα is probably bimodal. The rapid initial increase in t-PA level is most likely due to indirect stimulation of the acute release of t-PA. The continuously increased concentration of t-PA in the plasma is caused by an increase in t-PA synthesis. Simultaneously, there is a large increase in PAI-1 production observed two hours after TNFα- or LPS-administration, which far exceeds that of t-PA (Suffredini et al., 1989; van Hinsbergh et al., 1990b; van Deventer et al., 1990). This may result — after an initial rise in fibrinolytic activity — in a prolonged attenuation of the fibrinolysis process. It is generally believed that induction of PAI-1 by inflammatory mediators may contribute to the thrombotic complications in endotoxemia and sepsis. Nevertheless, during infusion of TNFα the fibrinolytic system was still sufficiently active to cause all fibrin formed to be immediately converted into fibrin degradation products, also during the period that a large excess of circulating PAI-1 was detected (van Hinsbergh et al., 1990b).

The inflammatory mediators TNFα, IL-1 and LPS also elicit another effect on the regulation of plasminogen activator production in endothelial cells. Simultaneous with the increase in PAI-1 and t-PA, these inflammatory mediators induce the synthesis of u-PA in human endothelial cells in vitro (van Hinsbergh et al., 1990a). Whereas u-PA is normally not found in endothelial cells in vivo, association of u-PA with the endothelium was observed in acute appendicitis (Grøndahl-Hansen et al., 1989) and in rheumatoid arthritis (Weinberg et al., 1991). Induction of endothelial u-PA by TNFα in vitro is associated by an increased degradation of matrix proteins (Niedbala and Stein Picarella, 1992). The enhanced secretion of u-PA occurs entirely towards the basolateral side of the cell, whereas the secretion of t-PA and PAI-1 proceeds equally to the luminal and basolateral sides of the cell (van Hinsbergh et al., 1990a). The polar secretion of u-PA suggests that u-PA may be involved in local remodeling of the basal membrane of the cell. u-PA activity is controlled in space by interaction of u-PA with its cellular receptor and by the inhibitor PAI-1. The increase of PAI-1 induced by inflammatory mediators may represent, in addition to a role in the modulation of fibrinolysis, a protective mechanism of the cell against uncontrolled u-PA activity.

The observation that TNFα also increases the production of matrix-degrading metalloproteinases (MMPs) by endothelial cells is consistent with a putative role for TNFα and IL-1 in inflammation-induced local pericellular proteolysis. In human microvascular and vein endothelial cells, TNFα increases the mRNA levels and the synthesis of interstitial collagenase (MMP-1), stromelysin-1 (MMP-3) and — if protein kinase C is also activated — gelatinase-B (MMP-9), whereas the mRNAs levels and synthesis of their physiological inhibitors TIMP-1 and TIMP-2 are not changed (Hanemaaijer et al., 1993; Cornelius et al., 1995). Furthermore, activation of gelatinase A (MMP-2) was observed after exposure of the cells to TNFα (Hanemaaijer et al., 1993). Recently, it has been shown that activation of gelatinase A depends on the activity of membrane-type MMP (MT-MMP) (Sato et al., 1994; Foda et al., 1996), and that MMP-2 interacts with $\alpha_v\beta_3$ integrin (Brooks et al., 1996). Interestingly, secretion of gelatinases by bovine endothelial cells occurs predominantly towards the basolateral side of the cells (Unemori et al., 1990) similarly to the TNFα-induced production of uPA (van Hinsbergh et al., 1990a). A role of MMPs in endothelial cell-matrix remodeling has indeed been shown in a three dimensional collagen matrix in vitro (Fisher et al., 1994). Furthermore, MMPs have been detected in vivo in proliferating endothelial cells during development (Karelina et al., 1995) and in endothelial cells present in atherosclerotic plaques and growing tumors (Galis et al., 1994; Nikkari et al., 1995; Rao et al., 1996).

The plasmin-plasminogen activator system and the matrix metalloproteinases cooperate in the degradation of extracellular matrix proteins (Liotta et al., 1991). Figure 5 depicts the interaction between the two systems. It should be noted, however, that this schematic picture is based on in vitro data, and that it is still uncertain whether all the

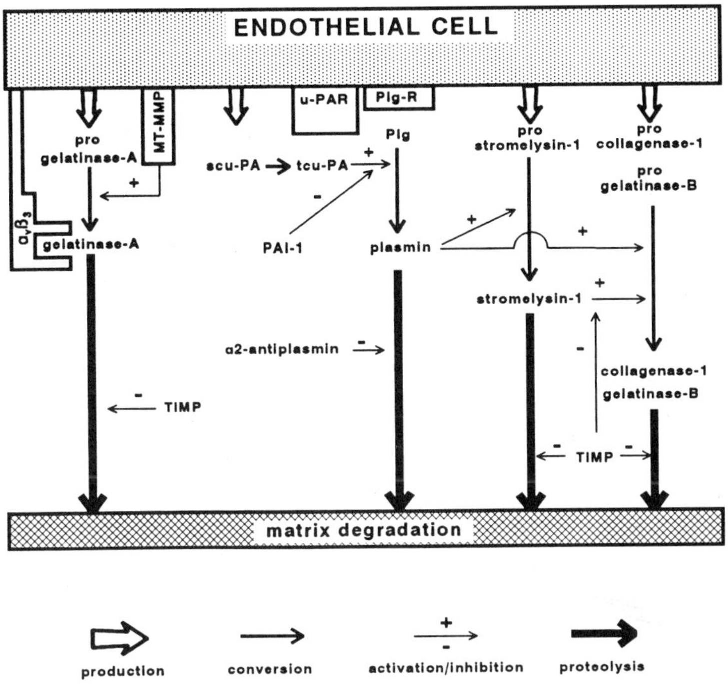

Figure 5. Schematic representation of the interaction between the u-PA-plasmin system and the presumed activation of matrix degrading metalloproteinases (MMPs). Abbreviations: PA: plasminogen activator; u-PA: urokinase-type PA; sc-u-PA: single-chain u-PA; tc-u-PA: two-chain u-PA; u-PAR: u-PA receptor; Plg: plasminogen; Plg-R: Plg receptor; PAI-1: PA inhibitor-1; MT-MMP: membrane-bound MMP; TIMP: tissue inhibitor of MMP. +: stimulation; –: inhibition.

depicted steps also act in vivo. Nevertheless, u-PA and MMP expression frequently coincide in time and location in pathological tissues. If these data are taken together, it will be clear that activation of endothelial cells by TNFα affects multiple sites in the proteolytic cascades involved in the degradation of matrix proteins to such an extent that it markedly enhances the breakdown and remodeling of the endothelial cell basal membrane.

5. INVOLVEMENT OF u-PA AND u-PA RECEPTOR IN CELL MIGRATION AND REALIGNMENT OF ENDOTHELIAL CELLS

Concentration of u-PA activity at the cellular protrusions of migrating or invading cells has been frequently observed (Danø et al., 1985; Estreicher et al., 1990). Blasi (1993) suggested that a continuous activation and removal of u-PA bound to the receptor could contribute to the formation and detachment of focal attachment sites and hence to locomotion of the cell. Indeed, migrating and invading cells, such as monocytes and tumor cells, express u-PA activity bound to u-PA receptors on their cellular protrusions (Estreicher et al., 1990) and on focal attachment sites (Pöllänen et al., 1988; Hébert and Baker, 1990). Receptor-bound u-PA activity is also thought to be involved in smooth muscle and endothelial cell migration and in the formation of new blood vessels (angiogenesis). Inhibition of plasminogen activation interferes with smooth muscle cell migration in vitro (Schleef and Birdwell, 1982; Quax et al., 1996) and affects smooth muscle migration and proliferation in vivo (Clowes et al., 1990). In mice lacking u-PA intimal hyperplasia of injured arteries is less pronounced than in wild-type or t-PA-deficient mice (Carmeliet and Collen, 1994). Moreover, PAI-1-deficient mice show an exacerbated intimal proliferation (Carmeliet and Collen, 1994). Animals made deficient for plasminogen show comparable pathological features as those with a combined deficiency of u-PA and t-PA (Ploplis et al., 1995; Bugge et al., 1995b). These data suggest that u-PA and plasminogen have a role in cell recruitment.

An involvement of u-PA and the u-PA receptor in the migration of bovine endothelial cells has been demonstrated by several investigators (Pepper et al., 1987, 1993; Sato et al., 1988). After the wounding of a monolayer of these endothelial cells, the cells that migrate into the wounded area express u-PA activity (Pepper et al., 1987) bound to the u-PA receptor (Pepper et al., 1993). The migration and expression of u-PA depended on the release of bFGF from the wounded area (Sato et al., 1988). bFGF is a potent inducer of plasminogen activator activity, in particular u-PA, in bovine endothelial cells (Saksela et al., 1987; Gualandris and Presta, 1995), but not in human endothelial cells (Koolwijk et al., 1996). In both species bFGF increases the number of u-PA receptors on endothelial cells (Mignatti et al., 1991; Pepper et al., 1993; Koolwijk et al., 1996).

Receptor-bound u-PA was demonstrated in focal adhesion sites of fibroblasts (Hébert and Baker, 1990) and endothelial cells (Conforti et al., 1994). It may act proteolytically on these structures and hence influence cell-matrix interactions and cell migration. Alternatively, it has been suggested that u-PA can act on cell migration without involvement of its proteolytic activity (Odekon et al., 1992), either by u-PA receptor-dependent signal transduction (Dumler et al., 1993; Rao et al., 1995) or by interaction of the occupied u-PA receptor with vitronectin (Wei et al., 1994). After exposure of endothelial cells to shear forces induced by fluid flow, the cells realign according to the direction of flow (Franke et al., 1984). This process is paralleled by a movement of the cellular focal contact sites of endothelial cells (Davies et al., 1994). Ponfoort et al. (1995) observed that the production of u-PA by human iliac vein endothelial cells was elevated several-fold after exposure of the cells to fluid flow. Interestingly, the realignment of iliac vein endothelial cells according to the direction of flow appeared to be related to cell-bound u-PA activity. Anti-u-PA antibodies, which inhibited u-PA activity, reduced the cellular realignment induced by fluid shear forces (Ponfoort et al., 1995).

6. ROLE OF THE PLASMIN/PLASMINOGEN ACTIVATOR SYSTEM IN THE FORMATION OF ENDOTHELIAL TUBES IN A FIBRIN MATRIX

The outgrowth of new blood vessels from existing ones, angiogenesis, is an essential process during development, but normally stops when the body becomes adult. The half-life of endothelial cells in the adult body varies between 100 to 10,000 days in normal tissues, whereas it is reduced to several days in placenta and tumors (Hobson and Denekamp, 1984). With the exception of the female reproductive system, angiogenesis in the adult is associated with tissue repair after injury by wounding or inflammation. Repair-associated angiogenesis in the adult is usually accompanied by the presence of fibrin and inflammatory cells or mediators, in contrast to developmental angiogenesis in the embryo. Fibrin is a temporary matrix, which is formed after the wounding of a blood vessel and when plasma leaks from blood vessels forming a fibrous exudate, often seen in areas of inflammation and in tumors (Dvorak, 1986; Dvorak et al., 1992). The fibrin matrix acts as a barrier preventing further blood loss, and provides a structure in which new microvessels can infiltrate during wound healing. A proper timing of the outgrowth of microvessels as well as the subsequent (partial) disappearance of these vessels is essential to ensure adequate wound healing and to prevent the formation of scar tissue. Although essential for the formation of granulation tissue and tissue repair, angiogenesis, once under the control of pathological stimuli, can contribute to a number of pathological conditions, such as tumor neovascularisation, pannus formation in rheumatoid arthritis, and diabetic retinopathy. Understanding the mechanisms involved in angiogenesis may provide clues to preventing pathological angiogenesis without seriously impairing tissue repair. A number of studies and reviews have focused on angiogenic factors and the formation of capillary-like structures (Folkman, 1995; Burgess and Maciag, 1989; Montesano, 1992; Aiello et al., 1994; Fong et al., 1995; Shalaby et al., 1995; Sato et al., 1995; Dvorak et al., 1995). Among them, Dvorak (1986) and Colvin (1986) have pointed to the importance of fibrin in angiogenesis. Furthermore, the work of Polverini and colleagues have demonstrated the involvement of monocytes and their products in the induction of angiogenesis (Polverini et al., 1989; Leibovitch et al., 1987; Fratèr-Schröder et al., 1987; Koch et al., 1992). Because of the specific roles of fibrin and inflammatory cells and mediators in pathological angiogenesis in the adult, we have focused our studies on the invasion of human endothelial cells into three-dimensional fibrin matrices and the role of endothelial plasminogen activators in this process. This model resembles recanalization of a fibrin clot by invading endothelial cells. This invasion is usually preceded by the infiltration of inflammatory cells, which interact with the vascular structures from which subsequently endothelial cells migrate into the fibrin clot (Kwaan, 1966).

A three-dimensional fibrin matrix model was used by Pepper and Montesano to demonstrate a direct correlation between the expression of PA activity and the formation of capillary sprouts by bovine microvascular endothelial cells in vitro (Pepper et al., 1990; Montesano, 1992). The outgrowth of tubular structures was increased by bFGF, which increases both u-PA activity and u-PA receptor in bovine endothelial cells. Interestingly, the extent of tube formation and the diameter of the formed tubes were reduced by the simultaneous presence of TGF-β (Pepper et al., 1990). The latter is a growth factor, which amongst others exerts a strong enhancement of PAI-1 synthesis in cultured endothelial cells, and thus inhibits PA activity (Saksela et al., 1987). In addition to bFGF, VEGF can also stimulate bovine endothelial cells to form tubular structures. It acts cooperatively with bFGF in this induction (Pepper et al., 1992).

Studies in human endothelial cells showed that no tubular structure are formed when a quiescent monolayer of human microvascular endothelial cells grown on a fibrin matrix is exposed to bFGF or VEGF. However, when bFGF or VEGF are added simultaneously with TNFα, a large number of capillary-like tubular structures are formed (Figure 6) (Koolwijk et al., 1996). The outgrowth of tubular structures requires u-PA activity and is completely reduced by anti-u-PA immunoglobulins but not by anti-t-PA antibodies (Figure 7). It is also reduced by inhibiting the interaction of u-PA with its receptor. Furthermore, proteolytic activation of plasminogen appears to be involved, because the plasmin inhibitor aprotinin largely inhibits the formation of tubular structures. These data agree with the data on bovine endothelial cells, except that in human endothelial cells a second mediator is required to induce u-PA synthesis. In animals bFGF and aFGF have been shown to induce neovascularisation (Thompson et al., 1988; Broadley et al., 1989). Because it is difficult to rule out the involvement of a limited number of leukocytes in

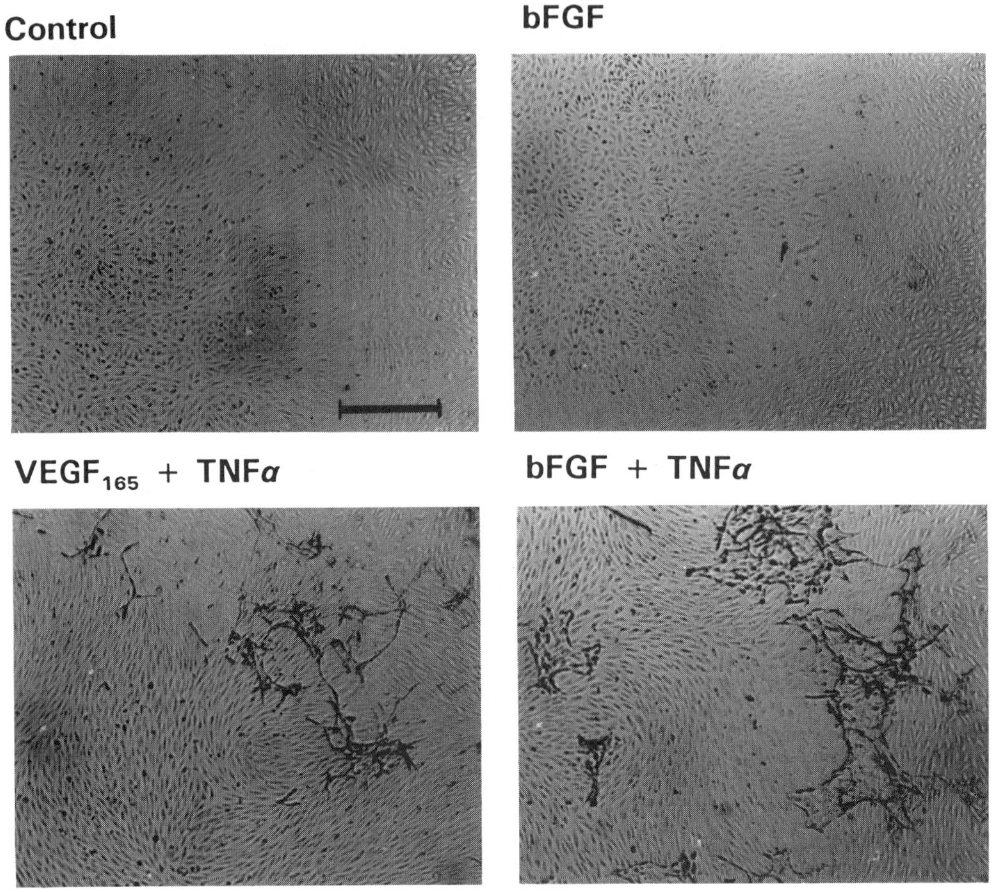

Figure 6. Formation of capillary-like tubular structures in a three dimensional fibrin matrix by human endothelial cells. Human microvascular endothelial cells grown under control conditions on top of a three-dimensional fibrin matrix (control). Formation of tubular structures is induced by the simultaneous addition of VEGF (100ng/mL) and TNFα (20ng/mL), or b-FGF (20ng/mL) and TNFα (20ng/mL). When b-FGF, VEGF or TNFα were added solely, no tubular structures were induced, as shown for 20ng/mL b-FGF. The bar represents 0.5mm.

these experiments in vivo, it remains to be elucidated whether these growth factors act in vivo in conjunction with inflammatory mediators or independently of the latter mediators.

Proteolysis of the basement membrane of endothelial cells and invasion of endothelial cells into the underlying matrix are prerequisites for angiogenesis (Ausprunk and Folkman, 1977). However, not only proteolysis, but also the formation of new cell attachment sites are important for the formation of tubular structures. It should be noted that fibrin contains cell binding domains for endothelial cells: a RGD sequence in its α-chains which binds to the vitronectin receptor, i.e. the $\alpha_v\beta_3$ integrin (Dejana et al., 1990; Thiagarajan et al., 1996); and another site in the β-chain (residues 5–42) (Bunce et al., 1992), which binds to an 130 kD receptor (Erban and Wagner, 1992). Brooks et al. (1994a,b) and Hammes et al. (1996) have shown that inhibition of the $\alpha_v\beta_3$ integrin reduces angiogenesis in several in vivo models. Friedlander et al. (1995) recently reported that both $\alpha_v\beta_3$ and $\alpha_v\beta_5$ integrin are involved in growth factor-stimulated angiogenesis in the rabbit cor-

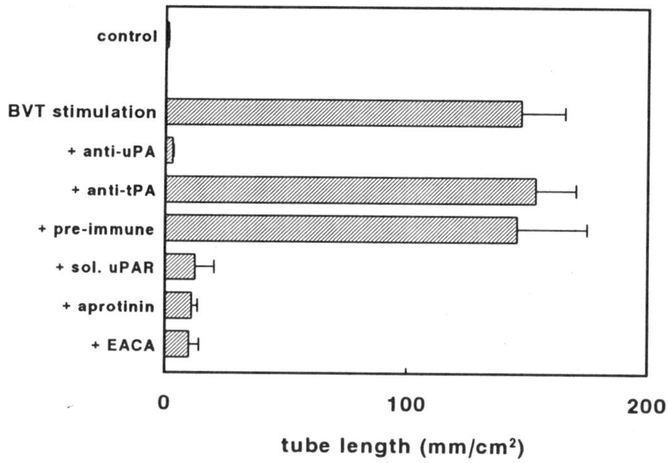

Figure 7. Effect of various inhibitors on the formation of tubular structures by human microvascular endothelial cells in a three-dimensional fibrin matrix *in vitro*.

nea, but via distinct mechanisms. It is of interest to note that bFGF- and TNFα-induced angiogenesis is inhibited by an antibody against the $\alpha_v\beta_3$ integrin, whereas angiogenesis induced by VEGF or by a protein kinase C activating phorbol ester required $\alpha_v\beta_5$ integrin. It remains to be established whether both $\alpha_v\beta_3$- and $\alpha_v\beta_5$-dependent mechanisms are active in the invasion of endothelial cells into a fibrin matrix. The involvement of $\alpha_v\beta_3$ integrin interaction with the fibrin matrix is likely (Chang et al., 1995), while the role of $\alpha_v\beta_5$ integrin, which more selectively interacts with vitronectin, has to be evaluated. Irrespective of the exact role of fibrin in stimulating angiogenesis, the fibrin structure has important consequences for wound healing, and proteolytic modification of fibrin, e.g., by leukocyte elastase, or interaction of fibrin with other matrix proteins, such as vitronectin and fibronectin, may affect cell invasion and angiogenesis and the success of wound healing. Furthermore, the recent observation that MMP-2 (gelatinase-A) binds to $\alpha_v\beta_3$ (Brooks et al., 1996) further points to the complexity of interactions that the $\alpha_v\beta_3$ integrin may have in fibrinous exudates, in which the matrix consists of a mixture of fibrin, collagens and other extracellular matrix components.

7. SUMMARY

The endothelial cell uses PAs for several functions (Figure 8). The endothelium is able to respond to emerging fibrin deposits in the blood stream by regulating the production of t-PA, the main fibrinolysis regulator in blood. In addition it can fine-tune fibrinolytic activity by the simultaneous production of PAI-1, so that recurrent bleeding of a wound is prevented. On the other side of the coin, the endothelium uses PAs, in particular u-PA, for proteolytically changing its interaction with its underlying matrix and for remodeling of its basement membrane, a process which is necessary for cell migration and angiogenesis. This process is limited in space and time by the interaction of u-PA with its specific receptor and by the presence of PAI-1. Locally generated u-PA and plasmin activities interact with those of matrix metalloproteinases, which degrade another spectrum of extracellular matrix proteins. The expression of u-PA, u-PA receptor and many

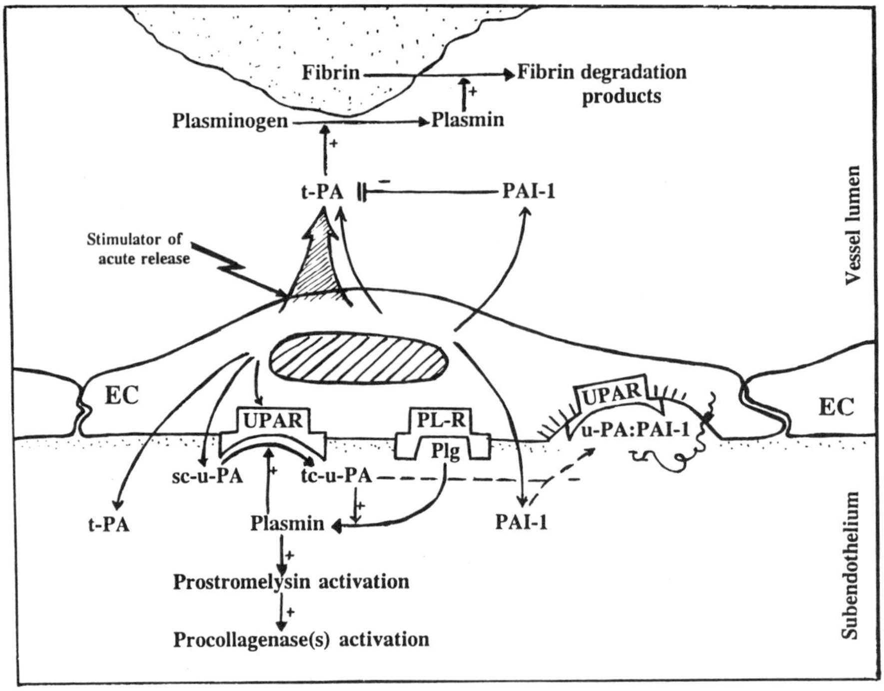

Figure 8. Schematic representation of postulated aspects of the involvement of endothelial cell plasminogen activators in fibrinolysis and local proteolysis. Abbreviations: PA: plasminogen activator; t-PA: tissue-type PA; u-PA: urokinase-type PA; sc-u-PA: single-chain u-PA; tc-u-PA: two-chain u-PA; Plg: plasminogen; UPAR: u-PA receptor; P1-R: plasminogen receptor; PAI-1: PA inhibitor-1; EC: endothelial cell. +: stimulation; –: inhibition.

matrix metalloproteinases in endothelial cells is under the control of the monocyte-derived cytokines TNFα and IL-1 and by angiogenic growth factors. Cell migration and invasion require not only proteolytic activity but also the formation of new cell-matrix contacts needed for the cell to pull itself ahead. Recent data point to a close interaction of proteases, matrix proteins and cellular receptors involved in cell-matrix interaction and cell surface-bound pericellular proteolysis.

ACKNOWLEDGMENTS

The financial support of the Netherlands Heart Foundation (grants M93.001 and 92.071) and the Praeventiefonds (grants 28-2621 and 28-2622) is gratefully acknowledged.

REFERENCES

Aiello, L.P., Avery, R.L., Arrigg, P.G., Keyt, B.A., Jampel, H.D., Shah, S.T., Pasquale, L.R. et al. Vascular endothelial cell growth factor in ocular fluid of patients with diabetic retinopathy and other retinal disorders. *N. Engl. J. Med.*, 331:1480–1487, 1994.

Appella, E., Robinson, E. A., Ullrich, S. J., Stoppelli, M. P., Corti, A., Cassani, G., and Blasi, F. The receptor-binding sequence of urokinase A biological function for the growth-factor module of proteases. *J. Biol. Chem.*, 262: 4437–4440, 1987.

Ausprunk, D. and Folkman, J. Migration and proliferation of endothelial cells in preformed and newly formed blood vessels during tumor angiogenesis. *Microvasc. Res.*, 14:52–65, 1977.

Bachmann, F. Fibrinolysis, in: *Thrombosis and Haemostasis 1987*, M. Verstraete, J. Vermylen, R. Lijnen, and J. Arnout, eds., pp. 227–265, Leuven University Press, Leuven, 1987.

Bachmann, F. The enigma PAI-2. Gene expression, evolutionary and functional aspects. *Thromb. Haemostas.*, 74:172–179, 1995.

Barnathan, E.S., Kuo, A., Van der Keyl, H., McCrae, K.R., Larsen, G.R. and Cines, D.B. Tissue-type plasminogen activator binding to human endothelial cells. Evidence for two distinct sites. *J. Biol. Chem.*, 263:7792–7799, 1988.

Barnathan, E. S. Characterization and regulation of the urokinase receptor of human endothelial cells. *Fibrinolysis*, 6:1–9, 1992.

Beebe, D. P. Binding of tissue plasminogen activators to human umbilical vein endothelial cells. *Thromb. Res.*, 46:241–254, 1987.

Blasi, F. Urokinase and urokinase receptor - A paracrine/autocrine system regulating cell migration and invasiveness. *Bioessays*, 15:105–111, 1993.

Blasi, F., Conese, M., Møller, L. B., Pedersen, N., Cavallaro, U., Cubellis, M., Fazioli, F., Hernandez-Marrero, L., Limongi, P., Munoz-Canoves, P., Resnati, M., Riittinen, L., Sidenius, N., Soravia, E., Soria, M., Stoppelli, M., Talarico, D., Teesalu, T., and Valcamonica, S. The urokinase receptor: Structure, regulation and inhibitor- mediated internalization. *Fibrinolysis*, 8:182–188, 1994.

Bonfanti, R., Furie, B. C., Furie, B., and Wagner, D. D. PADGEM (GMP140) is a component of Weibel-Palade bodies of human endothelial cells. *Blood*, 73:1109–1112, 1989.

Broadley, K.N., Aquino, A.M., Woodward, S.C., Buckley-Sturrock, A., Sato, Y., Rifkin, D. and Davidson, J.M. Monospecific antibodies implicate basic fibroblast growth factor in normal wound repair. *Lab. Invest.* 61: 571–575, 1989.

Brooks, P.C., Clark, R.A. and Cheresh, D.A. Requirement of vascular integrin $\alpha_v\beta_3$ for angiogenesis. *Science* 264:569–571, 1994a.

Brooks, P.C., Montgomery, A.M., Rosenfeld, M., Reisfeld, R.A., Hu, T., Klier, G and Cheresh, D.A. Intergrin $\alpha_v\beta_3$ antagonists promote tumor regression by inducing apoptosis of angiogenic blood vessels. *Cell*, 79:1157–1164, 1994b.

Brooks, P.C., Strömblad, S., Sanders, L.C., von Schalscha, T.L., Aimes, R.T., Stetler-Stevenson, W.G., Quigley, J.P., and Cheresh, D.A. Localization of matrix metalloproteinase MMP-2 to the surface of invasive cells by interaction with integrin $\alpha_v\beta_3$. *Cell*, 85:683–693, 1996.

Bu, G., Warshawsky, I., and Schwartz, A. L. Cellular receptors for the plasminogen activators. *Blood*, 83:3427–3436, 1994.

Bu, G., Williams, S., Strickland, D.K. and Schwartz, A.L. Low density lipoprotein receptor-related protein/α_2-macroglobulin receptor is an hepatic receptor for tissue-type plasminogen activator. *Proc. Natl. Acad. Sci. U.S.A.*, 89:7427–7431, 1992.

Bugge, T.H., Suh, T.T., Flick, M.J., Daugherty, C.C., Rømer, J., Solberg, H., Ellis, V., Danø, K. and Degen, J.J. The receptor for urokinase-type plasminogen activator is not essential for mouse development or fertility. *J.Biol. Chem.* 270:16886–16894, 1995a.

Bugge, T.H., Flick, M.J., Daugherty, C.C., and Degen, J.L. Plasminogen deficiency causes severe thrombosis but is compatible with development and reproduction. *Genes & Development*, 9:794–807, 1995b.

Bunce, L.A., Sporn, L.A. and Francis, C.W. Endothelial cell spreading on fibrin requires fibrinopeptide B cleavage and amino acid residues 15–42 of the β-chain. *J. Clin. Invest.*, 89:842–850, 1992.

Burgess, W.H., and Maciag, T. The heparin-binding (fibroblast) growth factor family of proteins. *Annu. Rev. Biochem.*, 58:575–606, 1989.

Carmeliet, P., Schoonjans, L., Kieckens, L., Ream, B., Degen, J., Bronson, R., De Vos, R., Van den Oord, J. J., Collen, D., and Mulligan, R. C. Physiological consequences of loss of plasminogen activator gene function in mice. *Nature*, 368:419–424, 1994.

Carmeliet, P. and Collen, D. Evaluation of the plasminogen/plasmin system in transgenic mice. *Fibrinolysis* 8:269–276, 1994.

Carmeliet, P., Stassen, J. M., Schoonjans, L., Ream, B., Van den Oord, J. J., De Mol, M., Mulligan, R. C., and Collen, D. Plasminogen activator inhibitor-1 gene-deficient mice. II. Effects on hemostasis, thrombosis, and thrombolysis. *J. Clin. Invest.*, 92:2756–2760, 1993.

Carmeliet, P., Kieckens, L., Schoonjans, L., Ream, B., Van Nuffelen, A., Prendergast, G., Cole, M., Bronson, R., Collen, D., and Mulligan, R. C. Plasminogen activator inhibitor-1 gene-deficient mice. I. Generation by homologues recombination and characterization. *J. Clin. Invest.*, 92:2746–2755, 1994.

Chang M-C., Wang, B-R. and Huang T-F. Characterization of endothelial cell differential attachment to fibrin and fibrinogen and its inhibition by Arg-Gly-Asp-containing peptides. *Thromb. Haemostas.*, 74:764–769, 1995.

Ciambrone, G. J., and McKeown-Longo, P. J. Vitronectin regulates the synthesis and localization of urokinase-type plasminogen activator in HT-1080 cells. *J. Biol. Chem.*, 267:13617–13622, 1992.

Clowes, A.W., Clowes, M.M., Au, Y.P.T., Reidy, M.A. and Belin, D. Smooth muscle cells express urokinase during mitogenesis and tissue-type plasminogen activator during migration in injured rat carotid artery. *Circ. Res.* 67:61–67, 1990.

Colucci, M., Paramo, J.A., and Collen, D., Generation in plasma of a fast-acting inhibitor of plasminogen activator in response to endotoxin stimulation. *J. Clin. Invest.*, 75:818–824, 1985.

Colvin R.B. Wound Healing Processes in Hemostasis and Thrombosis. in: *"Vascular Endothelium in Hemostasis and Thrombosis,"* M.A. Gimbrone Jr., ed., pp. 220–241, Churchill Livingstone, Edinburgh, 1986.

Conforti, G., Dominguez-Jimenez, C., Rønne, E., Høyer-Hansen, G., and Dejana, E. Cell-surface plasminogen activation causes a retraction of *in vitro* cultured human umbilical vein endothelial cell monolayer. *Blood*, 83:994–1005, 1994.

Cornelius, L. A., Nehring, L. C., Roby, J. D., Parks, W. C., and Welgus, H. G. Human dermal microvascular endothelial cells produce matrix metalloproteinases in response to angiogenic factors and migration. *J. Invest. Dermatol.*, 105:170–176, 1995.

Danø, K., Andreasen, P.A., Grøndahl-Hansen, J., Kristensen, P., Nielsen, L.S. and Skriver, L. Plasminogen activators in tissue degradation and cancer. *Adv. Cancer Res.* 44:139–264, 1985.

Danø, K., Behrendt, N., Brünner, N., Ellis, V., Ploug, M., and Pyke, C. The urokinase receptor Protein structure and role in plasminogen activation and cancer invasion. *Fibrinolysis*, 8:189–203, 1994.

Davies, P.F., Robotewsky, A., Griem, M.L., Quantitative studies of endothelial adhesions: directional modeling of focal adhesion sites in response to flow forces. *J. Clin. Invest.* 93:2031–2038, 1994.

Dejana E., Lampugnani, M.G., Giorgi, M., Gaboli, M., and Marchisio, P.C. Fibrinogen induces endothelial cell adhesion and spreading via the release of endogenous matrix proteins and the recruitment of more than one integrin receptor. *Blood*, 75:1509–1517, 1990.

Diamond, S.L., Eskin, S.G., and McIntire, L.V. Fluid flow stimulates tissue plasminogen activator secretion by cultured human endothelial cells. *Science*, 243:1483–1485,1989.

Dumler, I., Petri, T., and Schleuning, W-D. Interaction of urokinase-type plasminogen activator (u-PA) with its cellular receptor (u-PAR) induces phosphorylation on tyrosine of a 38 kDa protein. *FEBS Letters*, 322:37–40, 1993.

Dvorak, H.F. Tumors: wounds that do not heal: similarities between tumor stroma generation and wound healing. *N. Engl. J. Med.* 315:1650–1659, 1986.

Dvorak, H. F., Nagy, J. A., Berse, B., Brown, L. F., Yeo, K-T., Yeo, T-K., Dvorak, A. M, Van De Water, L., Sioussat, T. M, and Senger, D. R. Vascular permeability factor, fibrin, and the pathogenesis of tumor stroma formation. *Ann. N. Y. Acad. Sci.*, 667:101–111, 1992.

Dvorak, H. F, Brown, L. F., Detmar, M., and Dvorak, A. M. Vascular permeability factor/vascular endothelial growth factor, microvascular hyperpermeability, and angiogenesis. *Am. J. Pathol.*, 146:1029–1039, 1995.

Emeis, J. J. Regulation of the acute release of tissue-type plasminogen activator from the endothelium by coagulation activation products. *Ann. N. Y. Acad. Sci.*, 667:249–258, 1992.

Emeis, J.J., and Kooistra, T. Interleukin-1 and lipopolysaccharide induce a fast-acting inhibitor od tissue-type plasminogen activator in vivo and in cultured endothelial cells. *J. Exp. Med.* 163:1260–1266, 1986.

Emeis, J.J., and Kooistra, T. Animal models and experimental procedures to study the synthesis and acute release of tissue-type plasminogen activator, *Fibrinolysis* 7, Suppl.1:31–32, 1993.

Emeis, J.J., van den Eijnden-Schrauwen, Y., and Kooistra, T. Tissue-type plasminogen activator and the vessel wall: synthesis, storage and secretion. in: "*Vascular Control of Hemostasis*", V.W.M. van Hinsbergh ed., pp. 187–206, Harwood Acad. Publ., Amsterdam, 1996.

Erban J.K. and Wagner D.D. A 130-kDa protein on endothelial cells binds to amino acids 15–42 of the Bβ chain of fibrinogen. *J. Biol. Chem.* 267:2451–2458, 1992.

Estreicher, A., Mühlhauser, J., Carpentier, J-L., Orci, L., and Vassalli, J-D. The receptor for urokinase type plasminogen polarizes expression of the protease to the leading edge of migrating monocytes and promotes degradation of enzyme inhibitor complexes. *J. Cell Biol.*, 111:783–792, 1990.

Fearns, C., Samad, F., and Loskutoff, D.J. Synthesis and localization of PAI-1 in the vessel wall. in: "*Vascular Control of Hemostasis*", V.W.M. van Hinsbergh ed., pp. 207–226, Harwood Acad. Publ., Amsterdam, 1996.

Feng, P., Ohlsson, M., and Ny, T. The structure of the TATA-less rat tissue-type plasminogen activator gene. Species-specific sequence divergences in the promoter predict differences in regulation of gene expression, *J. Biol. Chem.*, 265:2022–2027, 1990.

Fisher, C., Gilbertsonbeadling, S., Powers, E.A., Petzold, G., Poorman, R., and Mitchell, M.A. Interstitial collagenase is required for angiogenesis *in vitro*. *Dev. Biol.* 162:499–510, 1994.

Foda, H.D., George, S., Conner, C., Drews, M., Tomkins, D.C., Zucker, S. Activation of human umbilical vein endothelial cell progelatinase A by phorbol myristate: a protein kinase C-dependent mechanism involving a membrane-type matrix metalloproteinase. *Lab. Invest.* 74:538–545, 1996.

Folkman, J. Tumor angiogenesis. in: "*The Molecular Basis of Cancer*", J. Mendelsohn, P.M. Howley, M.A. Israel, and L.A. Liotta, eds., pp. 206–232, W.B. Saunders Comp., Philadelphia, 1995.

Fong, G-H., Rossant, J., Gertsenstein, M. and Breitman, M.L. Role of the flt-1 receptor tyrosine kinase in regulating the assembly of vascular endothelium. *Nature*, 376:66–70, 1995.

Franke, R.P., Gräfe, M., Schnittler, H., Seiffge, D., Mittermayer, C., and Drenckhahn, D. Induction of human vascular endothelial stress fibres by fluid shear stress. *Nature*, 307:648–649, 1984.

Fràter-Schröder, M., Risau, W., Hallman, R., Gautschi, P., and Böhlen, P. Tumor necrosis factor type α, a potent inhibitor of endothelial cell growth in vitro, is angiogenic *in vivo*. *Proc. Natl. Acad. Sci. USA*, 84:5277–5281, 1987.

Friedlander, M., Brooks, P.C., Shaffer, R.W., Kincais, C.M., Varner, J.A. and Cheresh, D.A. Definition of two angiogenic pathways by distinct α_v integrins. *Science*, 270:1500–1502, 1995.

Galis Z.S., Sukhova, G.K., Lark, M.W., and Libby, P. Increased expression of matrix metalloproteinases and matrix degrading activity in vulnerable regions of human atherosclerotic plaques. *J. Clin. Invest.*, 94:2493–2503, 1994.

Grøndahl-Hansen, J., Kirkeby, L., Ralfkiær, Kristensen, P., Lund, L.R., Danø, K. Urokinase-type plasminogen activator in endothelial cells during acute inflammtion of the appendix. *Am. J. Pathol.* 135:631–636, 1989.

Gualandris, A., and Presta, M. Transcriptional and posttranscriptional regulation of urokinase-type plasminogen activator expression in endothelial cells by basic fibroblast growth factor. *J. Cell. Physiol.*, 162:400–409, 1995.

Haddock, R. C., Spell, M. L., Baker III, C. D., Grammer, J. R, Parks, J. M., Speidel, M., and Booyse, F. M. Urokinase binding and receptor identification in cultured endothelial cells. *J. Biol. Chem.*, 266:21466–21473, 1991.

Hajjar, K.A. The endothelial cell tissue plasminogen activator receptor. Specific interaction with plasminogen. *J. Biol. Chem.* 266:21962–21970, 1991.

Hajjar, K. A. Cellular receptors in the regulation of plasmin generation. *Thromb. Haemostas.*, 74:294–301, 1995.

Hajjar, K.A., and Nachman, Endothelial cell-mediated conversion of Glu-plasminogen to Lys-plasminogen. Further evidence for assembly of the fibrinolytic system on the endothelial cell surface. *J. Clin. Invest.* 82:1769–1778, 1988.

Hajjar, K. A., and Reynolds, C. α-Fucose-mediated binding and degradation of tissue-type plasminogen activator by HepG2 cells. *J. Clin. Invest.*, 93:703–710, 1994.

Hajjar, K. A., Hamel, N. M., Harpel, P. C., and Nachman, R. L. Binding of tissue plasminogen activator to cultured human endothelial cells. *J. Clin. Invest.*, 80:1712–1719, 1987.

Hajjar, K., Jacovina, A. and Chacko, J. An endothelial cell receptor for plasminogen tissue plasminogen activator 1 identity with annexin II. *J. Biol. Chem.*, 269:21191–21197, 1994.

Hammes, H.P., Brownlee, M., Jonczyk, A., Sutter, A., and Preissner, K. Subcutaneous injection of a cyclic peptide antagonist of vitronectin receptor-type integrins inhibits retinal neovascularization. *Nature Medicine*, 2:529–533, 1996.

Hanemaaijer, R., Koolwijk, P., Leclercq, L., De Vree, W. J. A., and Van Hinsbergh, V. W. M. (1993) Regulation of matrix metalloproteinase expression in human vein and microvascular endothelial cells - Effects of tumour necrosis factor-α, interleukin-1 and phorbol ester. *Biochem. J.*, 296: 803–809, 1993.

Hébert, C. A., and Baker, J. B. Linkage of extracellular plasminogen activator to fibroblast cytoskeleton: Colocalization of cell surface urokinase with vinculin. *J. Cell Biol.*, 105: 1241–1247, 1988.

Heegaard, C.W., Wiborg Simonsen, A.C., Oka, K., Kjøller, L., Christensen, A., Madsen, B., Ellgaard, L., Chan, L., and Andreasen, P.A. Very low density lipoprotein receptor binds and mediates endocytosis of urokinase-type plasminogen activator-type-1 plasminogen activator inhibitor complex. *J. Biol. Chem.*, 270:20855–20861, 1995.

Hobson B. and Denekamp J. Endothelial proliferation in tumours and normal tissues: Continuous labelling studies. *Br. J. Cancer*, 49:405–413, 1984.

Holmes, W.E., Nelles, L., Lijnen, H.R. and Collen, D. Primary structure of human α_2-antiplasmin, a serine protease inhibitor (Serpin). *J. Biol. Chem.*, 262:1659–1664, 1987.

Karelina, T.V., Goldberg, G.I., and Eisen, A.Z. Matrix metalloproteinases in blood vessel development in human fetal skin and in cutaneous tumors. *J. Invest. Dermatol.*, 105:411–417, 1995.

Keeton, M., Eguchi, Y., Swadey, M., Ahn, C. and Loskutoff, D. Cellular localization of type 1 plasminogen activator inhibitor messenger RNA and protein in murine renal tissue. *Am. J. Pathol.*, 142:59–70, 1993.

Koch, A.E. Polverini, P.J., Kunkel, S.L., Harlow, L.A., DiPietro, L.A., Elner, V.M., Elner S.G., and Strieter, R.M. Interleukin-8 as a macrophage-derived mediator of angiogenesis. *Science*, 258:1798–1801, 1991.

Kooistra, T., Schrauwen, Y., and Emeis, J. J. Regulation of endothelial cell t-PA synthesis and release. *Int. J. Hematol.*, 59:233–255, 1994.

Kooistra, T., Bosma, P.J., Toet, K., Cohen, L.H., Griffioen, M., Van den Berg, E., Le Clercq, L., and Van Hinsbergh, V.W.M. Role of protein kinase C and cyclic adenosine monophosphate in the regulation of tissue-type plasminogen activator, plasminogen activator inhibitor-1, and platelet-derived growth factor mRNA levels in human endothelial cells. Possible involvement of proto-oncogenes c-jun and c-fos. *Arterioscler. Thrombos.*, 11:1042–1052, 1991a.

Kooistra, T., Opdenberg, J.P., Toet, K., Hendriks, H.F.J., Van den Hoogen, R.M., and Emeis, J.J. Stimulation of tissue-type plasminogen activators synthesis by retinoids in cultured human endothelial cells and rat tissues *in vivo*. *Thromb. Haemostas.*, 65:565–572, 1991b.

Kooistra, T., Toet, K., Kluft, C., Von Voigtlander, P.F., Ennis, M.D., Aiken, J.W., Boadt, J.A. and Erickson, L.A. Triazolobenzodiazepines: a new class of stimulators of tissue-type plasminogen activator synthesis in human endothelial cells. *Biochem. Pharmacol.* 46:61–67, 1993.

Kooistra, T., Lansink, M., Arts, J., Sitter, T., and Toet, K. Involvement of retinoic acid receptor α in the stimulation of tissue-type plasminogen activator gene expression in human endothelial cells. *Eur. J. Biochem.*, 232:425–432, 1995.

Koolwijk, P., van Erck, M.G.M., de Vree, W.J.A., Vermeer, M.A., Weich, H.A., Hanemaaijer, R. and van Hinsbergh, V.W.M. Cooperative effect of TNFα, bFGF and VEGF on the formation of tubular structures of human microvascular endothelial cells in a fibrin matrix. Role of urokinase activity. *J. Cell Biol.*, 132:1177–1188, 1996.

Kwaan, H.C. Tissue fibrinolytic activity studied by a histochemical method. *Fed. Proc.*, 25:52–56, 1966.

Langer, D. J., Kuo, A., Kariko, K., Ahuja, M., Klugherz, B. D., Ivanics, K. M, Hoxie, J. A., Williams, W. V., Liang, B. T., Cines, D. B., and Barnathan, E. S. Regulation of the endothelial cell urokinase-type plasminogen activator receptor - Evidence for cyclic AMP-dependent and protein kinase-C dependent pathways. *Circ. Res.*, 72:330–340, 1993.

Lansink, M., and Kooistra, T. Stimulation of tissue-type plasminogen activator expression by retinoic acid in human endothelial cells requires retinoic receptor β2 induction. *Blood*, 88: 531–541, 1996.

Leibovich S.J., Polverini, P.J., Shepard, H.M., Wiseman, D.M., Shively, V. and Nuseir, N. Macrophage-induced angiogenesis is mediated by tumour necrosis factor-α. *Nature*, 329:630–632, 1987.

Levin, E.G., Marotti, K.R., and Santell, L. Protein kinase C and the stimulation of tissue plasminogen activator release from human endothelial cells. Dependence on the elevation of messenger RNA. *J. Biol. Chem.*, 264:16030–16036, 1989.

Liotta, L.A., Steeg, P.S., and Stetler-Stevenson, W.G. Cancer metastasis and angiogenesis - An imbalance of positive and negative regulation. *Cell*, 64:327–336, 1991.

Loskutoff, D.J. Regulation of PAI-1 gene expression. *Fibrinolysis*, 5:197–206, 1991.

Malek, A.M., Jackman, R., Rosenberg, R.D., Izumo, S. Endothelial expression of thrombomodulin is reversibly regulated by fluid shear stress. *Circ. Res.*, 74:852–860, 1994.

Mandriota, S., Seghezzi, G., Vassalli, J-D., Ferrara, N., Wasi, S., Mazzieri, R., Mignatti, P. and Pepper, M. Vascular endothelial growth factor increases urokinase receptor expression in vascular endothelial cells. *J. Biol. Chem.*, 270:9709–9716, 1995.

Mignatti, P., Mazzieri, R., and Rifkin, D. B. Expression of the urokinase receptor in vascular endothelial cells is stimulated by basic fibroblast growth factor. *J. Cell Biol.*, 113:1193–1201, 1991.

Miles, L.A., Levin, E.G., Plescia, J., Collen, D. and Plow, E.F. Plasminogen receptors, urokinase receptors, and their modulation on human endothelial cells. *Blood*, 72:628–635, 1988.

Miles, L. A., Fless, G. M., Levin, E. G., Scanu, A. M., and Plow, E. F. A potential basis for the thrombotic risks associated with lipoprotein(a). *Nature*, 399:301–303, 1989a.

Miles, L. A., Dahlberg, C. M., Levin, E. G., and Plow, E. F. Gangliosides interact directly with plasminogen and urokinase and may mediate binding of these fibrinolytic components to cells. *Biochemistry*, 28:9337–9343, 1989b.

Montesano, R. Regulation of angiogenesis *in vitro*. *Eur. J. Clin. Invest.*, 22:504–515, 1992.

Nachman, R.L. Thrombosis and atherogenesis: molecular connections. *Blood*, 79:1897–1906, 1992.

Niedbala, M. J., and Stein-Picarella, M. Tumor necrosis factor induction of endothelial cell urokinase-type plasminogen activator mediated proteolysis of extracellular matrix and its antagonism by γ-interferon. *Blood*, 79:678–687, 1992.

Nikkari S.T., O'Brien, K.D., Ferguson, M., Hatsukami, T., Welgus, H.G., Alpers, C.E., and Clowes, A.W. Interstitial collagenase (MMP-1) expression in human carotid atherosclerosis. *Circulation* 92:1393–1398, 1995.

Nykjær, A., Petersen, C. M., Møller, B., Jensen, P. H., Moestrup, S. K., Holtet, T. L., Etzerodt, M., Thogersen, H. C., Munch, M., Andreasen, P. A., and Gliemann, J. Purified α2-macroglobulin receptor/LDL receptor-related protein binds urokinase activator inhibitor type-1 complex - Evidence that the α2-macroglobulin receptor mediates cellular degradation of urokinase receptor-bound complexes. *J. Biol. Chem.*, 267:14543–14546, 1992.

Odekon, L. E., Sato, Y., and Rifkin, D. B. Urokinase-type plasminogen activator mediates basic fibroblast growth factor-induced bovine endothelial cell migration independent of its proteolytic activity. *J. Cell. Physiol.*, 150:258–263, 1992.

Olson, D., Pöllänen, J., Høyer-Hansen, G., Rønne, E., Sakaguchi, K., Wun, T-C., Appella, E., Danø, K. and Blasi, F. Internalization of the urokinase-plasminogen activator inhibitor type-1 complex is mediated by the urokinase receptor. *J. Biol. Chem.*, 267:9129–9133, 1992.

Orth, K., Madison, E. L., Gething, M-J., Sambrook, J. F., and Herz, J. Complexes of tissue-type plasminogen activator and its serpin inhibitor plasminogen-activator inhibitor type-1 are internalized by means of the low density lipoprotein receptor- related protein/α2-macroglobulin receptor. *Proc. Natl. Acad. Sci. USA*, 89:7422–7426, 1992.

Otter, M., Barrett-Bergshoef, M.M. and Rijken, D.C. Binding of tissue-type plasminogen activator by the mannose receptor. *J. Biol. Chem.* 266:13931–13935, 1991.

Pepper, M. S., Vassalli, J-D., Montesano, R., and Orci, L. Urokinase-type plasminogen activator is induced in migrating capillary endothelial cells. *J. Cell Biol.*, 105:2535–2541, 1987.

Pepper, M. S., Belin, D., Montesano, R., Orci, L., and Vassalli, J. Transforming growth factor-ß-1 modulates basic fibroblast growth factor induced proteolytic and angiogenic properties of endothelial cells *in vitro*. *J. Cell Biol.*, 111:743–755, 1990.

Pepper, M. S., Ferrara, N., Orci, L., and Montesano, R. Potent synergism between vascular endothelial growth factor and basic fibroblast growth factor in the induction of angiogenesis *in vitro*. *Biochem. Biophys. Res. Commun.*, 189:824–831, 1992.

Pepper, M. S, Sappino, A-P., Stocklin, R., Montesano, R., Orci, L., and Vassalli, J-D. Upregulation of urokinase receptor expression on migrating endothelial cells. *J. Cell Biol.*, 122:673–684, 1993.

Pepper, M. S., Vassalli, J-D., Wilks, J. W., Schweigerer, L., Orci, L., and Montesano, R. Modulation of bovine microvascular endothelial cell proteolytic properties of inhibitors of angiogenesis. *J. Cell. Biochem.*, 55:419–434, 1994.

Ploplis, V.A., Carmeliet, P., Vazirzadeh, S., Van Vlaenderen, I., Moons, L., Plow, E.F., and Collen, D. Effects of disruption of the plasminogen gene on thrombosis, growth, and health in mice. *Circulation*, 92:2585–2593, 1995.

Ploug, M., Behrendt, N., Lober, D., and Danø, K. Protein structure and membrane anchorage of the cellular receptor for urokinase-type plasminogen activator. *Seminars in Thromb. Haemostas.*, 17:183–193, 1991.

Plow, E.F., Felez, J. and Miles, L.A. Cellular regulation of fibrinolysis. *Thromb. Haemostas.*, 66:32–36, 1991.

Pöllänen, J., Hedman, K., Nielsen, L. S., Danø, K., and Vaheri, A. Ultrastructural localization of plasma membrane-associated urokinase-type plasminogen activator at focal contacts. *J. Cell Biol.*, 106:87–95, 1988.

Polverini, P. Macrophage-induced angiogenesis - A review. *Macrophage-Derived Cell Regulatory Factors*, 1:54–73, 1989.

Ponfoort, E.D., van Bockel, J.H., and van Hinsbergh, V.W.M. Shear forces induce the synthesis of urokinase-type plasminogen activator in human iliac vein endothelial cells *in vitro*. *Circulation* 92 (suppl.) abstract 3016, 1995.

Presta, M., Maier, J. A. M., and Ragnotti, G. The mitogenic signalling pathway but not the plasminogen activator-inducing pathway of basic fibroblast growth factor is mediated through protein kinase C in fetal bovine aortic endothelial cells. *J. Cell Biol.*, 109:1877–1884, 1989.

Quax, P. H. A., Van den Hoogen, C. R., Verheijen, J. H., Padro, T., Zeheb, R., Gelehrter, T. D., van Berkel, T. J. C., Kuiper, J., and Emeis, J. J. Endotoxin induction of plasminogen activator and plasminogen activator inhibitor type 1 mRNA in rat tissues *in vivo*. *J. Biol Chem.*, 265:15560–15563, 1990.

Quax, P.H.A., Koolwijk, P., Verheijen, J.H.M., and Van Hinsbergh, V.W.M. The role of plasminogen activators in vascular pericellular proteolysis. in *"Vascular Control of Hemostasis"*, V.W.M. van Hinsbergh, ed., pp. 227–245, Harwood Academic Publishers, Amsterdam, 1996.

Quigley, J. P., Gold, L. I., Schwimmer, R., and Sullivan, L. M. Limited cleavage of cellular fibronectin by plasmin activator purified from transformed cells. *Proc. Natl. Acad. Sci. USA*, 84:2776–2780, 1987.

Rao, N. K., Shi, G-P., and Chapman, H. A. Urokinase receptor is a multifunctional protein: Influence of receptor occupancy on macrophage gene expression. *J. Clin. Invest.*, 96:465–474, 1995.

Rao, J.S., Yamamoto, M., Mohaman, S., Gokaslan, Z.L., Stetler-Stevenson, W.G., Roa, V.H., Liotta, L.A., Nicolson, G.I.,and Sawaya, R.E. Expression and localization of 92 kD type IV collagenase genatinase B (MMP-9) in human gliomas. *Clin. Exp. Metastasis*, 14:12–18, 1996.

Redlitz, A., Tan, A. K., Eaton, D. L., and Plow, E. F. Plasma carboxypeptidases as regulators of plasminogen system. *J. Clin. Invest.*, 96:2534–2538, 1995.

Resnick N., and Gimbrone M.A. Jr. Hemodynamic forces are complex regulators of endothelial gene expression. *FASEB J.* 9:874–982, 1995.

Rijken, D. D., Wijngaards, G., and Welbergen, J. Relationship between tissue plasminogen activator and the activators in blood and vascular wall. *Thromb. Res.*, 18:815–830, 1980.

Saksela, O.D., Moscatelli, D. and Rifkin, D. The opposing effects of basic fibroblast growth factor and transforming growth factor beta on the regulation of plasminogen activator activity in capillary endothelial cells. *J. Cell. Biol.* 105: 957–963, 1987.

Sato, H., Takin, T., Okada, Y., Cao, J., Shinagawa,A., Yamamoto, E., and Seiki, M. A matrix metalloproteinase expressed in the surface of invasive tumour cells. *Nature,* 370:61–65, 1994.

Sato, T.N., Tozawa, Y., Deutsch, U., Wolburg-Buchholz, K., Fujiwara, Y., Gendron-Maguire, M., Gridley, T., Wolburg, H., Risau, W. and Qin, Y. Distinct roles of the receptor tyrosine kinases tie-1 and tie-2 in blood vessel formation. *Nature,* 376:70–74, 1995.

Sato, Y., and Rifkin, D. B. Autocrine activities of basic fibroblast growth factor: Regulation of endothelial cell movement, plasminogen activator synthesis, and DNA synthesis. *J. Cell Biol.*, 107:1199–1205, 1988.

Schleef, R. R., and Birdwell, C. R. The effect of proteases on endothelial cell migration *in vitro. Exp. Cell Res.*, 141:503–508, 1982.

Schleef, R. R., Bevilacqua, M. J., Sawdey, M., Gimbrone, M. A., and Loskutoff, D. J. Cytokine activation of vascular endothelium. Effect on tissue-type plasminogen activator and type 1 plasminogen activator inhibitor. *J. Biol. Chem.*, 263:5797–5803, 1988.

Schrauwen, Y., de Vries, R.E.M., Kooistra T. and Emeis, J.J. Acute release of tissue-type plasminogen activator (t-PA) from the endothelium; regulatory mechanisms and therapeutic target. *Fibrinolysis,* 8 (suppl.2): 8–12, 1994.

Shalaby, F., Rossant, J., Yamaguchi, T.P., Gertsenstein, M., Wu, X-F., Breitman, M.L. and Schuh, A.C. Failure of blood-island formation and vasculogenesis in flk-1-deficient mice. *Nature* 376:62–66, 1995.

Sprengers, E.D. and Kluft, C. Plasminogen activator inhibitors, *Blood,* 69:381–387, 1987.

Suffredini, A.F., Harpel, P.C. and Parrillo, J.E. Promotion and subsequent inhibition of plasminogen activation after administration of intravenous endotoxin to normal subjects. *N. Engl. J. Med.* 320:1165–1172, 1989.

Thiagarajan, P., Rippon, A.J., and Farrell, D.H. Alternative adhesion sites in human fibrinogen for vascular endothelial cells. *Biochemistry,* 35:4169–4175, 1996.

Thompson, E.A., Nelles, L., and Collen, D. Effect of retinoic acid on the synthesis of tissue-type plasminogen activator and plasminogen activator inhibitor-1 in human endothelial cells. *Eur. J. Biochem.* 201:627–632, 1991.

Thompson, J.A., Anderson, K.D., DiPietro, J.M., Zwiebel, J.A., Zametta, M., Anderson, W.F. and Maciag, T. Site-directed neovessel formation *in vivo. Science,* 241:1349–1352, 1988.

Unemori, E.N., Bouhana, K.S. and Werb, Z. Vectorial secretion of extracellular matrix proteins, matrix-degrading proteinases, and tissue inhibitor of metalloproteinases by endothelial cells. *J. Biol Chem.*, 265:445–451, 1990.

Van Bennekum, A.M., Emeis, J.J., Kooistra, T., and Hendriks, H.F.J. Modulation of tissue-type plasminogen activator by retinoids in rat plasma and tissues. *Am. J. Physiol.* 264:R931-R937, 1993.

Van den Eijnden-Schrauwen, Y., Kooistra, T., De Vries, R. E. M., and Emeis, J. J. Studies on the acute release of tissue-type plasminogen activator from human endothelial cells in vitro and in rats *in vivo*: Evidence for a dynamic storage pool. *Blood*, 85:3510–3517, 1995.

Van Deventer, S.J.H., Büller, H.R., Ten Cate, J.W., Aarden, L.A., Hack, E. and Sturk, A. Experimental endotoxemia in humans: analysis of cytokine release and coagulation, fibrinolytic, and complement pathways. *Blood,* 76:2520–2526, 1990.

Van Hinsbergh, V. W. M. Impact of endothelial activation on fibrinolysis and local proteolysis in tissue repair. *Ann. N. Y. Acad. Sci.*, 667:151–162, 1992.

Van Hinsbergh, V.W.M., Kooistra, T., Van den Berg, E.A., Princen, H.M.G., Fiers, W. and Emeis, J.J. Tumor necrosis factor increases the production of plasminogen activator inhibitor in human endothelial cells in vitro and in rats *in vivo. Blood*, 72:1467–1473, 1988.

Van Hinsbergh, V. W. M., van den Berg, E. A., Fiers, W., and Dooijewaard, G. Tumor necrosis factor induces the production of urokinase-type plasminogen activator by human endothelial cells. *Blood*, 75:1991–1998, 1990a.

Van Hinsbergh, V. W. M., Bauer, K. A., Kooistra, T., Kluft, C., Dooijewaard, G., Sherman, M. L., and Nieuwenhuizen, W. Progress of fibrinolysis during tumor necrosis factor infusions in humans Concomitant increase in tissue-type plasminogen activator, plasminogen activator inhibitor type-I, and fibrin(ogen) degradation products. *Blood*, 76:2284–2289, 1990b.

Van Hinsbergh, V.W.M., Vermeer, M., Koolwijk, P., Grimbergen, J., and Kooistra, T. Genistein reduces tumor necrosis factor α -induced plasminogen activator inhibitor-1 transcription but not urokinase expression in human endothelial cells. *Blood*, 84:2984–2991, 1994.

Vassalli, J-D. The urokinase receptor. *Fibrinolysis*, 8:172–181, 1994.

Wallén, P. Structure and Function of Tissue Plasminogen Activator and Urokinase, in: "*Fundamental and Clinical Fibrinolysis*," P.J. Castellino, P.J. Gaffney, M.M. Samama, and A. Takada, eds., pp. 1–18, Elsevier, Amsterdam, 1987.

Wei, Y., Waltz, D., Rao, N., Drummond, R., Rosenberg, S., and Chapman, H. Identification of the urokinase receptor as cell adhesion receptor for vitronectin. *J. Biol. Chem.*, 269:32380–32388, 1994.

Weinberg, J.B., Pippen, A.M.M and Greenberg, C.S. Extravascular fibrin formation and dissolution in synovial tissue of patients with osteoarthitis and rheumatoid arthritis. *Arthritis Rheum*. 34:996–1005, 1991.

Wohlwend, A., Belin, D. and Vassalli, J-D. Plasminogen activator-specific inhibitors produced by human monocytes/macrophages. *J. Exp. Med.* 165:320–339, 1987.

Wun, T-C. and Capuano, A. Spontaneous fibrinolysis in whole human plasma. Identification of tissue activator-related protein as the major plasminogen activator causing spontaneous activity *in vitro*, *J. Biol. Chem.* 260:5061–5066, 1985.

Wyne, K.L., Pathak, R.K., Seabra, M.C., and Hobbs, H.H. Expression of the VLDL receptor in endothelial cells. *Arterioscl. Thromb. Vasc. Biol.*, 16:407–415, 1996.

8

THE ANTITHROMBOTIC EFFECTS OF NITRIC OXIDE

Joseph Loscalzo

Whitaker Cardiovascular Institute and Evans Department of Medicine
Boston University School of Medicine
Boston, Massachusetts

Nitric oxide (NO) is a free radical product of the endothelial cell that maintains blood flow. These effects of NO are the direct consequence of its ability to relax vascular smooth muscle and to inhibit platelet activation and platelet-dependent thrombus formation. In concert with prostacyclin, the other important endothelial product that regulates vessel tone and inhibits platelet function, endothelial NO serves to optimize tissue perfusion under a variety of physiologic and pathophysiologic conditions.

1. MECHANISMS OF PLATELET INHIBITION BY NO

NO inhibits platelet adhesion, activation, and aggregation (Figure 1). Importantly, this profile of effects is different from that for prostacyclin, which does not inhibit adhesion. The mechanism(s) by which NO inhibits platelet function is complex and only partly understood at the current time. By binding to the prosthetic heme group of soluble guanylyl cyclase, NO increases platelet cyclic guanosine monophosphate (GMP) concentration in a concentration- and time-dependent manner[1]. Cyclic GMP activates a cyclic GMP-dependent protein kinase that phosphorylates myosin light chain kinase and, as a result, modulates actomyosin ATPase activity[2]. Phosphyorylated myosin light chain kinase has a reduced affinity for the calcium-calmodulin complex leading to a reduction in myosin light chain phosphorylation and, hence, a reduction in myosin ATPase activity[3]. In addition to this effect on actomyosin ATPase activity, activation of a cyclic GMP-dependent protein kinase can lead to phosphorylation of calcium transporter(s) with a resulting reduction in intracellular calcium concentration[4,1]. Recent evidence from our laboratory also suggests that NO can inhibit phosphoinositide 3'-kinase activity induced by thrombin activation, an effect that reduces the stability of the platelet thrombus by promoting dissociation of fibrinogen from its platelet surface receptor (*vide infra*)[5].

Vascular Endothelium, edited by Catravas *et al.*
Plenum Press, New York, 1998

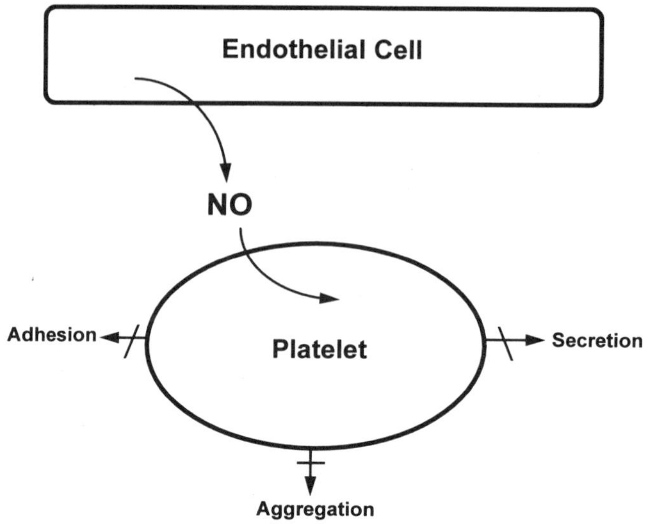

Figure 1. The antiplatelet effects of endothelial NO.

2. FUNCTIONAL CONSEQUENCES OF PLATELET INHIBITION BY NO

The functional consequences of these changes in intracellular signaling evoked by NO are several and include inhibition of platelet adhesion[6], inhibition of P-selectin expression[7], inhibition of secretion[8], and inhibition of fibrinogen binding to the integrin glycoprotein IIb/IIIa[1,6]. These changes lead to the physiologic effect of bleeding time prolongation[9,10] under normal conditions and inhibition of shear-dependent platelet activation under conditions of vascular injury[11,12].

3. ROLE OF PLATELET-DERIVED NO

While the vascular source of NO that inhibits platelet function is primarily derived from the endothelial cell, recent evidence suggests that the platelet itself also has the enzymatic machinery needed to synthesize the inhibitory radical. A constitutive isoform of NO synthase has been identified in the platelet, and both constitutive and inducible isoforms are present in human megakaryocytes[13,14]. Activation of platelet NO synthase appears to occur with platelet activation, and the generation of NO by the platelet serves to limit the recruitment of additional platelets to the growing thrombus[15].

4. THERAPEUTIC POTENTIAL OF PLATELET INHIBITION BY NO

Thiols potentiate the inhibition of platelet function by NO through the formation of S-nitrosothiols or thionitrites[16,17]. These derivatives of NO and biologic thiols are naturally occurring, and serve to stabilize NO by preventing its inactivation by oxygen-derived free radicals[18,19]. One can expoit these antiplatelet actions of S-nitrosothiols to limit platelet activation at sites of atherosclerotic plaque rupture[11,12], an especially relevant therapeutic aim owing to the relative lack of bioactive NO produced by dysfunctional endothelium in atherosclerotic individuals. By impairing adhesion and the release of growth factor-containing granules from the activated platelet, inhibiting platelet activation in this setting dramatically limits vascular smooth muscle migration and proliferation following vascular injury[20].

5. SUMMARY

The antiplatelet actions of NO are important for the maintenance of normal vascular function. By impairing platelet activation, NO maintains blood fluidity and tissue perfusion, particularly in the microvasculature. The importance of this action of NO can be appreciated both by the finding that inhibition of NO synthase can lead to intravascular platelet thrombus formation in the renal bed[21] and by our recent observation that a deficiency in bioactive NO caused by reduced extracellular glutathione peroxidase activity in plasma can lead to thrombotic stroke in children[22]. NO is, then, central to the regulation of hemostasis, thrombosis, and vascular homeostasis.

REFERENCES

1. Mendelsohn, M.E., O'Neill, S., George, D., and Loscalzo, J. Inhibition of fibrinogen binding to human platelets by S-nitroso-N-acetylcysteine. *J. Biol. Chem.*, 265:19028–19034, 1990.
2. Nishikawa, M., DeLanerolle, P., Lincoln, T.M., and Adelstein, R.S. Phosphorylation of mammalian myosin light chain kinases by the catalytic subunit of cyclic AMP-dependent protein kinase and by cyclic GMP-dependent protein kinase. *J. Biol. Chem.*, 259:8429–8436, 1984.
3. Ikebe, M. and Reardon, S. Phosphorylation of myosin light chain kinase by smooth muscle calcium/calmodulin-dependent multi-functional protein kinase. *J. Biol. Chem.*, 265:8975–8978, 1990.
4. Ignarro, U. and Kadowitz, P.J. A pharmacological and physiological role of cyclic GMP in vascular smooth muscle relaxation. *Annu. Rev. Pharmacol. Toxicol.*, 25:171–191, 1985.
5. Pigazzi, A., Heydrick, S., Folli, F., and Loscalzo, J. Thrombin stimulates phosphoinositide 3-kinase-associated protein kinase activity in human platelets. *J. Invest. Med.*, 44:214A, 1996.
6. Radomski, M.W., Palmer, R.M.J., and Moncada, S. Endogenous nitric oxide inhibits human platelet adhesion to the vascular endothelium. *Lancet*, 2:1057–1058, 1987.
7. Michelson, A.D., Benoit, S.E., Furman, M.I., Breckwoldt, W.L., Rohrer, M.J., Barnard, M.R., and Loscalzo J. Effects of endothelium-derived relaxing factor/nitric oxide on platelet surface glycoproteins. *Am. J. Physiol.*, 270:H1640–H1648, 1996.
8. Lieberman, E.H., O'Neill, S., and Mendelsohn, M.E. S-nitrosocysteine inhibition of human platelet secretion is correlated with increases in platelet cGMP levels. *Circ. Res.*, 68:1722–1728, 1991.
9. Keaney, J.F., Stamler, J.S., Simon, D.I., Jaraki, O., and Loscalzo, J. S-nitros-bovine serum albumin: An NO adduct with EDRF-like properties. *J. Clin. Invest.*, 91:1582–1589, 1993.
10. Simon, D.I., Stamler, J.S., Loh, E., Loscalzo, J., Francis, S.A., and Creager, M.A. Effect of nitric oxide synthase inhibition on bleeding time in humans. *J. Cardiovasc. Pharm.*, 26:339–342, 1995.
11. Folts, J.D., Stamler, J., and Loscalzo, J. Intravenous nitroglycerin infusion inhibits cyclic blood flow responses caused by periodic platelet thrombus formation in stenosed canine coronary arteries. *Circulation*, 83:2122–2127, 1991.
12. Keaney J.F., Stamler, J.S., Scharfstein, J., Folts, J.D., and Loscalzo, J. NO forms a stable adduct with serum albumin that has potent antiplatelet properties *in vivo*. *Clin. Res.*, 40:194A, 1992.
13. Pigazzi, A., Fabian, A., Johnson, J., Upchurch, G.R., Loscalzo, J. Identification of nitric oxide synthases in human megakaryocytes and platelets. *Circulation*, 92:1–365, 1995.
14. Lelchuk, R., Radomski, M.W., Martin, J.F., and Moncada, S. Constitutive and inducible nitric oxide synthases in human megakaryoblastic cells. *J. Pharmacol. Exp. Therap.*, 262:1220–1224, 1992.

15. Freedman, J.E., Alpert, C., Piedad, B., and Loscalzo, J. Nitric oxide production by platelets alters platelet recruitment to the platelet thrombus. *J. Invest. Med.*, 44:328A, 1996.
16. Stamler, J., Mendelsohn, M., Amarante, P., Smick, D., George, D., and Loscalzo, J. N-acetylcysteine potentiates platelet inhibition by endothelium-derived relaxing factor. *Circ. Res.*, 65:789–795, 1989.
17. Loscalzo, J. N-acetylcysteine markedly potentiates the inhibition of platelet aggregation by nitroglycerin. *J. Clin. Invest.*, 76:703–708, 1985.
18. Stamler, J.S., Jaraki, O., Osborne, J., Keaney Jr., J.F., Vita, J.A., Singel, D.J., Valeri, C.R., and Loscalzo, J. Nitric oxide circulates in mammalian plasma primarily as an S-nitroso adduct of serum albumin. *Proc. Natl. Acad. Sci. (USA)*, 89:7674–7677, 1992.
19. Stamler, J.S., Singel, D.J., Loscalzo, J. Biochemistry of nitric oxide and its redoxactived forms. *Science*, 251:1898–1901, 1992.
20. Marks, D.S., Vita, J.A., Folts, J.D., Keaney Jr., J.F., Welch, G.N., and Loscalzo, J. Inhibition of neointimal proliferation in rabbits following vascular injury by a single treatment with a protein adduct of nitric oxide. *J. Clin. Invest.*, 96:2630–2638, 1996.
21. Shultz, P.H. and Raij, L. Endogenously synthesized nitric oxide prevents endotoxin-induced gomerular thrombosis. *J. Clin. Invest.*, 90:1718–1725, 1992.
22. Freedman, J.E., Loscalzo, J., Benoit, S.E., Valeri, C.R., Barnard, M.R., and Michelson, A.D. Decreased platelet inhibition by nitric oxide in two brothers with a history of arterial thrombosis. *J. Clin. Invest.*, 97:979–987, 1996.

IV. INFLAMMATION/ANGIOGENESIS

9

CYTOKINE REGULATION OF ENDOTHELIAL CELL FUNCTION

Alberto Mantovani, Paola Allavena, Annunciata Vecchi, Elisabetta Dejana, Silvano Sozzani, and Martino Introna

Instituto di Ricerche Farmacologiche "Mario Negri"
via Eritrea 62
20157 Milano, Italy

1. INTRODUCTION

Endothelial cells (EC) have long considered a "passive" lining of blood vessels, endowed with negative properties, the most important one being that of representing a non-thrombogenic substrate for blood. As such, EC were viewed to participate in tissue reactions essentially as targets for injurious agents. The possibility of isolating and culturing EC from various tissues gave the tools for studying their complex reactions to a variety of activating stimuli. EC have, in this way emerged as active participants in many physiological and pathological processes. It is now evident that hemostasis, inflammatory reactions, and immunity involve close interactions between immunocompetent cells and vascular endothelium. In particular, the ontogeny and function of white blood cells require an intimate relationship with vascular EC. Cytokines are mediators of these complex bi-directional interactions between leukocytes and vascular elements (for review see Mantovani and Dejana, 1989, Mantovani, Bussolino, and Dejana, 1992).

EC are strategically located at the interface between blood and tissues. In addition to responding rapidly (seconds to minutes) to agonists such as histamine and thrombin, vascular cells, upon exposure to cytokines, undergo profound alterations of function that involve gene expression and protein synthesis, require hours to develop, and are relatively long-lasting. This functional reprogramming of endothelial cells has been referred to as activation, in analogy with the terminology long in use for the monocyte-macrophage lineage.

Vascular cells are both a target for and a source of cytokines, and these soluble polypeptide mediators serve as communication signals with leukocytes (primary producers of these molecules) as well as with diverse tissues and organs. Here we will concisely review the response elicited by cytokines in vascular cells and analyze how vessel wall elements participate in cytokine cascades by producing copious amounts of some of these polypeptide mediators. The framework for this contribution can be found in Mantovani

Vascular Endothelium, edited by Catravas *et al.*
Plenum Press, New York, 1998

and Dejana (1989); Pober and Cotran (1990); and Libby and Hansson (1991). The reader is referred to these for more comprehensive referencing.

2. EC RESPONSES TO CYTOKINES

The spectrum of responses elicited in endothelial cells by different cytokines is extremely vast and varied. Different cytokines activate distinct, largely non-overlapping, sets of functions which can be grouped into programs of activation/differentiation (Figure 1). The phenomenology of endothelial cell activation by cytokines was reviewed in detail previously (Mantovani, Bussolino and Dejana, 1992). Here we will update these previous descriptions and emphasize production and action of relatively novel mediators (e.g. chemokines) as well as the molecular basis for EC activation.

2.1. IL-1 and TNF

2.1.1. Activation. As previously described in detail, IL-1 and TNF induce the expression of a functional program related to thrombosis and inflammation (Figure 1). Briefly, these cytokines facilitate thrombus formation by inducing procoagulant activity, inhibiting the thrombomodulin/protein C anticoagulation pathway and blocking fibrin dissolution via stimulation of the type 1 inhibitor of plasminogen activator.

IL-1 and TNF induce augmented production of the vasodilatory mediators PG12 and NO (Mantovani, Bussolino and Dejana, 1992; Rosenkranz Weiss et al., 1994; and Rossi et al., 1985). Cytokines and autocoids constitute interlaced systems of soluble mediators that participate in microvasculature homeostasis during host defense. A well established example is the production of platelet activating factor or prostanoids by EC stimulated with TNF or IL-1. The PAF synthesis is the neosynthesis of an elastase like protease (Bussolino et al., 1994a) which is the final effector of the synthesis of PAF. PAF expressed on EC surface cooperates with adhesion molecules (E-selectin) in the transmigration of leukocytes. A little amount of PAF is also secreted and can prime or activate circulating cells or EC, and triggers a vicious circle of activation leading to a recruitment of more inflammatory cells (revised in Bussolino and Camussi, 1995). Furthermore, PAF produced by ECs is a secondary mediator of angiogenesis induced by inflammatory cytokine (Montrucchio et al., 1994). Prostanoids synthesis is strictly dependent by the induction of the pivotal enzymes phospholipase A2 and cyclooxygenase 1 and 2 (Jackson et al., 1993; Camacho et al., 1995). Furthermore, IL-1 induces the transcription of prostaglandin G/H synthase 2 and in combination with aspirin increases the synthesis of a new class of lipoxins during endothelial cell-leukocyte interaction having inhibitory effect on neutrophil adhesion (Claria and Serhan, 1995).

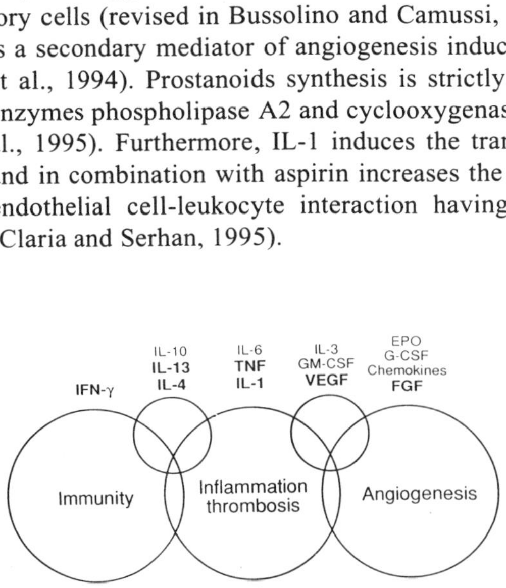

Figure 1. Cytokines activate district, but overlapping, sets of functions in EC (modified from ref. Mantovani, Bussolino, and Dejana, 1992).

NO is a representative of a new class of autacoids serving a variety of functions in several different tissues including vasculature and acting through an independent receptor mechanism (revised in Nathan and Xie, 1994). Nitric oxide synthase (NOS) type II (inducible, Ca++-independent isoform) and type III (constitutive, endothelium cell-isoform) are present in ECs, and catalyze the conversion of arginine into citrulline and NO. The former continuously produces "quanta" of NO, which especially participates in the regulation of vascular tone (Nathan and Xie, 1994). In man, experiments done on ECs from large vessels demonstrate that inflammatory cytokines increase the activity of NOS III and the synthesis of NO (Rosenkrnaz Weiss et al., 1994; Werner Felmayer et al., 1993). The effect of inflammatory cytokines on metabolic pathways leading to NO formation is characterized by a decrease of stability of NOS III mRNA (Rosenkranz Weiss et al., 1994), by an increase of GTP cyclohydrolase I, the rate-limiting enzyme of tetrahydrobiopterin synthesis, and of tetrahydrobiopterin, a cofactor for full activity of NOS III (Rosenkranz Weiss et al. 1994; Werner Felmayer et al., 1993). However, the increase of tetrahydrobiopterin is compensated for by a decrease in stability of NOSIII mRNA (Yoshizumi et al., 1993) necessary to counteract the toxic effect of NO and its metabolite peroxinitrite on EC themselves.

Results obtained in several laboratories have demonstrated that the enhancement of NO production in murine, rat, porcine and bovine ECs observed upon cytokine challenge is mediated by transcription in NOSII gene (Lamas et al., 1992; Radomski, Palmer, and Moncada, 1990; Suschek et al., 1993; Balligand et al., 1995). Of interest the promoter of this gene contains cis-acting DNA elements that are targets of cytokine-activated transcriptions factors NK-κB/Rel and IRF (Martin, Nathan, and Xie, 1994; Kamijo et al., 1994), while recent data point to a role of SP1 and GATA factors for the endothelial, basal transcriptions of the NOSIII gene (Zhang, Min, and Sessa, 1995). At present, unequivocal information on the role of this NOS isoform in human ECs is lacking, since experiments have been only done on EC from large vessels. However, we can speculate that NOSII is operative in human microvasculature, after induction by inflammatory cytokines because it has been detected in EC lining the vessels localized in synovia of patients with rheumatoid arthritis (Sakurai et al., 1995) and in brain tumors (Cobbs et al., 1995).

The ability of different cytokines alone or in combination to induce NO production is in dispute. In human EC, the combination of at least three cytokines (IL-1, TNF and INF) is required to enhance NO synthesis (Rosenkranz Weiss et al., 1994). In other animal species only one cytokine is sufficient to elicit NO synthesis (Lamas et al., 1992; Balligand et al., 1995). The analysis of these data could indicate that in human a complex and refined mechanism for NO synthesis in EC has been developed during evolution.

The complexity of this phenomenon is also stressed by the opposite effect of one cytokine in different animal species or on different NOS isoforms. For instance IFN-γ, is required for NO synthesis in human EC (Rosenkranz Weiss et al., 1994), but is an inhibitor in bovine EC stimulated with TNF (Lamas et al., 1992). Furthermore, TGFβ inhibits NOSII (Murata et al., 1995), but up-regulates NOSIII (Inoue et al., 1995).

NO produced by EC acts in autocrine and paracrine manner. For example, alteration of NO levels in ECs has a primary target EC themselves and modifies their biological properties. The inhibition of basal production of NO by competitive analogues of arginine increase the adhesion of leukocytes to EC surface, by a mechanism in part dependent to the up-regulation of VCAM-1 and the PAF synthesis (Niu, Smith, and Kubes, 1994; De Caterina et al., 1995). Furthermore, the treatment of EC with NO donors reduce the expression of VCAM-1 induced by IL-1, IL-4, TNF or LPS (De Caterina et al., 1995). NO also decreases the expression of other adhesion molecules (ICAM-1 and E-selectin) and secretable cytokine (IL-6, MCP-1, IL-8, and M-CSF) (De Caterina et al., 1995, Peng et al., 1995, Zeiher et al., 1995). These effects of NO do not appear to involve stimulation of guanylate cyclase, the classical intracellular target of NO, but are mediated by an inhibition of NF-κB through the induction and stabilization of NF-κB inhibitor (De Caterina et al., 1995; Peng, Libby, and Liao, 1995). The finding that inhibition of endogenous, cytokine-independent NO production by arginine analogues could activate NF-κB suggests that constitutively produced NO may play an important physiologic role in tonically inhibiting the expression of NF-κB-dependent pro-inflammatory genes (Niu, Smith, and Kubes, 1994; De Caterina et al., 1995; Peng, Libby, and Liao, 1995). Moreover, NO produced by cytokine stimulation attenuates the inflammatory effects of these molecules.

However, NO is not only an intracellular regulator of EC function, but is instrumental in EC response. For example, a NO-dependent killing of Schistosoma mansoni has been described in murine EC stimulated with IFN-γ, IL-1 or TNF (Oswald et al., 1994). NO participates in a long-term morphogenetic program of vasculature as in delayed hypersensitivity response, chronic inflammation or in atherogenesis. It enhances the motility of EC, inhibits the proliferations of vascular smooth cells, is angiogenic *in vivo* and reduces neointima formation (Ziche et al., 1994; Fukuo et al., 1995; von der Leyen et al., 1995). Finally in septic shock, NO produced by EC cooperates in the pathogenesis of the depression in contractile function of myocardium (Schulz et al., 1995; Sato et al., 1995).

Inflammatory cytokines promote leukocyte extravasation by altering the rheology of microcirculation, by induction of adhesion molecules (E selectin, vascular cell adhesion molecule-1, VCAM-1, and intercellular adhesion molecule-1, ICAM-1) and of chemotatic cytokines. The regulated expression of adhesion molecules in the multistep process of recruitment has been the object of recent extensive reviews and chemokine production is discussed below. TNF and IFNγ cause redistribution of CD31 on the EC surface (Romer et al., 1995) and this could be important given the role of this molecule in transmigration. IL-1 and TNF cause production of platelet activating factor (PAF) in EC (Mantovani, Bussolino and Dejana, 1992). A serine protease is involved in the synthesis of PAF in activated EC (Bussolino et al., 1994a; Heller et al., 1992). PAF can act on leukocytes as well as on EC themselves (Bussolino et al., 1994c). PAF is not important for leukocyte adhesion, but primes leukocytes and facilitates transmigration (Hill et al., 1994).

IL-1 and TNF induce production by EC of copious amounts of IL-6 which acts locally and systemically (Sironi et al., 1989). IL-6 does not activate EC, but was recently reported to induce c-sis expression in this cell type (Calderon et al., 1992).

The membrane attack complex of complement augments TNF induced expression of adhesion molecules (Kilgore et al., 1995) and induces tissue factor by inducing IL-1α (Saadi et al., 1995). Interestingly, in turn, IL-1 augments production of C3 and factor B by EC (Coulpier et al., 1995) NO, in turn, inhibited ctyokine induction of adhesion molecules and cytokines (De Caterina et al., 1995; Peng et al., 1995), with controversy about IL-8 (Villarete and Remick, 1995).

TNF selectively down regulates protein S secretion in EC but not in hepatocytes (Hooper et al., 1994).

2.1.2. Gene Induction during Endothelial Activation. Endothelial cells have been the object of several systematic studies aimed at cloning genes induced by exposure to different signals, generally by differential screening approaches. The general assumption of such studies was to clone members of the "immediate early inducible genes" (i.e. in the absence of protein synthesis) in order to elucidate the crucial steps that occur in the setting of different metabolic programs in the EC. As expected, many such genes have been cloned by different groups because they are also present in "immediate early genes" programs induced by different signals in other cell types (see point 2) (e.g., myc, fos, etc.).

In particular, two groups have explored the proinflammatory prothrombotic program". Dixit and coworkers cloned several genes induced in EC by TNF exposure (Dixit et al., 1990), while we concentrated on the IL-1 effect (Introna et al., 1993). As expected, several genes have been cloned by both groups, which had been already characterized as part of the IL-1 response, including E-selectin, IL-8, MCP-1, PA-I-1, groα, and groβ (Dixit et al., 1990; Introna et al., 1993).

More interestingly, new genes have been characterized in both systems which include a novel zinc finger transcription factor, A20, which subsequently was shown to be able to protect from the TNF induced cytotoxicity in a variety of cell types (Opipari et al., 1992), the B12, a novel gene differentially expressed in the heart and liver during mouse embryogenesis and rapidly degradated (Wolf et al., 1992) C-193, which localizes to the nucleus C-28, and the B94 which after its initial cloning (Sarma et al., 1992), was subsequently further characterized as being expressed in a complex modulated way during vasculogenesis and other developmental programs including hematopoietic differentiation and spermatogenesis (Wolf et al., 1994). Very recently yet another cloned the secreted B61 protein (Holzman, Marks, and Dixit, 1990), was shown to b a chemoattractant *in vitro* for endothelial cells and an angiogenetic factor *in vivo* (Pandey et al., 1995). Its proposed function as the ligand for the Eck receptor may provide a clue for one circuit of angiogenesis during the inflammation through the production of TNF and subsequently the production and release of B61 (Pandey et al., 1995).

We cloned a gene (PTX3), both from human and mouse cells (Breviario et al., 1992; Introna et al., 1996) and found that a partial sequence, including several sequencing errors, was already in the EMBL data base library and identified as TSG-14, cloned from another group of "gene-searchers," (Lee et al., 1990). PTX3 is a secreted protein of 381 aminoacids composed of two halves, the NH2 terminal domain showing no precise sequence similarity information and the COOH domain being perfectly alignable with all the known members of the acute phase "pentraxin" gene family, including C-Reactive-Protein (CRP) and Serum Amyloid P Component (SAP) from many different species (Pepys and Baltz, 1983). While CRP and SAP are expressed mainly in the liver *in vitro* and *in vivo* following IL-6 stimulation (Pepys and Baltz, 1983), PTX3 shows a much wider range of responder cells *in vitro*, i.e. endothelial cells, fibroblasts, hepatocytes and monocytes (Brevario et al., 1992; Vidal Alles et al., 1994). *In vivo* PTX3 is inducible in several organs, particularly the heart and the muscle, upon i.v. injection of LPS, and may well represent, therefore, a new acute phase marker with novel tissue specificity (Introna et al., 1996). The secondary structure prediction, based on the SAP crystal (Emsley et al., 1994), while confirming its belonging to the pentraxin domain of the family, indicates the presence of many α-helices in the NH2 domain; recently, two other pentraxins have been cloned from guinea pig testis (Reid and Blobel, 1994; Noland et al., 1994) and from brain (Shclimgen et al., 1995) which also have "long" NH2 domains, although the alignment amongst them is very poor; PTX3 may represent the first cloned member of a "long pentraxins" family, whose expression is mainly confined at extrahepatic sites (Introna et al., 1996).

When a similar cloning strategy was applied to IL-1 and TNF long term stimulated endothelial cells, the isolation of a nuclear protein, C-193, was reported, which behaves as a rapidly induced and protein synthesis independent gene (Chu et al., 1995).

With a quite different perspective, another group searched for immediate early genes induced by exposure to PMA (which is able to induce a phenotyopic transition from a non polar cobblestone monolayer in polar elongated fibroblasts like phenotype) and one gene was isolated and called edg-1 which is a novel member of the G-protein coupled receptor family (Hla and Maciag, 1990).

2.2. Bacterial Products and Toxins

Bacterial lipopolysaccharides (LPS) elicit a spectrum of EC responses similar to that of IL-1 and TNF (Mantovani, Bussolino and Dejana, 1992). It has recently been shown that EC, which are CD14$^-$, interact with soluble CD14 and LPs-binding protein, both present in plasma (von Asmuth et al., 1993; Pugin et al., 1993; Haziot et al., 1993; Read et at., 1993; Goldblum et al., 1994). While direct activation of EC by LPS does occur, the indirect pathway involving mononuclear phagocyte activation and production of inflammatory cytokines is more effective in stimulation of EC by this bacterial product (Pugin, Ulevitch, and Tobias, 1993).

The Lexosceles venom toxin causes selective expression of IL-8, granulocyte-macrophage colony stimulating factor (GM-CSF), and E-selectin but not of IL-6 or ICAM-1 (Patel et al., 1994). S. aureus, a gram + bacterium, interact with EC and after internalization induces IL-1 and IL-6 production (Yao et al., 1995).

Infection with E. coli producing verocytotoxin (also known as Shiga toxin), is casually related to subsequent development of hemolytic uremic syndrome (Remuzzi and Ruggenenti, 1995). Cowdria rumirantum infects EC and elicits IL-1, IL-8, and IL-6 (Bourdoulous et al., 1995). EBV infects EC, and EBV-cell lines induce IL-6 productions (Jones et al., 1995). Rickettsia conroii, the agent of Mediterranean spotted fever, induces IL-6 and IL-8 production via IL-1 (Kaplanski et al., 1995). Various viruses including CMV, for instance, alter cytokine production (Woodroffe, Garnett, and Layton, 1994; Almeida et al., 1994). Verocytotoxin directly activates EC in a prothrombotic-proinflammatory sense and synergizes with IL-1 and TNF (van de Kar et al., 1995; van de Kar et al., 1992). These cytokines have recently been shown to augment the expression of receptors for verocytotoxin by enhancing galactosyltransferase and augmenting globotriaosylceramide (Gb3) (van de Kar et al., 1995), thus providing a molecular basis for synergism (van de Kar et al., 1992).

2.3. Hematopoietic Growth Factors

The hematopoietic growth factors GM-CSF, G-CSF, IL-3, and erytropoietin were shown to affect EC (for recent review Bussolino et al., 1994b). By and large, these factors are relatively weak agonists which affect migration and proliferation and amplify EC responsiveness to other signals (Bussolino et al., 1994b). In the same studies, M-CSF, IL-6 and IL-6 were inactive. However, in two studies M-CSF induced expression of monocyte chemotactic protein-1 (MCP-1) in endothelial cells (Shyy et al., 1993) and IL-6 induced c-sis expression (Calderon et al., 1992). We failed to detect c-fms mRNA, encoding the M-CSF receptor, in human umbilical vein EC (unpublished). Analysis of receptor expression revealed that EC express c-kit and the β chain common to the GM-CSF, IL-3, and IL-5 receptors. Moreover, they also have detectable mRNA for αIL-3 and αGM, but not for αIL-5 (Bussolino F. unpublished). It has been speculated that the ability of various hemopoietic cytokines to affect EC is a memoir of common ancestors, i.e. a reflection of the common ontogenetic origin of hematopoietic and endothelial elements in blood islands (Bussolino et al., 1994b). Endothelial cells express the IL-1Rα chain and TNF and IFN augment its expression (Korpelainen et al., 1995) IL-13 and TNF synergize in terms of induction of IL-8 and adhesion molecules and the same occurs with IFNγ in terms of class II MHC.

2.4. Interferons and IL-12

Interferon γ (IFNγ) was the first molecularly identified cytokine shown to affect EC (Pober and Cotran, 1990). IFNγ induces EC expression of MHC class II antigens and of the invariant chain and augments class I. It also amplifies responses to TNF, slowly stimulates ICAM-1 expression and augments LPs-induced production of IL-1. EC express CD40 and IFNγ, as well as IFNβ, IL-1 and TNF, augment its expression (Hollenbaugh et al., 1995; Yellin et al., 1995; Karmann et al., 1995). Engagement of CD40 amplifies induction of adhesion molecules (Karmann et al., 1995). Thus, ping pong reciprocal stimulation may occur when CD40L expressing cells interact with EC.

IL-12 is a heterodimeric cytokine active on T cells and NK cells with antitumor activity against a variety of murine tumors. Recently IL-12 was shown to have potent anti-angiogenic activity *in vivo* (Voest et al., 1995). In a mouse opportunistic vascular, IL-12 is the most effective single agent studied to date (Vecchi et al., unpublished). Circumstantial evidence suggests that IL-12 does not act per se on EC, but rather via induction of IFNγ (Voest et al., 1995). IFNs have long been known to have antiangiogenic activity and IFNα is used in the treatment of human hemangiomas (Ezekowitz, Mulliken, and Folkman, 1992). IFNγ, in turn, may act on EC via induction of the C-X-C chemokine IP-10 (see point 2.7). Thus, IL-12 may set in motion a cytokine cascade involving IFNγ and IP10, which eventually results in inhibition of angiogenesis.

2.5. IL-4 and IL-13

IL-4 has growth factor activity and induces uPA in micro, but not in macrovascular EC (for review, Mantovani, Bussolino, and Dejana, 1992; Wojta et al., 1993). It is intriguing that IL-4 and IL-13, but not IL-10, inhibited induction of RANTES expression in EC stimulated by IFNγ and TNF (Marfaingkoka et al., 1995). IL-4 selectively induces VCAM-1 and E-selectin expression. IL-4 is a weak inducer of IL-6 and MCP-1 in EC and amplifies production of these mediators in concert with other stimuli (for review, Mantovani, Bussolino, and Dejana, 1992). Recently it was shown that IL-13 has similar activities on EC (Sironi et al., 1994). Sharing of receptor components may underlie the similarity of action of IL-4 and IL-13 in EC. It is intriguing that these cytokines have divergent effects on monocytes (where they inhibit), and EC (where the amplify production of certain cytokines). Stimulation of certain EC functions may be important in the induction and local expression of TH2 type responses, by, for instance, inducing recruitment of eosinophils and basophils and favoring the transition to the late phase reaction.

2.6. IL-10

Information on the interaction of IL-10 with vascular endothelium is scanty and fragmentary. This cytokine was a weak stimulus for expression of chemokines and IL-6 in mouse Polyoma middle T (PmT) immortalized EC lines and amplified the action of IL-1 and TNF (Sironi et al., 1993). This effect was associated with prolongation of mRNA half line. Stimulation of HUVEC was variable and not reproducible. Under the same conditions, it inhibited cytokine production in human and murine mononuclear phagocytes. This dissociation of effects (inhibition versus costimulation) is reminiscent of the activity of IL-4 and

IL-13 (see above). IL-10 was reported to inhibit antigen presentation by human dermal microvascular EC (Vora et al., 1994) and induction of tetrahydrobiopterine in HUVEC (Schoedon et al., 1993). The effect of IL-10 on IL-8 production have been the object of conflicting reports (Chen and Manning, 1996; Debeaux et al., 1995). IL-10 inhibited the capacity of LPS to amplify apoptosis of EC by irradiation (Eissner et al., 1995). In one study (Krakauer, 1995), IL-10 inhibited induction of adhesion molecules by IL-1α.

2.7. Chemokines

By and large, the spectrum of action of chemokines is restricted to leukocytes, but recent evidence suggests that some members of the superfamily of inflammatory mediators may affect EC function. IL-8, groα and other C-X-C chemokines were reported to induce EC migration and proliferation *in vitro* and to be angiogenic *in vivo* (Koch et al., 1992; Strieter et al., 1992; Wen, Rowland, and Derynck, 1989). The expression of high affinity receptors and responsiveness to IL-8 of EC has however been the object of conflicting results (Schonbect et al., 1995; Petzelbauer et al., 1995).

Platelets factor 4, a C-X-C chemokine contained in the a granules of platelets, inhibits growth factor induced proliferation of EC and angiogenesis (Maione et al., 1990). Also interferon g-inducible protein 10 (IP 10) was shown to have angiostatic properties *in vivo*, though conflicting results have been obtained as to its capacity to inhibit bFGF-induced proliferation of HUVEC *in vitro* (Luster, Greengerg, and Leder, 1995; Angiolillo et al., 1995). IP-10 induced via IFNγ may represent the ultimate mediator of the anti-angiogenic activity of IL-12 (Voest et al., 1995).

A 3 aminoacidic motif (Glu-Leu-Arg) (ELR), is highly conserved in all members of C-X-C family which activate neutrophils. Recent results, including the action of ELR^+ versus ELR^- molecules and the activity of IL-8 muteins, suggest that the presence or absence of an ILR motif dictate whether C-X-C chemokines induce or inhibit angiogenesis (Strieter et al., 1995). However, the observation that groβ inhibits angiogenesis is not consistent with this model of function (Cao et al., 1995).

The C-C chemokines monocyte chemotactic protein-1 (MCP-1), -2 and -3 do not affect EC migration and proliferation.

Three types of chemokine binding sites have been identified (Mantovani et al., 1996). Chemokines activate leukocytes via 7 transmembrane domain, G-protein coupled receptors. A promiscuous chemokine receptor identical to the Duffy blood group antigen has been identified on erythrocytes where it may act as a sink for chemokines leaking from inflamed tissues. This promiscuous receptor is expressed by EC at post capillary venules *in vivo*, but not by endothelial cells *in vitro* (Mantovani et al., 1996; Peiper et al., 1995). The structure of the promiscuous chemokine receptor is that of a 7 transmembrane-type molecule, but there is no clear evidence of signaling activity. Finally, chemokines bind heparin and heparin-like proteoglycans and these molecules on endothelial cells present at least certain chemokines to leukocytes in the multistop process of recruitment. The molecular basis underlying the pro and anti-angiogenic has not been clearly established. The expression of classic chemokine serpentine receptors in endothelial cells is controversial (Petzelbauer et al., 1995; Maione et al., 1990). The anti-angiogenic chemokine IP-10 was suggested in one study to inhibit bFGF-induced HUVEC proliferation via interaction with proteoglican structures (Angiolillo et al., 1995), but did not affect HUVEC proliferation in another (Luster, Greenberg, and Leder, 1995). In the same vein, groα affected EC proliferation in one study (Wen, Rowland, and Derynck, 1989) but others failed to observe binding and activity of this chemokine on EC (Rot, 1992). The mouse C-C chemokine FIC has been shown to bind EC (Heinrich et al., 1993).

3. THE CYTOKINE REPERTOIRE OF ENDOTHELIAL CELLS

EC are important producers of various polypeptide mediators that regulate the hematopoietic system, the differentiation of T and B cells, and the extravasation of leukocytes at inflammatory sites. Here we will briefly summarize available information on the production of EC of cytokines that affect leukocytes. TGFβ, a product and a regulator of EC e.g. (Mantovani and Dejana, 1989; Sporn et al., 1987; Gamble and Vadas, 1991), is not considered in this review.

3.1. IL-1, TNF, and IL-6

IL-1, a cytokine that profoundly affects vascular cells (see point 2.1), is also a product of vessel elements (for review, see ref. Mantovani and Dejana, 1989; Pober and Cotran, 1990; Libby and Hansson, 1991). Inducers of IL-1 production by vascular cells include LPS, TNF, and IL-1 itself. *In vitro* passage of EC results in spontaneous expression of IL-1α message and refractoriness to activation by exogenous IL-1 (see above). IL-1 gene expression and protein were also detected *in vivo* after inoculation of LPS and feeding an atherogenic diet (Moyer et al., 1991; Clinton et al., 1991). Endothelial cells produce IL-1, but efforts to demonstrate expression of the IL-1 receptor antagonist (IL-1ra) umbilical vein cells yielded negative results (Bertini et al., 1992). IL-1ra blocks the action of IL-1 on EC, and these cells may represent a major target for the therapeutic activity of this molecule *in vivo*.

Upon exposure to inflammatory signals and superinduction by protein synthesis inhibitors, vascular SMC, but not EC, can express TNF mRNA and protein (Warner and Libby, 1989).

SMC and EC produce high levels of IL-6 (Sironi et al., 1989). Variable levels of spontaneous IL-6 production can usually be detected and these are markedly augmented by exposure to IL-1, TNF and LPS. IL-1 acts on T cells, at least in part, by inducing IL-6. Because IL-6 does not affect the proinflammatory and prothrombotic functions of IL-1 on EC, and anti-IL-6 antibodies do not affect the activity of IL-1, it can be concluded that IL-6 is not the ultimate mediator of the effects of IL-1 on vascular cells (Sironi et al., 1989). Production of IL-6 is amplified by costimulation with IL-4 or IL-13 (Rollins and Pober; 1991; Howell et al., 1991; Colotta et al., 1992). IL-6 is also induced by IL-17 (Fossiez et al., 1996). Hypoxia induces IL-6 production in EC and NFIL-6 plays a central role in this phenomenon (Yan et al., 1995). IL-6 is a pleiotropic cytokine that affects T and B lymphocytes and the production of acute-phase proteins by liver. The production of IL-6 by EC integrates these cells in immunological circuits and in the regulation of the acute-phase response. IL-1 also induces production of leukemia inhibitory factor (LIF) (Grosser et al., 1995), but not of IL-11 (Suen et al., 1994).

3.2. CSFs

ECs are an important source of CSFs. Hematopoietic growth factors produced by EC include stem cell factor (Meininger et al., 1995) G-CSF, GM-CSF, and M-CSF. CSF production is induced or augmented by a variety of stimuli including LPS, IL-1 and TNF (see refs. Mantovani and Dejana, 1989; Pober and Cotran, 1990; Libby and Hansson, 1991) and minimally modified low-density lipoproteins (MM-LDL) Rajavashisth et al., 1990). By producing CSFs as well as IL-1 and IKL-6, which have CSF activity, EC participate in the regulation of hematopoiesis. Because EC respond to certain CSF molecules (G-CSF, GM-CSF, and erythropoietin; see above), autocrine and paracrine circuits involving EC and CSFs could be important in the generation and maintenance of an appropriate bone marrow mileu.

3.3. Chemokines: *In Vitro* Studies

3.3.1. Endothelial Cells. EC produce various chemokines in response to signals representative of inflammatory reactions, immunity and thrombosis (Mantovani, Bussolino, and Dejana, 1992). Inflammatory cytokines (IL-1 and TNF) and bacterial LPS induce expression and release of IL-8 and groα (Strieter et al., 1989; Schroder and Christophers, 1989; Sica et al., 1990a; Dixit et al., 1990; Sica et al., 1990b; Rollins et al., 1990; Gimbrone et al., 1989; Wen, Rowland, and Derynck, 1989). Monocyte-asscociated IL-1α induces IL-8 expression in EC via a juxatacrine mechanism (Kaplanski et al., 1994). Induction of IL-8 expression is associated with and depends on gene transcription (Sica et al., 1990a). IL-4 and IL-13 are weak inducers of IL-8 expression and amplify induction by inflammatory cytokines (Rollins and Pober, 1991; Howell et al., 1991; Colotta et al., 1992; Korpelainen et al., 1993; Sironi et al., 1994). Histamine induces IL-8 production in EC (Jeanin et al., 1994). Hypoxia has recently been shown to induce IL-8 and MCP-1 expression in EC, a finding potentially relevant for pathological conditions in which activation and recruitment of leukocytes may amplify tissue damage (Kaplanski et al., 1993b; Karakurum et al., 1994). Platelets contain IL-1 and when they interact with vascular EC, induce IL-8 gene expression (Kaplanski et al., 1993a). Fibrin induces IL-8 in EC (Qi and Kreutzer, 1995).

As a result of proteolytic cleavage, IL-8 versions with a different NH_2 terminus and length can be produced (Baggiolini, Dewald, and Moser, 1994). It has been suggested that EC release predominantly a 77aa version of IL-8, which is a less active species at activating leukocytes than the most common 73 residue form (Gimbrone et al., 1989; Hebert et al., 1990). The proteolytic conversion to smaller versions of the molecule can be catalyzed by thrombin (Hebert et al., 1990).

The influence of IL-8 on the interaction of polymorphonuclear cells with vascular EC has been the object of seemingly conflicting observations, which seem now to reflect different experimental protocols and, most interestingly, different functions exerted by this cytokine under different pathophysiological conditions. IL-8 increased the adhesiveness of normal PMN for normal EC (Carveth et al., 1989). In apparent contrast with these findings, EC-derived IL-8 was reported to inhibit binding of the PMN to activated EC (Gimbrone et al., 1989). Although it elicits PMN extravasation when given locally, IL-8 inhibits recruitment if administered systemically by the i.v. route (Hechtman et al., 1991; Ley et al., 1993). The seemingly paradoxic anti-inflammatory effects of high levels of systemic IL-8, possibly dependent upon the action of a reverse chemotactic gradient and leukocyte deactivation, may represent a feedback mechanism to control tissue damage. Chemoattractants cause rapid release from leukocytes of the p75 TNF receptor and of the type II decoy IL-1 receptor (Colotta et al., 1995). These molecules are likely to play an important role in inhibition of distant inflammatory reactions by chemokines.

The role of IL-8 produced locally by vascular cells was reexamined using reconstructed vessel wall models (Huber et al., 1991). Unequivocal evidence was obtained for the importance of IL-8 in transendothelial migration induced by inflammatory cytokines (Huber et al., 1991; Smith et al., 1993).

EC of postcapillary venules bind IL-8, possibly via heparin-like molecules (Rot, 1992). EC of postcapillary venules in kidney and other tissues express the promiscuous chemokine receptor present also on erythrocytes and known as Duffy antigen (Hadley et al., 1994; Peiper et al., 1995). This receptor is present on EC of both Duffy+ and - individuals and may serve to present chemokines to circulating leukocytes. Solid phase IL-8 elicits haptotactic migration (Wang et al., 1990). Thus, locally produced IL-8 may be retained on the surface of EC and activate adhesive interactions and migration (Rot, 1992).

EC activated *in vitro* by inflammatory cytokines express groα, which according to one report, could in turn act on EC (Wen, Rowland and Derynck, 1989). It has been suggested that EC-bound groα may promote monocyte adhesion (Schwartz et al., 1994).

IP10, a member of the C-X-C family, but unique in that it attracts monocytes, is expressed in certain endothelia of mice exposed *in vivo* to IFNγ or to LPS (Narumi et al., 1992; Gómez-Chiarri et al., 1993). There are no reports on *in vitro* expression of this chemokine in EC.

EC produces substantial amounts of the C-C chemokine MCP-1 (Sica et al., 1990b; Rollins et al., 1990). The proinflammatory signals IL-1, TNF and to a lesser extent, LPS are potent stimuli for MCP-1 production (Sica et al., 1990b; Rollins et al., 1990). Under the same conditions, MCP-2 was undetectable in EC supernatants (J. Van Damme, personal communication). IL-4 and IL-13 are active, though less potent inducers of MCP-1 expression (Rolling and Pober, 1991; Howell et al., 1991; Colotta et al., 1992). IFNγ was recently shown to induce MCP-1 in human microvascular EC (Brown et al., 1994). M-CSF was reported to induce MCP-1, though we did not detect the M-CSF receptor c-fms in HUVEC by Northern analysis (Shyy et al., 1993). Given the role that lipids and monocytes play in the natural history of atherosclerosis, it is of interest that minimally modified low density lipoproteins (MM-LDL) induce MCP-1 production in EC and smooth muscle cells (Crushing et al., 1990). Thrombin was recently found to induce expression of MCP-1 in monocytes and, less prominently, in EC (Colotta et al., 1994). The C-C chemokine RANTES was produced by EC exposed to TNF and IFNγ (Marfaingk-oka et al., 1995).

The molecular basis of stimulation of chemokine expression in EC has been studied to a limited extent. Induction by inflammatory signals and thrombin is protein synthesis independent in EC, but, interestingly, not in monocytes (Colotta et al., 1994). Direct demonstration of enhanced gene transcription was obtained for MCP-1 and IL-8 by nuclear run off analysis (Sica.et al., 1990b; Sica et al., 1990a).

3.3.2. SMC and Mesangial Cells. Already in early studies on MCP-1, it had been described that vascular SMC secreted a monocyte chemoattractant (Valente et al., 1984; Valente et al., 1988) which resembled a similar factor produced by tumor lines (Botazzi et al., 1983). The SMC-derived factor was subsequently identified as MCP-1. Indeed IL-1, TNF, and MM-LDL induce expression of MCP-1 in SMC with mechanisms and kinetics resembling those observed in EC (Wang et al., 1991; Cushing et al., 1990). IL-1 and TNF also stimulate expression and release of IL-8 in SMC (Wang et al., 1991).

Kidney mesangial cells may have an ontogenetic relationship with SMC. They express and release MCP-1 and IL-8 in response to IL-1 and TNF (Zoja et al., 19911; Rovin, Yoshimura, and Tan, 1992; Satriano et al., 1993b; Satriano et al., 1993a). Of special importance in relation to pathology is induction of MCP-1 and IP10 by aggregated immunoglobulins (Sariano et al., 193b; Satriano et al., 1993a). Reactive oxygen intermediates act as second messengers in induction of MCP-1 by aggregated Ig (Satriano et al., 1993a). These results may have considerable relevance in kidney pathology involving mononuclear phagocyte infiltration.

3.4. Chemokines: *In Vivo* Studies

EC at sites of delayed type hypersensitivity reactions and kidney allograft rejection express the C-C chemokine RANTES (Devergne et al., 1994; Pattison et al., 1994). *In vivo* studies on chemokines in vessel wall pathology have largely been restricted to atherosclerosis. MCP-1 expression has been detected in atheromatous lesions of rabbits, primates, and man (Takeya et al., 1993; Yu et al., 1992; Yla Herttuala et al., 1991; Nelken et al., 1991). IL-8 and MCP-1 mRNA have been detected in increased amounts in aortic aneurysms (Koch et al., 1993). Chemokine gene expression has been detected in various cellular elements, including SMC, EC and mononuclear phagocytes, with somewhat different results in different studies. In the only study with mAb (Takeya et al., 1993), cell populations positive for MCP-1 were different in lesions representative of different stages of the natural history of atherosclerosis. EC staining was prominent in diffuse intimal thickening and in fatty streaks, whereas it was weak in atheromatous lesions. Subendothelial macrophages were strongly positive for MCP-1 in fatty streak lesions and in atherosclerotic plaques. In plaques, a few intimal SMC stained for MCP-1. These results suggest that EC and macrophages are the major source of MCP-1 in early atherosclerotic lesions (Takeya et al., 1993). Monocyte adhesion and infiltration is an early event in the natural history of atherosclerosis (Libby and Hansson, 1991; Ross, 1993). Mononuclear phagocyte infiltration is also a prominent feature of vasculitis (Fauci, Haynes, and Katz, 1978). Locally produced MCP-1 may play an important role in regulating extravasation of leukocytes, of monocytes in particular, in vessel wall pathology.

4. EC HETEROGENEITY

Vascular endothelia are remarkably heterogeneous in terms of morphology, marker expression (e.g. von Willebrand factor), and function. Heterogeneity is evident at multiple levels and emerges from analysis of macro- vs. microvascular EC, of the vascular bed from different organs, and of EC obtained from different compartments in the same organ. However, studies of EC heterogeneity in response to and production of cytokines are scanty and fragmentary. It has been shown that, unlike large vessel EC, microvascular EC express PDGF receptors (Beitz et al., 1991) and respond with growth and uPA production to IL-4 (Streiter et al., 1989; Wojta et al., 1993). When EC obtained from different levels of the lung (pulmonary artery and microvessels) were investigated for responsiveness to TNF, only the former were damaged by this cytokine, which increased PG12 production in both (Meyrick, Christman, and Jesmok, 1991).

In various studies human microvascular EC of dermal or brain origin have been studied. These differed from HUVEC in terms of expression of adhesion molecules in response to inflammatory cytokines or endothelins (Rott et al., 1993; McCarron et al., 1995; Petzelbauer et al., 1993), antigen presentation regulated by IFNγ and IL-10 (Vora et al., 1994), and tetrahydrobiopterine production regulated by various mediators (Schoedon et al., 1993). As discussed above, production of chemokines can be differentially regulated in different EC populations (Brown et al., 1994). All in the phenomenology of EC heterogeneity is growingly complex. However its molecular and cellular basis as well as the formulation of a conceptual framework for this descriptive phenomenology, remains elusive.

5. MOLECULAR BASIS FOR ACTIVATION OF THE PROINFLAMMATORY PROGRAM

In this section, the molecular basis for endothelial cell activation by IL-1 and TNF will be reviewed in some detail. Attention will be focused on these molecules because among cytokines active on EC, IL-1 and TNF are the ones extensively studied and elicit the wildest spectrum of responses. These include regulation of genes whose expression is largely (e.g. VCAM-1) or absolutely (e.g. E-selectin) restricted to EC, thus raising the largely unsolved issue of the molecular basis of endothelial specificity.

5.1. Receptors and Second Messengers

EC express both the p75 and the p55 TNF receptor. Following the binding to the receptors, the importance of a subsequent step of internalization has been recently shown for the induction of several genes, showing that most of the internalized receptor is the p75 (which is the most abundant in the membrane) and is then transported to the lysosomes via coated pits and coated vesicles (Bradley, Thiru, and Pober, 1995; Bradley, Johnson, and Pober, 1993).

On the contrary, the p55 receptor is expressed at much lower levels in the membrane but is more abundant overall and it is detectable mainly in the Golgi apparatus and in the cytoplasmatic vacuoles (Bradley, Thiru, and Pober, 1995).

TNF muteins (Barbara et al., 1994) and agonistic or antagonistic antibodies (Mackay et al., 1993) have been used to define the relative importance of p55 versus p75 in EC activation. By and large, these studies indicate that TNF activates EC predominantly via the p55 receptors (Paleolog et al., 1994; Slowik et al., 1993). However, there is also evidence for a contribution of the p75 receptor (van de Kar et al., 1995). The contribution of the p75 receptor is best observed at low TNF concentrations, consistently with the "ligand passing" model of function of this molecule (Tartaglia, Pennica, and Goeddel, 1993). The transmembrane form of TNF is the prime ligand of p75 and it may play an important role in fuxtacrine interactions between EC and monocytes (Grell et al., 1995; Schmid et al., 1995; Lukacs et al., 1995).

EC express only the type I IL-1 receptor (Colotta et al., 1993b). Under resting conditions or upon exposure to inducers (e.g. IL-4) active on myelomonocytic cells, EC do not express the type II decoy receptor (Colotta et al., 1993b), a molecule without a demonstrable role in signaling, capable of blocking IL 1 (Colotta et al., 1993a).

Blocking of the type I receptor inhibited the activation of EC by IL-1 (Colotta et al., 1993b). IL-1α has been suggested to act intracellularly in the regulation of EC life span (Maier et al., 1990) and to localize to the nucleus (Maier, Statuto, and Ragnotti, 1994). The role, if any, of intracellular receptors in these putative intracellular activities of IL-1 remains to be elucidated.

Second messengers involved in EC activation by IL-1 or TNF have not been systematically investigated. A role of PKA-mediated phosphorylation was studied by forskolin mediated activation of the adenylate cyclase in HUVEC (Ghersa et al., 1994) and suprisingly, a reduction in the cytokine-induced E-selectin and VCAM-1 expression was observed concomitantly with a stimulation in the ICAM-1 expression (Ghersa et al., 1994). The data on E-selectin were further studied at molecular level showing a direct reduction of phosphorylation of some component of the protein complexes, but without affecting the binding of NF-κB (see below).

Somewhat similar conclusions were also reached by others (Pober et al., 1993) who affirm that cAMP is neither an intracellular mediator nor a physiological regulator of TNF induced E-selectin and VCAM-1 expression, but pharmacological elevations of cAMP in

EC could have a role because of their inhibitory effects on the TNF-induced synthesis of E-selectin and VCAM-1 (Pober et al., 1993).

Interestingly, IL-4 has been hypothesized to induce a yet uncharacterized adhesive molecule on the EC for PBL (different from VCAM-1) through a cAMP dependent pathway (Galea et al., 1993).

Apart from still controversial results on the distribution of the various PKC isoforms in the different endothelial cell populations (Mattila et al., 1994 and ref. therein), several observations agree that PKC does not play a major role in the TNF inducible expression of several genes, such as E-selectin (Ritchie et al., 1991; Niedbala and Stein Picarella, 1993a; van de Kar et al., 1995), although PMA can indeed recapitulate the adhesive phenotype; recently one interesting exception was suggesting a role for PKC in uPA (but not uPAR) reporter expression (Niedbala and Stein Picarella, 1993).

Finally, few reports have explored the role of tyrosine phosphorylation in cytokine activation of EC. van Hinsbergh et al reported that TNF induced signal transduction pathway of PAI-1 transcription involves a genistein-sensitive step that is not involved in the induction of u-PA by TNF (van Hinsbergh et al., 1994). TNF induction of E-selectin and VCAM-1 via NF-κB involves activation of tyrosine kinase (Weber et al., 1995). With the use of a vast array of tyrosine protein kinase inhibitors and phosphatase inhibitors they could show that the effect of genistein was quite selective, irrespective of the signal used (IL-4, thrombin, and TNF mutants) (van Hinsbergh et al., 1994). VCAM-1 expression induced by IL-4 seems to require a tyrosine phosphorylation step, including the phosphorylation of its own receptor (Palmercrocker and Pober, 1995), similarly to what observed with Oncostatin M, which phosphorylates its own receptor and requires tyrosine phosphorylation to induce IL-6 secretion (Schieven et al., 1992).

EC express TGFβ receptors *in vitro* and *in vivo* (Dickson et al., 1995; Sankar et al., 1995). The type I and type IIR have been identified and transfer type III increases responsiveness of bovine EC to TGFβ2 (Sankar et al., 1995).

All in all, second messenger studies implying some phosphorylation event have to be considered in conjunction with the following observations on gene expression by NF-κB molecules, because the final activation step for NF-κB activity is phosphorylation of the I-κB inhibitor (see below).

Interestingly, many data suggest a role for reactive oxygen intermediates (ROI) as a common and critical denominator for various activating signals leading to NF-κB activation (see below). In particular, many different antioxidants (such as N-acetyl-L-cysteine (NAC), dithiocarbamates, vitamin E derivatives, metal chelators, and pyrolidinedithiocarbamate (PDTC)) have been shown to effectively block the NF-κB activation and subsequent gene induction. This issue has been scarcely explored so far in endothelial cells: α-tocopherol was shown to inhibit the cytokine-induced expression of E-selectin but the authors failed to detect any change in NF-κB activity on the promoter (Faruqi, de la Motte, and DiCorleto, 1994). Other agents like PDTC and NAC were shown, on the contrary, to inhibit the expression of VCAM-1, possibly through a reduction in the binding of NF-κB proteins to promoter elements, but interestingly, in the same study, no effect was reported in respect to E-selectin or ICAM (Marui et al., 1993).

5.2. Transcription Factors

IL-1 and TNF regulate expression of a large and complex set of genes in EC, as revealed by investigations aimed at defining the molecular basis of functional responses and by "fishing expeditions" (see above). An analysis of regulatory elements of this vast array of genes is beyond the scope of this review. Here we will focus on regulation of genes whose expression is, at least to some extent, restricted to EC, such as E-selectin, VCAM-1, and TM. In most cases, both from studies with deletion mutants of the different promoters and from gel retardation assays, molecules of the NF-κB family have been implicated.

The best studied gene so far is E-selectin, whose promoter has been extensively characterized in EC. A fragment of 233 bp was found essential but not sufficient to confer cytokine inducibility and defined one NF-κB binding site (Whelan et al., 1991). With transfection assays in EC or nuclear extracts from cells of endothelial origin, at least six distinct elements exist in the minimal promoter which are al required for maximal inducibility, but none of them is sufficient (Figure 2) Whitley et al., 1994; Schindler and Baichwal, 1994). Canonical NF-κB proteins (dimers p50/p65) seem to be able to bind at least three such elements, which share the canonical consensus (5′-GGGPuNNPyPyCC-3′) although the possible presence of alternative NF-κB dimers has not yet been completely clarified. Many reports suggest that the dimers sitting on the promoter in resting conditions are in dynamic equilibrium with other, and possibly different dimers that are mobilized upon activation (Hooft van Huijsduijnen, Pescini, and DeLamarter, 1993). A fourth element, NF-ELAM-1 is a CRE/ATF-like binding site, recognized by ATFa, ATF2, and ATF3 homodimers as well as ATFa/ATF2/c-jun heterodimers (Pescini et al., 1994; Kaszubska et al., 1993), while the fourth element, NF-ELAM-2 binds a still unknown protein. In most of these binding sites, there is yet another protein member of the high-mobility group I (Y), HMG I (Y) (Lewis et al., 1994), which facilitates the binding of NF-κB (and possibly also of the ATF) proteins to the DNA and has also a binding site on its own (Figure 2). Finally a role for the relative helix orientation of ATF and one NF-κB element, more than the distance between them, has been reported as crucial in determining a correctly oriented transcription factor induced DNA bending, which is pivotal for transcription to occur in endothelial cells (Meacock et al., 1994). Cooperativity between different families of transcription factors is increasingly recognized as a crucial element in gene expression. Only accurate and consistent studies in the same cellular context and with a few selected stimuli will eventually lead to the understanding of how a relatively small number of transcription factors can be used to achieve the high level of specificity required to regulate the complex pattern of gene expression in higher eukaryotes (Tijan and Maniatis, 1994). Recent findings of interaction between NF-κB and other transcription factors, as in the case of the E-selectin and VCAM promoter, greatly expands the possibilities for combinatorial control of promoters (Siebenlist, Franzoso, and Brown, 1994; Neish et al., 1995).

A further recently recognized element of complexity for the regulation of the NF-κB activity in endothelial cells, is the family of the NF-κB inhibitors known as IkB (Thanos and Maniatis, 1995). In resting cells IkB proteins bind and sequester NF-κB in the cytoplasm. Stimulation rapidly degrades IkB (Grimm and Baeuerle, 1993) and releases NF-κB which translocates to the nucleus. Recently, it was suggested that TNF induced phosphorylation of Ik-Bα is a signal for its degradation which occurs while it is still complexed with NF-κB (Miyaloto et al., 1994). In any case, phosphorylation is a key regulatory element upstream to the NF-κB activation (Miyalota et al., 1994).

It was observed that immediately after TNF stimulation, protein levels of IkBα fall rapidly (Read et al., 1994; de Martin et al., 1993) while in parallel RNA levels are dramatically upregulated (d'Aniello et al., 1993). Interestingly enough, the Ik-Bα clone was one of the IL-1 immediate early inducible genes cloned from HUVEC and described to be rapidly upregulated even in the absence of protein synthesis (Introna et al., 1993; d'Aniella et al., 1993). It then became clear that the transcriptional activation of the IkBα gene was itself under the control of the NF-κB activity (Read et al., 1994; de Martin et al.,

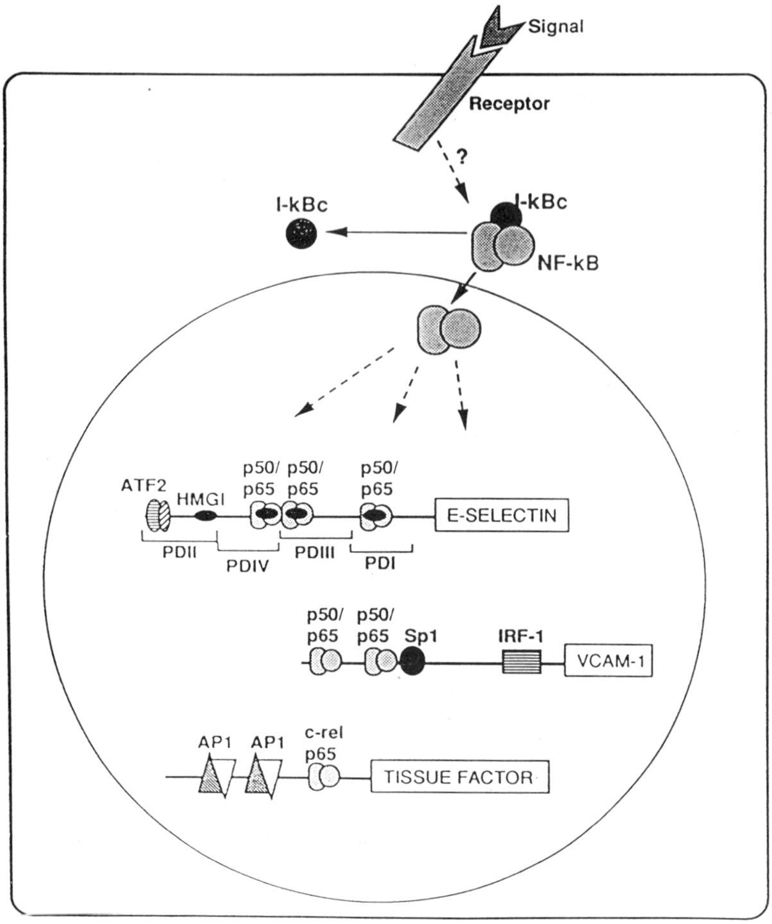

Figure 2. Promoters of cytokine inducible genes in endothelial cells. IL-1 = Interleukin-1; IL-1R = Interleukin-1 receptor; IkB-a = α inhibitor of NF-κB. 1 = protected site previously defined as NF ELAM-1 bound by ATF-2 (−153/−146 nucleotides upstream the transcriptional start site). 2 = site bound by HMG I (high mobility group protein) (−140/−135). 3 = protected site previously defined as NF ELAM-3 or NF-κB or as site 1 bound by NF-κB (−126/117). 4 = protected site previously identified as site 2 bound by NF-κB (−116/−106) this site includes also a binding site for HMG-1 (−113/−110). 5 = protected site identified as site 2 NF-ELAM 2 the bound protein has not been yet characterized (around −100). 6 = protected site previously identified as NF-κB 1 or site 3 bound by NF-κB (−94/−83) the site includes also a binding site for HMG-1 (−90/−86).

1993). This circuit may explain the transient induction in the endothelial cells of many cytokine inducible genes (e.g. ELAM-1).

VCAM-1 promoter was found to contain two canonical NF-κB binding sites (Neish et al., 1992; Lademarco et al., 1992) which activate the gene when bound by the heterodimers p65/p50 or by the p65 homodimers (Shu et al., 1993). In addition, two GATA sites were described which were found necessary for the induction of the full responsiveness to the cytokines (Neish et al., 1992). Very recently a functional interferon regulatory factor 1 (IRF-1) sequence motif was demonstrated in the VCAM-1 promoter and evidence was produced of slow cooperative DNA binding and transactivation properties between NF-κB and IRF-1 (Neish et al., 1995).

Interestingly, like for the E-selectin promoter, the HMGI (Y) protein appears to facilitate the NF-κB and IRF-1 activation (Neish et al., 1995).

The ICAM-1 promoter has been recently studied in endothelial cells and it was shown that transcription induced by TNFα critically depends on a single atypical NF-κB site, which mediates possibly also the IL-1 and LPS inducibility (Ledebur and Parks, 1995). Even if p65 homodimers and p65/p50 heterodimers can bind *in vitro* to this element, functional studies implicate the p65 homodimer as the key transactivator on this site (Ledebur and Parks, 1995). This information may have a more general implication when compared with a previous report on *in vitro* studies with nuclear extracts from IL-1 stimulated HUVEC with several fragments derived from many IL-1 inducible genes, which showed that while one canonical NF-κB sequence from E-selectin promoter could bind the canonical p65/p50 heterodimer, the other genes all contain variant NF-κB sites, including IL-8, GM-CSF, ICAM-1 and TF and specifically bound c-Rel/p65 heterodimers (Parry and Mackman, 1994). One possible conclusion is that all the NF-κB sequences described so far, in spite of different decameric sequences, if possess the triplet of guanine residues at the 5′ end may bind and be activated by the p65/p50 heterodimer (like ELAM-1 and VCAM-1,) while the variant may be activated by different dimers (like ICAM-1).

A different mechanism seems to be involved in the transcriptional activation of the LDL receptor gene, which in endothelial cells shows a binding of a SP-1 transcription factor and does not have any NF-kB sites. (Hamanaka et al., 1992). Induction of tissue factor involves members of the b2lp and NF-kB family (Bierhaus et al., 1995).

Finally, the promoter of TM gene is repressed rather than induced by TNF. Both the basal activity and the TNF-mediated repression, can be reconducted to the same element which binds an ets-like factor, not further characterized (von der Ahe et al., 1993).

The results discussed here provide a basis for a new level of comprehension of EC activation by IL-1 and TNF, by identifying transcription factors and regulatory elements involved in the induction or inhibition of gene expression and its time-course. However, the central issue of specificity to endothelium of expression of certain genes remains unsolved. For instance, E-selectin is exclusively expressed in cytokine activated EC and considerable progress has been made in defining the molecular basis of induction and ensuing inhibition (see above): yet no genetic clues as to the endothelial specificity of this molecule are available. The difficulty in transfecting primary EC and the lack of appropriate EC lines may have hampered solving this critical question, which may ultimately require *in vivo* studies in transgenic mice.

6. PATHOLOGY

Diseases involving a wide range of cellular and organ systems involve EC as targets or active participants. Here we will focus on selected disorders in which the interplay between EC and cytokines plays a primary role.

6.1. Infectious Diseases

Viral, bacterial or protozoan infections directly or indirectly involve EC. Viruses infect or interact with EC and induce or modify cytokine production. For instance Epstein Bar virus and cytomegalovirus (CMV) infect EC and these agents as well as EBV cell lines induce IL-6 production (Jones et al., 1995; Almeida et al., 1994). We recently found high levels of MCP-1 in the cerebrospinal fluid of patients with HIV infection, specifically associated with CMV encephalitis (Bernasconi et al., unpublished): immunoreactive MCP-1 was detected in brain EC (Vago, unpublished).

EC cytokine networks have assumed an increased relevance in HIV-1 infection and associated clinical disorders, including Kaposi's sarcoma, bacillary angiomatosis, vasculitis, psoriasis and localization of non-Hodgkins lymphomas in central nervous system. The interaction between infected monocytes with EC increases the virus replication. This response appears to be via both cell contact and EC-derived cytokines (IL-6, GM-CSF and IL-1) (Fan, Hsia and Edgington,1994). Furthermore, an unbalance of inflammatory cytokine production has been described in AIDS patients and precedes the appearance of Kaposi's sarcoma (Fan, Bass and Fahey, 1993). A combination of inflammatory cytokines (IFNγ, IL-1 and TNF) induce EC *in vitro* to acquire phenotypic and functional features closely resembling cultured and in situ Kaposi's sarcoma spindle cells. These include spindle morphology, the responsiveness to HIV-Tat protein and the neo-expression or up-regulation of adhesion molecules such as ICAM-l, VCAM and E-selectin. Finally these ECs become angiogenic in nude mice (Samaniego et al., 1995; Fiorelli et al., 1995). These data suggest that in vivo hyperplasia of EC observed in Kaposi's sarcoma and angiogenesis may be due to a chronic exposure to inflammatory cytokines.

Ricketsia ricketsiae and R. conorii cause spotted fevers characterized by prominent, generalized vascular inflammation. Ricketsiae enter and proliferate in EC. EC function affected include permeability, expression of adhesion molecules, release of von Willebrand factor. R. conorii was shown to induce IL-1α in EC and cell associated IL-1α was in turn responsible for induction of IL-8 and IL-6 (Kaplanski et al., 1995). Therefore a cytokine cascade initiated by ricketsial infection of EC may play an important role is spotted fever-associated vasculitis.

Gram+ and gram- bacteria and their products interact with EC, induce cytokine production and this interaction accounts for some of the systemic manifestations of sepsis as well as for located reactions. S. aureus, a gram+ bacterium, interacts with EC and after internalization induces IL-1 and IL-6 production (Yao et al., 1995). EC are a major determinant of systemic inflammatory response syndrome (SIRS)/toxic shock. Hypotension and coagulation abnormalities are likely primarily dependent upon EC activation by LPS or LPS induced IL-1 and TNF (review Bradley, Wilks and Rubenstein, 1994). Bacterial lipopolysaccharides (LPS) elicits a spectrum of EC responses similar to that of IL-1 and TNF (Mantovani, Bussolino and Dejana, 1992). It has recently been shown that EC, which are $CD14^-$, interact with soluble CD14 and LPS-binding protein, both present in plasma (von Asmuth et al., 1993; Pugin et al., 1993; Haziot et al., 1993; Read et al., 1993; Goldblum et al., 1994). While direct activation of EC by LPS does occur, the indirect pathway involving mononuclear phagocyte activation and production of inflammatory cytokines, is more effective in stimulation of EC by this bacterial product (Pugin, Ulevitch and Tobias, 1993).

Among parasitic disorders vascular involvement is a prominent feature of cerebral malaria, a fatal complication of Plasmodium falciparum infection in man. Circumstantial evidence in man and mouse model suggest that cytokines, TNF in particular; and alterations of endothelial cells play a pivotal role in arrest of erythrocytes in the brain and subsequent vascular damage (Willimann et al., 1995; Chirivi et al. 1994). The glycosyphosphatidylinositol toxin induces VCAM-1, ICAM-1, E-selectin, and NO synthase in EC via a tyrosine kinase and PKC dependent mechanism (Schofield et al., 1996; Tachado et al., 1996). The mechanisms responsible for selective involvement of the brain vascular bed remain undefined. Hemolytic uremic syndrome (HUS), the most common cause of acute renal failure in infants and small children, is caused by infection with verotoxin (VT, or Shiga-like toxin) producing

E. coli (Remuzzi and Ruggenenti, 1995). VT is toxic for EC, activates them in a pro-inflammatory/prothrombotic sense and synergizes with IL-1 and TNF (van de Kar et al., 1995; van de Kar et al., 1992). These cytokines enhance galactosyltransferase and consequently surface expression of globotriaosylceramide (Gb3) (van de Kar et al., 1995). The VT receptor is expressed in greater amounts on EC of certain vascular beds, such as the kidney. Therefore, augmentation of Gb3 underlies synergism between inflammatory cytokines and VT and explains the selective organ involvement in HUS.

6.2. Neoplasia

Formation of new blood vessels is a limiting determinant of tumor growth and progression. An analysis of pro and anti-angiogenic factors is beyond the scope of this review. IFNs, IL-12 and chemokines were discussed above and here selected aspects of IL-1 and TNF, mainly related to hemorrhagic necrosis and promotion of metastasis, will be discussed (Chirivi et al., 1994).

IL-1 and TNF regulate the lifespan of EC and angiogenesis with seemingly contradictory actions. Endogenous IL-1α has been suggested to reduce the lifespan of EC by acting intracellularly and localizing to the nucleus (Maier et al., 1990). IL-1 and TNF are not growth factors for EC, but TNF induces EC migration *in vivo* and angiogenesis *in vivo* (Mantovani, Bussolino and Dejana, 1992). These functions may be indirect. TNF (and IL-1) induce production of the secreted protein B.61 which functions as an EC chemoattractant and angiogenic factor by acting as the ligand for the eck tyrosine receptor (Holzman, Marks and Dixit, 1990).

Other molecules cloned as IL-1/TNF inducible genes in EC may be relevant to vasculogenesis or angiogenesis. A20 is a zincfinger transcription factor which may be important in protection of EC and other cell types against TNF toxicity. B24 is a gene of unknown function expressed in a complex way during vasculogenesis (Sarma et al., 1992). Finally, IL-1 and TNF induce chemokines in EC, which act as pro or anti-angiogenic factors (see above).

At least certain tumors utilize the same molecular pathways as leukocytes for interacting with endothelial cells and seeding at distant anatomical sites. Cytokines expressed constitutively by cancer cells or produced locally augment expression of adhesion molecules recognized by tumor cell counter-receptors (e.g. VLA4-VCAM-1 in melanoma) (for review Chirivi et al., 1994). In the same vein, EC-derived chemokines may play a role in cancer seeding at specific sites (Wakabayashi, Cavanaugh and Nicolson, 1995).

The eponymous function of TNF shared by other inflammatory cytokines is to cause hemorrhagic necrosis of neoplastic as well as of appropriately conditioned normal tissues (Schwartzman reaction). Involvement and damage of the vascular bed is a prominent feature of hemorrhagic necrosis (Mantovani, Bussolino and Dejana, 1992). TNF is not cytotoxic for confluent EC. The mechanism responsible for the capacity of TNF to selectively damage the vasculature of murine and human (Lienard et al., 1992) tumors remains unclear.

6.3. Atherosclerosis and Cardiovascular Pathology

Monocyte recruitment is the first recognizable event in the natural history of atherosclerosis (Libby and Hansson, 1991). EC and smooth muscle cells produce chemokines, MCP-1 in particular, in response to pathophysiologically relevant stimuli (see above). Various studies have shown expression of MCP-l in atherosclerotic lesions (Mantovani, Bussolino and Dejana, 1992). In particular, EC staining for MCP-1 was prominent in the early lesions constituted by diffuse intimal thickening and fatty streaks (Takeya et al., 1993). IL-1α was also detected in atherosclerosis (Clinton et al., 1991).

Cardiovascular disorders, such as myocardial infarction, are late consequences of atherosclerosis, associated with increased levels of inflammatory cytokines (e.g. TNF). The levels of the acute phase pentraxin C reactive protein have prognostic significance in ischemic heart disease (Liuzzo et al., 1994). We recently found that the long pentraxin PTX3 is preferentially expressed by heart EC (Introna et al., 1996) and is elevated in myocardial infarction (Peri G. unpublished). Its significance as a direct indicator of myocardial EC function remains to be established.

6.4. Transplantation and Autoimmunity

EC are a major target for immune reactions directed against allo or xenogeneic transplanted organs (Bach et al., 1994) or self antigens. Circulating anti-endothelial cell antibodies (AECA) have been detected in a variety of autoimmune diseases with vascular involvement.

AECA are a common occurrence in a variety of autoimmune diseases including sclerederma, SLE and Wegener granulomatosis. Antigens recognized include proteins (e.g. protease 3), phospholipids and protein-PL complexes. The presence of AECA is frequently associated with severe clinical symptoms, most notably thrombosis. AECA are generally not cytotoxic for EC, except for those found in Kawasaki disease. AECA activate expression of adhesion molecules, procoagulant activity and cytokine production, and endogenous IL-1 is the ultimate mediator of at least some of these actions (Carvalho et al., 1996; Simantov et al., 1995; Del Papa et al., 1996). Thus, EC activation is likely a major pathogenetic factor of clinical manifestations associated with AECA.

7. CONCLUDING REMARKS

EC, long considered a layer of nucleated cellophane, play an important, active role in the onset and regulation of inflammatory and immune reactions. The varied spectrum of responses elicited by cytokines is a dramatic demonstration of EC plasticity. The vast phenomenology of EC responses can be rationalized and represented in terms of cytokine-induced activation/differentiation programs. Recent work summarized here has begun to unravel the molecular basis of EC activation by cytokines. However central questions related to the mechanisms underlying expression restricted to EC and the phenomenology of EC heterogeneity remain without answer.

REFERENCES

Brown, Z., M.E. Gerritsen, W.W. Carley, R.M. Strieter, S.L. Kunkel, and J. Westwick. Chemokine gene expression and secretion by cytokine- activated human microvascular endothelial cells - Differential regulation of monocyte chemoattractant protein-1 and interleukin-8 in response to interferon- gamma. *Am. J. Pathol.*, 145:913–921, 1994.

Bussolino, F., M. Arese, L. Silvestro, R. Soldi, E. Benfanati, F. Sanavio, M. Aglietta, A. Bosia, and G. Camussi. Involvement of a serine protease in the synthesis of platelet-activating factor by endothelial cells stimulated by tumor necrosis factor-alpha or interleukin-1 alpha. *Eur. J. Immunol.*, 24:3131–3139, 1994a.

Bussolino, F., E. Bocchietto, F. Silvagno, R. Soldi, M. Arese, and A. Mantovani. Actions of molecules which regulate hemopeiesis on endothelial cells: memoirs of common ancestors? *Path. Res. Pract.*, 190:834–839, 1994b.

Bussolino, F., F. Silvagno, G. Garbarino, C. Costamagna, F. Sanavio, M. Arese, R. Soldi, M. Aglietta, G. Pescarmona, G. Camussi, et al. Human endothelial cells are targets for platelet-activating factor (PAF). Activation of alpha and beta protein kinase C isozymes in endothelial cells stimulated by PAF. *J. Biol. Chem.*, 269:2877–2886, 1994c.

Bussolino, F. And G. Camussi. Platelet-activating factor produced by endothelial cells. A molecule with autocrine and paracrine properties. *Eur. J. Biochem.*, 229:327–337, 1995.

Calderon, T.M., J. Sherman, H. Wilkerson, V.B. Hatcher, and J.W. Berman. Interleukin 6 modulates c-sis gene expression in cultured human endothelial cells. *Cell Immunol.*, 143:118–126, 1992.

Camacho, M., Godessart, R. Anton, M. Garcia, and L. Vila. Interleukin-1 enhances the ability of cultured human umbilical vein endothelial cells to oxidize linoleic acid. *J. Biol. Chem.*, 270:17279–17286, 1995.

Cao, Y.H., Chen, J.A. Weatherbee, M. Tsang, and J. Folkman. gro-beta, a -C-X-C-chemokine, is an angiogenesis inhibitor that suppresses the growth of Lewis lung carcinoma in mice. *J. Exp. Med.*, 182:2069–2077, 1995.

Carvalho, D., C.O.S. Savage, C.M. Blanck, and J.D. Pearson. IgG antiendothelial cell autoantibodies from scleroderma patients induce leukocyte adhesion to human vascular endothelial cells *in vitro*. *J. Clin. Invest.*, 97:111–119, 1996.

Caverth, H.J., J.F. Bohnsack, T.M. McIntyre, M. Baggiolini, S.M. Prescott, and G.A. Zimmerman. Neutrophil activating factor (NAF) induces polymorphonuclear leukocyte adherence to endothelial cells and to subendothelial matrix proteins. *Biochem. Biophys. Res. Comm.*, 162:387–393, 1989.

Chen, C.C. and A.M. Manning. TGF-beta 1, IL-10, and IL-4 differentially modulate the cytokine-induced expression of the IL-6 and IL-8 in human endothelial cells. *Cytokine*, 8:58–65, 1996.

Chirivi, R.G., M.I. Nicoletti, A. Remuzzi, and R. Giavazzi. Cytokines and cell adhesion molecules in tumor-endothelial cell interaction and metastasis. *Cell Adhes. Commun.*, 2:219–224, 1994.

Chu, W., D.K. Burns, R.A. Swerlick, and D.H. Presky. Identification and characterization of a novel cytokine-inducible nuclear protein from human endothelial cells. *J. Biol. Chem.*, 270:10236–10245, 1995.

Claria, J. and C.N. Serhan. Aspirin triggers previously undescribed bioactive eicosanoids by human endothelial cell-leukocyte interactions. *Proc. Natl. Acad. Sci. USA*, 92:9475–9479, 1995.

Clinton, S.K., J.C. Fleet, H. Loppnow, R.N. Salomon, B.D. Clark, J.G. Cannon, A.R. Shaw, C.A. Dinarello, and P. Libby. Interleukin-1 gene expression in rabbit vascular tissue *in vivo*. *Am. J. Pathol.*, 138:1005–10014, 1991.

Cobbs, C.S., J.E. Brenman, K.D. Aldape, D.S. Bredt, and M.A. Israel. Expression of nitric oxide synthase in human central nervous system tumors. *Cancer Res.*, 55;727–730, 1995.

Colotta, F., M. Sironi, A. Borre, W. Luini, F. Maddalena, and A. Mantovani. Interleukin 4 amplifies monocyte chemotactic protein and interleukin 6 production by endothelial cells. *Cytokine*, 4;24–28, 1992.

Colotta, F., F. Re, M. Muzio, R. Bertini, N. Polentarutti, M. Sironi, J.G. Giri, S.K. Dower, J.E. Sims, and A. Mantovani. Interleukin-1 type II receptor: a decoy target for IL-1 that is regulated by IL-4. *Science*, 261:472–475, 1993a.

Colotta, F., M. Sironi, A. Borre, T. Pollicino, S. Bernasconi, D. Boraschi, and A. Mantovani. Type II interleukin-1 receptor is not expressed in cultured edothelial cells and is not involved in endothelial cell activation. *Blood*, 81:1347–1357, 1993b.

Colotta, F., F.L. Sciacca, M. Sironi, W. Luini, M.J. Rabiet, and A. Mantovani. Expression of monocyte chemotactic protein-1 by monocytes and endothelial cells exposed to thrombin. *Am. J. Pathol.*, 144:975–985, 1994.

Colotta, F., S. Orlando, E.J. Fadlon, S. Sozzani, C. Matteucci, and A. Mantovani. Chemoattractants induce rapid release of the interleukin 1 type II decoy receptor in human polymorphonuclear cells. *J. Exp. Med.*, 181:2181–2188, 1995.

Coulpier, M., S. Andreev, C. Lemercier, H. Dauchel, O. Lees, M. Fontaine, and J. Ripoche. Activation of the endothelium by IL-1 alpha and glucocorticoids results in major increase of complement C3 and factor B production and generation of C3a. *Clin. Exp. Immunol.*, 101:142–149, 1995.

Cushing, S.D., J.A. Berliner, A.J. Valente, M.C. Territo, M. Navab. F. Parhami, R. Gerrity, C.J. Schwartz, and A.M. Fogelman. Minimally modified low density lipoprotein induces monocyte chemtactic protein 1 in human endothelial cells and smooth muscle cells. *Proc. Natl. Acad. Sci. USA*, 87:5134–5138, 1990.

d'Aniello, E.M., F. Brevario, I. Martin Padura, M.G. Lampugnani, E. Dejana, A. Mantovani, and M. Introna. Interleukin-1 and tumor necrosis factor induce transient expression of an inhibitor of nuclear factor kB in endothelial cells. *Endothlelium*, 1:161–165, 1993.

De Caterina, R., P. Libby, H.B. Peng, V.J. Thannickal, T.B. Rajavashisth, M.A.J. Gimbrone, W.S. Shin, and J.K. Liao. Nitric oxide decreases cytokine-induced endothelial activation. Nitric oxide selectivity reduces endothelial expression of adhesion molecules and proinflammatory cytokines. *J. Clin. Invest.*, 96:60–68, 1995.

de Martin, R., B. Vanhove, Q. Cheng, E. Hofer, V. Csizmadia, H. Winkler, and F.H. Bach. Cytokine-inducible expression in endothelial cells of and I kappa B alpha-like gene is regulated by NF kappa B. *EMBO J.*, 12;2773–2779, 1993.

Debeaux, A.C., J.P. Maingay, J.A. Ross, K.C.H. Fearon, and D.C. Carter. Interleukin-4 and interleukin-10 increase endotoxin-stimulated human umbilical vein endothelial cell interleukin-8 release. *J. Interferon. Cytokine Res.*, 15:441–445, 1995.

Del Papa, N., L. Guidali, M. Sironi, Y. Shoenfeld, A. Mantovani, A. Tincani, G. Balestrier, A. Radice, R.A. Sinico, and P.L. Meroni. Anti-endothelial IgG antibodies from Wegener's granulomatosis bind to endothelial cells *in vitro* and induce adhesion molecule expression and cytokine secretion. *Arthritis Rheum*, (In Press).

Devergne, O., A. Marfaingkoka, T.J. Schall, M.B. Legerravet, M. Sadick, M. Peuchmaur, M.C. Crevon, K.J. Kim, P. Galanaud, and D. Emilie. Production of the RANTES chemokine in delayed-type hypersensitivity reactions: Involvement of macrophages and endothelial cells (vol 180, pg 1689, 1994), *J. Exp. Med.*, 180:775, 1994.

Dickson, K., A. Philip, H. Warshawsky, M. O'Connor McCourt, and J.J. Bergeron. Specific binding of endocrine transforming growth factor-beta 1 to vascular endothelium. *J. Clin. Invest.*, 95:2539–2554, 1995.

Dixit, V.M., S. Green, V. Sarma, L.B. Holzman, F.W. Wolf, K. O'Rourke, P.A. Ward, E.V. Prochownik, and R.M. Marks. Tumor necrosis factor-alpha induction of novel gene products in human endothelial cells including a macrophage-specific chemotaxin. *J. Biol. Chem.*, 265:2973–2978, 1990.

Eissner, G., F. Kohlhuber, M. Grell, M. Ueffing, P. Scheurich, A. Hieke, G. Multhoff, G.M. Bornkamm, and E. Holler. Critical involvement of transmembrane tumor necrosis factor-alpha in endothelial programmed cell death mediated by ionizing radiation and bacterial endoxin. *Blood*, 86:4184–4193, 1995.

Emsley, J., H.E. White, B.P. O'Hara, G. Oliva, N. Srinivasan, I.J. Tickle, T.L. Blundell, M.B. Pepys, and S.P. Wood. Structure of pentameric human serum amyloid P component. *Nature*, 367:338–345, 1994.

Ezekowitz, R.A., J.B. Mulliken, and J. Folkman. Interferon alfa-2a therapy for life-threatening hemangiomas of infancy [see comments] [published errata appear in N. Engl. J. Med., 1994, Jan 27:330(4):300 and 1995 Aug 31:333(9):595–6]. *N. Engl. J. Med.*, 326:1456–1463, 1992.

Fan, J., H.Z. Bass, and J.L. Fahey. Elevated IFN-gamma and decreased IL-2 gene expression are associated with HIV infection. *J. Immunol.*, 151:5031–5040, 1993.

Fan, S.T., K. Hsia, and T.S. Edgington. Upregulation of human immunodeficiency virus-1 in chronically infected monocytic cell line by both contact with endothelial cells and cytokines. *Blood*, 84:1567–1572, 1994.

Faruqi, R., C. de la Motte, and P.E. DiCorleto. Alpha-tocopherol inhibits agonist-induced monocytic cell adhesion to cultured human endothelial cells. *J. Clin. Invest.*, 94:592–600, 1994.

Fauci, A.S., B.F. Haynes, and P. Katz. The spectrum of vasculitis. *Ann. Intern. Med.*, 89(Part 1):660–676, 1978.

Fiorelli, V., R. Gendelman, F. Samaniego, P.D. Markham, and B. Ensoli. Cytokines from activated T cells induce normal endothelial cells to acquire the phenotypic and functional features of AIDS-Kaposi's sarcoma spindle cells. *J. Clin. Invest.*, 95:1723–1734, 1995.

Fossiez, F., O. Djossou, P. Chomarat, L. Flores-Romo, S. Ait-Yahia, C. Maat, J.J. Pin, P. Garrone, E. Garcia, S. Saelalnd, D. Blanchard, C. Gaillard, B.D. Mahapatra, E. Rouvier, P. Goldstein, J. Banchereau, and S. Lebecque. T-cell IL-17 induces stromal cells to produce proinflammatory and hematopoietic cytokines. *J. Exp. Med.*, (In Press).

Fukuo, K., T. Inoue, S. Morimoto, T. Nakaashi, O. Yasuda, S. Kitano, R. Sasada, and T. Ogihara. Nitric oxide mediates cytotoxicity and basic fibroblast growth factor release in cultured vascular smooth muscle cells. A possible mechanism of neovascularization in atherosclerotic plaques. *J. Clin. Invest.*, 95:669–676, 1995.

Galea, P., G. Thibault, M. Lacord, P. Bardos, and Y. Lebranchu. IL-4, but not tumor necrosis factor-alpha, increases endothelial cell adhesiveness for lymphocytes by activating a cAMP-dependent pathway. *J. Immunol.*, 151:588–596, 1993.

Gamble, J.R. and M.A. Vadas. Endothelial cell adhesiveness for human T lymphocytes is inhibited by transforming growth factor beta-1. *J. Immunol.*, 146:1149–1154, 1991.

Ghersa, P., R. Hooft van Huijsduijnen, J. Whelan, Y. Cambet, R. Pescini, and J.F. DeLamarter. Inhibition of E-selectin gene transcription through a cAMP-dependent protein kinase pathway. *J. Biol. Chem.*, 269:29129–29137, 1994.

Gimbrone, M.A.J., M.S. Obin, A.F. Brock, E.A. Luis, P.E. Hass, C.A. Hebert, Y.K. Yip, D.W. Leung, D.G. Lowe, W.J. Kohr, W.C. Darbonne, K.B. Bechtol, and J.B. Baker. Endothelial interleukin-8: a novel inhibitor of leukocyte-endothelial interactions. *Science*, 246:1601–1603, 1989.

Goldblum, S.E., T. W. Brann, X. Ding, J. Pugin, and P.S. Tobias. Lipopolysaccharide (LPS)-binding protein and soluble CD14 function as accessory molecules for LPS-induced changes in endothelial barrier function, *in vitro. J. Clin. Invest.*, 93:692–702, 1994.

Grell, M., E. Douni, H. Wajant, M. Löheden, M. Clauss, B. Maxeiner, S. Georgopoulos, W. Lesslauer, G. Kolias, K. Pfizenmaier, and P. Scheurich. The transmembrane form of tumor necrosis factor is the prime activating ligand of the 80 kDa tumor necrosis factor receptor. *Cell*, 83:793–802, 1995.

Grimm, S. and P.A. Baeuerle. The inducible transcription factor NF-kappa B: structure-function relationship of its protein subunits. *Biochem. J.*, 290:297–308, 1993.

Grosset, C., B. Jazwiec, J.L. Taupin, H. Liu, S. Richard, F.X. Mahon, J. Reiffers, J.F. Moreau, and J. Ripoche. *In vitro* biosynthesis of leukemia inhibitory factor/human interleukin for DA cells by human endothelial cells: differential regulation by interleukin-1 alpha and glucocorticoids. *Blood*, 86:3763–3770, 1995.

Gómez-Chiarri, M., T.A. Hamilton, J. Egido, and S.N. Emancipator. Expression of IP-10, a lipopolysaccharide- and interferon-gamma-inducible protein, in murine mesangial cells in culture. *Am. J. Pathol.*, 142:433–439, 1993.

Hadley, T.J., Z.H. Lu, K. Wasniowska, A.W. Martin, S.C. Peiper, J. Hesselgesser, and R. Horuk. Postcapillary venule endothelial cells in kidney express a multispecific chemokine receptor that is structurally and functionally identical to the erythroid isoform, which is the Duffy blood group antigen. *J. Clin. Invest.*, 94:985–991, 1994.

Hamanaka, R., K. Kono, T. Seguchi, K. Okamura, A. Morimoto, M. Ono, J. Ogata, and M. Kuwano. Induction of low density lipoprotein receptor and a transcription factor SP-1 by tumor necrosis in human microvascular endothelial cells. *J. Biol. Chem.*, 267:13160–13165, 1992.

Haziot, A., G.W. Rong, J. Silver, and S.M. Goyert. Recombinant soluble CD14 mediates the activation of endothelial cells by lipopolysaccharide. *J. Immunol.*, 151:1500–1507, 1993.

Hebert, C.A., F.W. Luscinskas, J.M. Keily, E.A. Luis, W.C. Darbonne, G.L. Bennet, C.C. Liu, M.S. Obin, M.A.J. Gimbrone, and J.B. Baker. Endothelial and leukocyte forms of IL-8. Conversion by thrombin and interactions with neutrophils. *J. Immunol.*, 145:3033–3040, 1990.

Hechtman, D.H., M.I. Cybulsky, H.J. Fuchs, J.B. Baker, and M.A.J. Gimbrone. Intravascular IL-8. Inhibitor of polymorphonuclear leukocyte accumulation at sites of acute inflammation. *J. Immunol.*, 147:883–892, 1991.

Heinrich, J.N., R.P. Ryseck, H. Macdonald-Bravo, and R. Bravo. The product of a novel growth factor-activated gene, fic, is a biologically active C-C-type cytokine. *Mol. Cell Biol.*, 13:2020–2030, 1993.

Heller, R., F. Bussolino, D. Ghigo, G. Garbarino, G. Pescarmona, U. Till, and A. Bosia. Human endothelial cells are target for platelet-activating factor II. Platelet activating factor induces platelet-activating factor synthesis in human umbilical vein endothelial cells. *J. Immunol.*, 149:3682–3688, 1992.

Hill, M.E., I.N. Bird, R.H. Daniels, M.A. Elmore, and M.J. Finnen. Endothelial cell-associated platelet-activating factor primes neutrophils for enhanced superoxide production and arachidonic acid release during adhesion to but not transmigration across IL-1 beta-treated endothelial monolayers. *J. Immunol.*, 153:3673–3683, 1994.

Hla, T. and T. Maciag. An abundant transcript induced in differentiating human endothelial cells encodes a polypeptide with structural similarities to G-protein-coupled receptors. *J. Biol. Chem.*, 265:9308–9313, 1990.

Hollenbaugh, D., N. Mischel Petty, C.P. Edwards, J.C. Simon, R.W. Denfeld, P.A. Kiener, and A. Aruffo. Expression of functional CD40 by vascular endothelial cells. *J. Exp. Med.*, 182:33–40, 1995.

Holzman, L.B., R.M. Marks, and V.M. Dixit. A novel immediate-early response gene of endothelium is induced by cytokines and encodes a secreted protein. *Mol. Cell Biol.*, 10:5830–5838, 1990.

Hooft van Huijsduijnen, R., R. Pescini, and J.F. DeLamarter. Two distinct NF-kappa B complexes differing in their larger subunit bind the E-selectin promoter kappa B element. *Nucleic Acids Res.*, 21:3711–3717, 1993.

Hooper, W.C., D.J. Phillips, M.J. Ribeiro, J.M. Benson, V.G. George, E.W. Ades and B.L. Evatt. Tumor necrosis factor-alpha downregulates protein S secretion in human microvascular and umbilical vein endothelial cells but not in the HepG-2 hepatoma cell line. *Blood*, 84:483–489, 1994.

Howell, G., P. Pham, D. Taylor, B. Foxwell, and M. Feldmann. Interleukin 4 induces interleukin 6 production by endothelial cells: synergy with interferon- gamma. *Eur. J. Immunol.*, 21:91–101, 1991.

Huber, A.R., S.L. Kunkel, R.F. Todd, and S.J. Weiss. Regulation of transendothelial neutrophil migration by endogenous inerleukin-8. *Science*, 254:99–102, 1991.

Iademarco, M.F., J.J. McQuillan, G.D. Rosen, and D.C. Dean. Characterization of the promoter for vascular cell adhesion molecule-1 (VCAM-1). *J. Biol. Chem.*, 267:16323–16329, 1992.

Inoue, N., R.C. Venema, H.S. Sayegh, Y. Ohara, T.J. Murphy, and D.G. Harrison. Molecular regulation of the bovine endothelial cell nitric oxide synthase by transforming growth factor-beta 1. *Arterioscl. Thromb. Vasc. Biol.*, 15:1255–1261, 1995.

Introna, M., F. Breviario, E. D'Aniello, J. Golay, E. Dejana, and A. Mantovani. Il-1 inducible genes in human umbilical vein endothelial cells. *Eur. Heart J.*, 14(suppl K):78–81, 1993.

Introna, M., V. Vidal Alles, M. Castellano, G. Picardi, L. De Gioia, B. Bottazzi, G. Peri, F. Breviario, M. Salmona, L. De Gregorio, T.A. Dragani, N. Srinivasan, T.L. Blundell, T.A. Hamilton, and A. Mantovani. Cloning of mouse PTX3, a new member of the pentraxin gene family expressed at extrahepatic sites. *Blood*, 87:1862–1872, 1996.

Jackson, B.A., R.H. Goldstein, R. Roy, M. Cozzani, L. Taylor, and P. Polgar. Effects of transforming growth factor beta and interleukin-1 beta on expression of cyclooxygenase 1 and 2 phospholipase A2 mRNA in lung fibroblasts and endothelial cells in culture. *Biochem. Biophys. Res. Commun.*, 197:1465–1474, 1993.

Jeannin, P., Y. Delneste, P. Gosset, S. Molet, P. Lassale, Q. Hamid, A. Tsicopoulos, and A.B. Tonnel. Histamine induces interleukin-8 secretion by endothelial cells. *Blood*, 84:2229–2233, 1994.

Jones, K., C. Rivera, C. Sgadar, J. Franklin, E.E. Max, K. Bhatia, and G. Tosato. Infection of human endothelial cells with Epstein-Barr virus. *J. Exp. Med.*, 182:1213–1221, 1995.

Kamijo, R., H. Harada, T. Matsuyama, M. Bosland, J. Gerecitano, D. Shapiro, J. Le, S.I. Koh, T. Kimura, S.J. Green, et al. Requirement for transcription factor IRF-1 in NO synthase induction in macrophages. *Science*, 263:1612–1615, 1994.

Kaplanski, G., R. Porat, K. Aiura, J.K. Erban, J.A. Gelfand, and C.A. Dinarello. Activated platelets induce endothelial secretion of interleukin-8 *in vitro* via an interleukin-1-mediated event. *Blood*, 81:2492–2495, 1993b.

Kaplanski, G., C. Farnarier, S. Kaplanski, R. Porat, L. Shapiro, P. Bongrand, and C.A. Dinarello. Interleukin-1 induces interleukin-8 secretion from endothelial cells by juxtacrine mechanism. *Blood*, 84:4242–4248, 1994.

Kaplanski, G., N. Teysseire, C. Farnarier, S. Kaplanski, J.C. Lissitzky, J.M. Durand, J. Soubeyrand, C. Dinarello, and P. Bongrand. IL-6 and IL-8 production from cultured human endothelial cells stimulated by infection with Rickettsia conorii a cell-associated IL-1-alpha-dependent pathway. *J. Clin. Invest.*, 96:2839–2844, 1995.

Karakurum, M., R. Shreeniwas, J. Chen, D. Pinsky, S.D. Yan, M. Anderson, K. Sunouchi, J. Major, T. Hamilton, K. Kuwabara, A. Rot, P.R. Nowygrod, and D. Stern. Hypoxic induction of interleukin-8 gene expression in human endothelial cells. *J. Clin. Invest.*, 93:1564–1570, 1994.

Karmann, K., C.C. Hughes, J. Schechner, W.C. Fanslow, and J.S. Pober. CD40 on human endothelial cells: inducibility by cytokines and functional regulation of adhesion molecule expression. *Proc. Natl. Acad. Sci. USA*, 92:4342–4346, 1995.

Kaszubska, W., R.H. van Huijsduijnen, P. Ghersa, A.M. DeRaemy Schenk, B.P. Chen, T. Hai, J.F. DeLamarter, and J. Whelan. Cyclic AMP-independent ATF family members interact with NF-kappa B and function in the activation of the E-selectin promoter in response to cytokines. *Mol. Cell Biol.*, 13:7180–7190, 1993.

Kilgore, K.S., J.P. Shen, B.F. Miller, P.A. Ward, and J.S. Warren. Ehancement by the complement membrane attack complex of tumor necrosis factor-alpha-induced endothelial cell expression of E-selectin and ICAM-1. *J. Immunol.*, 155:1434–1441, 1995.

Koch, A.E., P.J. Polverini, S.L. Lunkel, L.A. Harlow, L.A. DiPietro, V.M. Elner, S.G. Elner, and R.M. Strieter. Interleukin-8 as a macrophage-derived mediator of angiogenesis. *Science*, 258:1798–1801, 1992.

Korpelainen, E.I., J.R. Gamble, W.B. Smith, G.J. Goodall, S. Qiyu, J.M. Woodcock, M. Dottore, M.A. Vadas, and A.F. Lopez. The receptor for interleukin 3 is selectively induced in human endothelial cells by tumor necrosis factor alpha and potentiates interleukin 8 secretion and neutrophil transmigration. *Proc. Natl. Acad. Sci. USA*, 90:11137–11141, 1993.

Korpelainen, E.I., J.R. Gamble, W.B. Smith, M. Dottore, M.A. Vadas, and A.F. Lopez. Interferon-gamma upregulates interleukin-3 (IL-3) receptor expression in human endothelial cells and synergizes with IL-3 in stimulating major histocompatability complex class II expression and cytokine production. *Blood*, 86:176–182, 1995.

Krakauer, T. IL-10 inhibits the adhesion of leukocytic cells to IL-1-activated human endothelial cells. *Immunol. Lett.*, 45:61–65, 1995.

Lamas, S., T. Michel, T. Collins, B.M. Brenner, and P.A. Marsden. Effects of interferon-gamma on nitric oxide synthase activity and endothelial-production by vascular endothelial cells. *J. Clin. Invest.*, 90:879–887, 1992.

Ledebur, H.C. and T.P. Parks. Transcriptional regulation of the intercellular adhesion molecule-1 gene by inflammatory cytokines in human endothelial cells - Essential roles of a variant NF-kappa B site and p65 homodimers. *J. Biol. Chem.*, 270:933–943, 1995.

Lee, T.H., G.W. Lee, E.B. Ziff, and J. Vilcek. Isolation and characterization of eight tumor necrosis factor-induced gene sequences from human fibroblasts. *Mol. Cell Biol.*, 10:1982–1988, 1990.

Lewis, H., W. Kaszubska, J.F. DeLamarter, and J. Whelan. Cooperatively between two NF-kappa B complexes, mediated by high-mobility-group protein I(Y), is essential for cytokine-induced expression of the E-selectin promoter. *Mol. Cell Biol.*, 14:5701–5709, 1994.

Ley, K. J.B. Baker, M.I. Cybulsky, M.A. Gimbrone, and F.W. Luscinskas. Intravenous interleukin-8 inhibits granulocyte emigration from rabbit messenteric venules without altering L-selectin expression or leukocyte rolling. *J. Immunol.*, 151:6347–6357, 1993.

Libby, P. and G.K. Hannson. Biology of Disease. Involvement of the immune system in human atherogenesis: current knowledge and unanswered questions. *Lab. Invest.*, 64:5–15, 1991.

Lienard, D., P. Ewalenko, J.J. Delmotte, N. Renard, and F.J. Lejeune. High-dose recombinant tumor necrosis factor alpha in combination with interferon gamma and melphalan in isolation perfusion of the limbs for melanoma and sarcoma. *J. Clin. Oncol.*, 10:52–60, 1992.

Liuzzo, G., L.M. Biasucci, J.R. Gallimore, R.L. Grillo, A.G. Rebuzzi, M.B. Pepys, and A. Maseri. The prognostic value of C-reactive protein and serum amyloid a protein in severe unstable angine [see comments]. *N. Engl. J. Med.*, 1994.

Lukacs, N.W., R.M. Strieter, V. Elner, H.L. Evanoff, M.D. Burdick, and S.L. Kunkel. Production of chemokines, interleukin-8 and monocyte chemoattractant protein-1, during monocyte endothelial cell interactions. *Blood*, 86:2767–2773, 1995.

Luster, A.D., S.M. Greenberg, and P. Leder. The IP-10 chemokine binds to a specific cell surface heparan sulfate site shared with platelet factor 4 and inhibits endothelial cell proliferation. *J. Exp. Med.*, 182:219–231, 1995.

Mackay, F., H. Loetscher, D. Stueber, G. Gehr, and W. Lesslauer. Tumor necrosis factor alpha (TNF-alpha)-induced cell adhesion to human endothelial cell is under dominant control of one TNF receptor type, TNF-R55. *J. Exp. Med.*, 177:1277–1286, 1993.

Maier, J.A., P. Voulalas, D. Roeder, and T. Maciag. Extension of the life-span of human endothelial cells by an interleukin-1 alpha antisense oligomer. *Science*, 249:1570–1574, 1990.

Maier, J.A., M. Statuto, and G. Ragnotti. Endogenous interleukin-1 alpha must be transported to the nucleus to exert its activity in human endothelial cells. *Mol. Cell Biol.*, 14:1845–1851, 1994.

Maione, T.E., G.S. Gray, J. Petro, A.J. Hunt, A.L. Donner, S.I. Bauer, H.F. Carson, and R.J. Sharpe. Inhibition of angiogenesis by recombinant human platelet factor-4 and related peptides. *Science*, 247:77–79, 1990.

Mantovani, A., P. Allavena, F. Collotta, and S. Sozzani. Chemokines in vascular pathophysiology. In: *Immune functions of the vassal wall.* G.K. Hansson and P. Libby, eds, Harwood Academic Publishers, 1996.

Mantovani, A., F. Bussolino, and E. Dejana. Cytokine regulation of endothelial cell function. *FASEB J.*, 6:2591–2599, 1992.

Mantovani, A. and E. Dejana. Cytokines as communication signals between leukocytes and endothelial cells. *Immunol. Today*, 10:370–375, 1989.

Marfaingkoka, A., O. Devergne, G. Gorgone, A. Portier, T.J. Schall, P. Galanaud, and D. Emilie. Regulation of the production of the RANTES chemokine by endothelial cells - Synergistic induction by IFN-gamma plus TNF-alpha and inhibition by IL-4 and IL-13. *J. Immunol.*, 154:1870–1878, 1995.

Martin, E. C. Nathan, and Q.W. Xie. Role of interferon regulatory factor 1 in induction of nitric oxide synthase. *J. Exp. Med.*, 180:977–984, 1994.

Marui, N., M.K. Offermann, R. Swerlick, C. Kunsch, C.A. Rosen, M. Ahmad, R.W. Alexander, and R.M. Medford. Vascular cell adhesion molecule-1 (VCAM-1) gene transcription and expression are regulated through and antioxidant-sensitive mechanism in human vascular endothelial cells. *J. Clin. Invest.*, 92:1866–1874, 1993.

Mattila, P., M.L. Majuri, S. Tiisala, and R. Renkonen. Expression of six protein kinase C isotypes in endothelial cells. *Life Sci.*, 55:1253–1260, 1994.

McCarron, R.M., L. Wang, D.B. Stanimirovic, and M. Spatz. Differential regulation of adhesion molecule expression by human cerebrovascular and umbilical vein endothelial cells. *Endothelium*, 2;339–346, 1995.

Meacock, S., R. Pescini Gobert, J.F. DeLamarter, and R. Hooft van Huijsduijnen. Transcription factor induced, phased bending of the E-selectin promoter. *J. Biol. Chem.*, 269:31756–31762, 1994.

Meininger, C.J., S.E. Brightman, K.A. Kelly, and B.R. Zetter. Increased stem cell factor release by hemangioma-derived endothelial cells. *Lab Invest.*, 72:166–173, 1995.

Meyrick, B., B. Christman, and G. Jesmok. Effects of recombinant tumor necrosis factor-alpha on cultured pulmonary artery and lung microvascular endothelial monolayers. *Am. J. Pathol.*, 138;93–101, 1991.

Miyaloto, S. M. Maki, M.J. Schmitt, M. Hatanaka, and I.M. Verna. Tumor necrosis factor alpha-induced phosphorylation of IkB alpha is a signal for its degradatioin but not dissociation from NF-kB. *Proc. Natl. Acad. Sci. USA*, 91:12740–12744, 1994.

Montrucchio, G., E. Lupia, E. Battaglia, G. Passerini, F. Bussolino, G. Emanuelli, and G. Camussi. Tumor necrosis factor alpha-induced angiogenesis depends on *in situ* platelet-activating factor biosynthesis. *J. Exp. Med.*, 180:377–382, 1994.

Moyer, C.F., D. Sajuthi, H. Tulli, and J.K. Williams. Synthesis of IL-1 alpha and IL-1 beta by arterial cells in atherosclerosis. *Am. J. Pathol.*, 138:951–960, 1991.

Murata, J., S.B. Corradin, E. Felley Bosco, and L. Juillerat Jeanneret. Involvement of a transforming-growth-factor-beta-like molecule in tumor-cell-derived inhibition of nitric oxide synthesis in cerebral endothelial cells. *Int. J. Cancer*, 62:743–748, 1995.

Narumi, S., L.M. Wyner, M.H. Stoler, C.S. Tannebaum, and T.A. Hamilton. Tissue-specific expression of murine IP-10 mRNA following systematic treatment with interferon-gamma. *J. Leukoc. Biol.*, 52:27–33, 1992.

Nathan, C. and Q.W. Xie. Nitric oxide synthases: roles, tolls, and controls. *Cell*, 78:915–918, 1994.

Neish, A.S., A.J. Williams, H.J. Palmer, M.Z. Whitley, and T. Collins. Functional analysis of the human vascular cell adhesion molecule 1 promoter. *J. Exp. Med.*, 176:1583–1593, 1992.

Neish, A.S., M.A. Read, D. Thanos, R. Pine, T. Maniatis, and T. Collins. Endothelial interferon regulatory factor 1 cooperates with NF-kB as a transcriptional activator of vascular cell adhesion molecule 1. *Mol. Cell Biol.*, 15:2558–2569, 1995.

Nelken, N.A., S.R. Coughlin, D. Gordon, and J.N. Wilcox. Monocyte chemoattractant protein-1 in human atheromatous plaques. *J. Clin. Invest.*, 88:1121–1127, 1991.

Niedbala, M.J. and M. Stein Picarella. Role of protein kinase C in tumor necrosis factor induction of endothelial cell urokinase-type plasminogen activator. *Blood*, 81:2608–2617, 1993.

Niu, X.F., C.W. Smith, and P. Kubes. Intracellular oxidative stress induced by nitric oxide synthesis inhibition increases endothelial cell adhesion to neutrophils. *Circ. Res.*, 74:113–1140, 1994.

Noland, T.D., B.B. Friday, M.T. Maulit, and G.L. Gerton. The sperm acrosomal matrix contains a novel member of the pentraxin family of calcuim-dependent binding proteins. *J. Biol. Chem.*, 269:32607–32614, 1994.

Opipari, A.W.J., H.M. Hu, R. Yabkowitz, and V.M. Dixit. The A20 zinc finger protein protects cells from tumor necrosis factor cytotooxicity. *J. Biol. Chem.*, 267:12424–12427, 1992.

Oswald, I.P., I. Eltoum, T.A. Wynn, B. Schwartz, P. Caspar, D. Paulin, A. Sher, and S.L. James. Endothelial cells are activated by cytokine treatment to kill an intravascular parasite, Schistosoma mansoni, through the production of nitric oxide. *Proc. Natl. Acad. Sci. USA*, 91:999–1003, 1994.

Paleolog, E.M., S.A. Delasalle, W.A. Buurman, and M. Feldmann. Functional activities of receptors for tumor necrosis factor-alpha on human vascular endothelial cells. *Blood*, 84:2578–2590, 1994.

Palmercrocker, R.L. and J.S. Pober. IL-4 induction of VCAM-1 on endothelial cells involves activation of a protein tyrosine kinase. *J. Immunol.*, 154:2838–2845, 1995.

Pandey, A., H. Shao, R.M. Marks, P.J. Polverini, and V.M. Dixit. Role of B61, the ligand for the Eck tyrosine kinase in TNF-alpha-induced angiogenesis. *Science*, 268,567–569, 1995.

Parry, G.C. and N. Mackman. A set of inducible genes expressed by activated human monolytic and endothelial cells contain kappa B-like sites that specifically bind c-Rel-p65 heterodimers. *J. Biol. Chem.*, 269:20823–20825, 1994.

Patel, K.D., V. Modur, G.A. Zimmerman, S.M. Prescott, and T.M. McIntyre. The necrotic venom of the brown recluse spider induces dysregulated endothelial cell-dependent neutrophil activation - Differential induction of GM-CSF, IL-8, and E-selectin expression. *J. Clin. Invest.*, 94:631–642, 1994.

Pattison, J., P.J. Nelson, P. Huie, I. Vonleuttichau, G. Farshid, R.K. Sibley, and A.M. Kresky. RANTES chemokine expression in cell-mediated transplant rejection of the kidney. *Lancet*, 343:209–211, 1994.

Peiper, S.C., Z.X. Wang, K. Neote, A.W. Martin, H.J. Showell, M.J. Conklyn, K. Ogborne, T.J. Hadley, Z.H. Lu, J. Hesselgesser, and R. Horuk. The Duffy antigen receptor for chemokines (DARC) is expressed in endothelial cells of Duffy negative individuals who lack the erythrocyte receptor. *J. Exp. Med.*, 181:1311–1317, 1995.

Peng, H.B., P. Libby, and J.K. Liao. Induction and stabilization of I kappa B alpha by nitric oxide mediates inhibition of NF-kappa B. *J. Biol. Chem.*, 270:14214–14219, 1995.

Peng, H.B., T.B. Rajavashisth, P. Libby, and J.K. Liao. Nitric oxide inhibits macrophage-colony stimulating factor gene transcription in vascular endothelial cells. *J. Biol. Chem.*, 270:17050–17055, 1995.

Pepys, M.B. and M.L. Baltz. Actuc phase proteins with special reference to C-reactive protein and related proteins (pentaxis) and serum amyloid A protein. *Adv. Immunol.*, 34:141–160, 1983.

Pescini, R., W. Kaszubska, J. Whelan, J.F. DeLamarter, and R. Hooft van Huijsduijnen. ATF-a0, a novel variant of the ATF/CREB transcription factor family, forms a dominant transcription inhibitor in ATF-a heterodimers. *J. Biol. Chem.*, 269:1159–1165, 1994.

Petzelbauer, P., J.R. Bender, J. Wilson, and J.S. Pober. Heterogeneity of dermal microvascular endothelial cell antigen expression and cytokine responsiveness *in situ* and in cell culture. *J. Immunol.*, 151:5062–5072, 1993.

Petzelbauer, P., C.A. Watson, S.E. Pfau, and J.S. Pober. Heterogeneity of dermal microvascular endothelial cell antigen expression and cytokine responsiveness *in situ* and in cell culture. *J. Immunol.*, 151:5062–5072, 1993.

Pober, J. and R.S. Cotran. Cytokines and endothelial cell biology. *Physiol. Rev.*, 70:427–451, 1990.

Pober, J.S. M.R. Slowik, L.G. De Luca, and A.J. Ritchie. Elevated cyclic AMP inhibits endothelial cell synthesis and expression of TNF-induced endothelial leukocyte adhesion molecule-1, and vascular cell adhesion molecule-1, but not intercellular adhesion molecule-1. *J. Immunol.*, 150:5114–5123, 1993.

Pugin, J., C.C. Schurer Maly, D. Leturcq, A. Moriarty, R.J. Ulevitch, and P.S. Tobias. Lipopolysaccharide activation of human endothelial and epithelial cells is medicated by lipopolysaccharide-binding protein and soluble CD14. *Proc. Natl. Acad. Sci. USA*, 90:2744–2478, 1993.

Pugin, J., R.J. Ulevitch, and P.S. Tobias. A critical role for monocytes and CD14 in endotoxin-induced endothelial cell activation. *J. Exp. Med.*, 178:2193–2200, 1993.

Qi, J.F. and D.L. Kreutzer. Fibrin activation of vascular endothelial cells - Induction of IL-8 expression. *J. Immunol.*, 155:867–876, 1995.

Radomski, M.W., R.M. Palmer, and S. Moncada. Glucocorticoids inhibit the expression of an inducible, but not the constitutive, nitric oxide synthase in vascular endothelial cells. *Proc. Natl. Acad. Sci. USA*, 87:10043–10047, 1990.

Rajavashisth, T.B., A. Andalibi, M.C. Territo, J.A. Berliner, M. Navab, A.M. Fogelman, and A.J. Lusis. Induction of endothelial cell expression of granulocyte and macrophage colony-stimulating factors by modified low-density lipoproteins. *Nature*, 344:254–257, 1990.

Read, M.A., S.R. Cordle, R.A. Veach, C.D. Carlisle, and J. Hawiger. Cell-free pool of CD14 mediates activation of transcription factor NF-kappa B by lipopolysaccharide in human endothelial cells. *Proc. Natl. Acad. Sci. USA*, 90:9887–9891, 1993.

Read, M.A., M.Z. Whitley, A.J. Williams, and T. Collins. NF-kappa B and I kappa B alpha: an inducible regulatory system in endothelial activation. *J. Exp. Med.*, 179:503–512, 1994.

Reid, M.S. and C.P. Blobel. Apexin, and acrosomal pentaxin. *J. Biol. Chem.*, 269:32615–32620, 1994.

Remzzi, G. and P. Ruggenenti. The hemolytic uremic syndrome. *Kidney Int.*, 48:2–19, 1995.

Ritchie, A.J., D.R. Johnson, B.M. Ewenstein, and J.S. Pober. Tumor necrosis factor induction of endothelial cell surface antigens is independent of protein kinase C activation or inactivation. Studies with phorbol myristate acetate and staurosporine. *J. Immunol.*, 146:3056–3062, 1991.

Rollins, B.J., T. Yoshimura, E.J. Leonard, and J.S. Pober. Cytokine-activated human endothelial cells synthesize and secrete a monocyte chemoattractant, MCP-1/JE. *Am. J. Pathol.*, 136:1229–1233, 1990.

Rollins, B.J. and J.S. Pober. Interleukin-4 induces the synthesis and secretion of MCP-1JE by human endothelial cells. *Am. J. Pathol.*, 138:1315–1319, 1991.

Romer, L.H., N.V. McLean, H.C. Yan, M. Daise, J. Sun, and H.M. DeLisser. IFN-gamma and TNF-alpha induce redistribution of PECAM-1 (CD31) on human endothelial cells. *J. Immunol.*, 154:6582–6592. 1995.

Rosenkrantz Weiss, P., W.C. Sessa, S. Milstien, S. Kaufman, C.A. Watson, and J.S. Pober. Regulation of nitric oxide synthesis by proinflammatory cytokines in human umbilical vein endothelial cells. Elevations in tetrahydrobiopterin levels enhance endothelial nitric oxide synthase specific activity [see comments]. *J. Clin. Invest.*, 93:2236–2243, 1994.

Ross, R. The pathogenesis of atherosclerosis: a perspective for the 1990s. *Nature*, 362:801–809, 1993.

Rossi, V., F. Brevario, P. Ghezzi, E. Dejana, and A. Mantovani. Prostacyclin synthesis induced in vascular cells by ineterleukin-1. *Science*, 229:174–176, 1985.

Rot, A. Endothelial cell binding of NAP-1/IL-8: role of neutrophil emigration. *Immunol. Today*, 13:291–294, 1992.

Rott, O., U. Tontsch, B. Fleischer, and E. Cash. Interleukin-6 production in "normal" and HTLV-1 tax-epressing brain specific endothelial cells. *Eur. J. Immunol.*, 23:1987–1991, 1993.

Rovin, B.H., T. Yoshimura, and L. Tan. Cytokine-induced production of monocyte chemoattractant protein-1 by cultured human mesangial cells. *J. Immunol.*, 148:2148–2153, 1992.

Saadi, S., R.A. Holznecht, C. Patte, D.M. Stern, and J.L. Platt. Complement-mediated regulation of tissue factor activity in endothelium. *J. Exp. Med.*, 182:1807–1814, 1995.

Sakurai, H., Kohsaka, M.F. Liu, H. Higashiyama, Y. Hirata, K. Kanno, I. Saito, and N. Miyasaka. Nitric oxide production and inducible nitric oxide synthase expression in inflammatory arthritides. *J. Clin. Invest.*, 96: 2357–2363, 1995.

Samaniego, F., P.D. Markham, R.C. Gallo, and B. Ensoli. Inflammatory cytokines induce AIDS-Kaposi's sarcoma-derived spindle cells to produce and release basic fibroblast growth factor and enhance Kaposi's sarcoma-like lesion formation in nude mice. *J. Immunol.*, 154:3582–3592, 1995.

Sankar, S., N. Mahooti Brooks, M. Centrella, T.L. McCarthy, and J.A. Madri. Expression of transforming growth factor type III receptor in vascular endothelial cells increases their responsiveness to transforming growth factor beta 2. *J. Biol. Chem.*, 270:13567–13572, 1995.

Sarma, V., F.W. Wolf, R.M. Marks, T.B. Shows, and V.M. Dixit. Cloning of a novel tumor necrosis factor-alpha-inducible primary response gene that is differentially expressed in development and capillary tube-like formation *in vitro*. *J. Immunol.*, 148:3302–3312, 1992.

Sato, K., K. Miyakawa, M. Takeya, R. Hattori, Y. Yui, M. Sunamoto, Y. Ichimori, Y. Ushio, and K. Takahashi. Immunohistochemical expression of inducible nitric oxide synthase (iNOS) in reversible endotoxic shock studied by a novel monoclonal antibody against rat iNOS. *J. Leukoc. Biol.*, 57:36–44, 1995.

Satriano, J.A., K. Hora, Z. Shan, E.R. Stanley, T. Mori, and D. Schlondorff. Regulation of monocyte chemoatttarctant protein-1 and macrophage colony-stimulating factor-1 by IFN-gamma, tumor necrosis factor-alpha, IgG aggregates, and cAMP in mouse mesangial cells. *J. Immunol.*, 150:1971–1978, 1993a.

Satriano, J.A., M. Shuldiner, K. Hora, Y. Xing, Z. Shan, and D. Schlondorff. Oxygen radicals as second messengers for expression of the monocyte chemoattractant protein, JE/MCP-1, and the monocyte colony-stimulating factor, CSF-1, in response to tumor necrosis factor-alpha and immunoglobulin-G- evidence for involvement of reduced nicotinamide adenine dinucleotide phosphate (NADPH)-dependent oxidase. *J. Clin. Invest.*, 92:1564–1571, 1993b.

Schieven, G.G., J.C. Kallestad, T.J. Brown, J.A. Ledbetter, and P.S. Linsley. Oncostatin M induces tyrosine phosphorylation in endothelial cells and activation of p62yes tyrosine kinase. *J. Immunol.*, 149:1676–1682, 1992.

Schindler, U. and V.R. Baichwal. Three NF-kappa B binding sites in the human E-selectin gene required for maximal tumor necrosis factor alpha-induced expression. *Mol. Cell Biol.*, 14:5820–5831, 1994.

Schlimgen, A.K., J.A. Helms, H. Vogel, and M.S. Perin. Neuronal pentraxin, a secreted protein with homology to acute phase proteins of the immune system. *Neuron.*, 14:519–526, 1995.

Schmid, E.F., K. Binder, M. Grell, P. Scheurich, and K. Pfizenmaier. Both tumor necrosis factor receptors, TNFR60 andTNFR80 are involved in a signaling endothelial tissue factor expression by juxtacrine tumor necrosis factor alpha. *Blood*, 86:1836–1841, 1995.

Schoedon, G., M. Schneemann, N. Blau, C.J. Edgell, and A. Schaffner. Modulation of human endothelial cell tetrahydrobiopterin synthesis by activating and deactivation cytokines: new perspectives on endothelium-derived relaxing factor. *Biochem. Biophys. Res. Commun.*, 196:1343–1348, 1993.

Schofield, L., S. Novakovic, P. Gerold, R.T. Schwartz, M.J. McConville, and S.D. Tachado. Glycosylphosphatidylinositol toxin of plasmodium upregulates intercellular adhesion molecule-1, vascular cell adhesion molecule-1, and E-selectin expression in vascular endothelial cells and increases leukocyte and parasite cytoadherence via tyrosine kinase-dependent signal transduction. *J. Immunol.*, 156:1886–1896, 1996.

Schonbeck, U., E. Brandt, F. Petersen, H.D. Flad, and H. Loppnow. IL-8 specifically binds to endothelial but not to smooth muscle cells. *J. Immunol.*, 154:2375–2383, 1995.

Schroder, J.M. and E. Christophers. Secretion of novel and homologous neutrophil-activating peptides by LPS-stimulated human endothelial cells. *J. Immunol.*, 142:244–251, 1989.

Schultz, R., D.L. Panas, R. Catena, S. Moncada, P.M. Olley, and G.D. Lopaschuk. The role of nitric oxide in cardiac depression induced by interleukin-1 beta and tumor necrosis factor-alpha. *Br. J. Pharmacol.*, 114:27–34, 1995.

Schwartz, D., A. Andalibi, L. Chaverrialmada, J.A. Berliner, T. Kirchgessner, Z.T. Fang, P. Tekampolson, A.J. Lusis, C. Gallegos, A.M. Fogelman, and M.C. Territo. Role of the GRO family of chemokines in monocyte adhesion to MM-LDL-stimulated endothelium. *J. Clin. Invest.*, 94:1968–1973, 1994.

Shu, H.B., A.B. Agranoff, E.G. Nabel, K. Leung, C.S. Duckett, A.S. Neish, T. Collins, and G.J. Nabel. Differential regulation of vascular cell adhesion molecule 1 gene expression by specific NF-kappa B subunits in endothelial and epithelial cells. *Mol. Cell Biol.*, 13:6283–6289, 1993.

Shyy, Y.J., L.L. Wickman, J.P. Hagan, H.J. Hsieh, Y.L. Hu, S.H. Telian, A.J. Valente, K.L.P. Sung, and S. Chien. Human monocyte colony-stimulating factor stimulates the gene expression of monocyte chemotactic protein-1 and increases the adhesion of monocytes to endothelial monolayers. *J. Clin. Invest.*, 92:1745–1751, 1993.

Sica, A., K. Matsushima, J. Van Damme, J.M. Wang, N. Polentarutti, E. Dejana, F. Colotta, and A. Mantovani. IL-1 transcriptionally activates the neutrophil chemotactic factor/IL-8 gene in endothelial cells. *Immunology*, 69:548–553, 1990a.

Sica, A., J.M. Wang, F. Colotta, E. Dejana, A. Mantovani, J.J. Oppenheim, C.G. Larsen, C.O. Zachariae, and K. Matushima. Monocyte chemotactic and activating factor gene expression induced in endothelial cells by IL-1 and tumor necrosis factor. *J. Immunol.*, 144:3034–3038, 1990b.

Siebenlist, U., G. Franzoso, and K. Brown. Structure, regulation and function of NF-kB. *Annu. Rev. Cell Biol.*, 10:405–455, 1994.

Simantov, R., J.M. LaSala, S.K. Lo, A.E. Gharavi, L.R. Sammaritano, and J.E. Salmon. Activation of cultured vascular endothelial cells by antiphospholipid antibodies. *J. Clin. Invest.*, 96:2211–2219, 1995.

Sironi, M., F. Breviario, P. Proserpio, A. Biondi, A. Vecchi, J. Van Damme, E. Dejana, and A. Mantovani. IL-1 stimulates IL-6 production in endothelial cells. *J. Immunol.*, 142:549–553, 1989.

Sironi, M. C. Munoz, T. Pollicino, A. Siboni, F.L. Sciacca, S. Bernasconi, A. Vecchi, F. Colotta, and A. Mantovani. Divergent effects of interleukin-10 on cytokine production by mononuclear phagocytes and endothelial cells. *Eur. J. Immunol.*, 23:2692–2695, 1993.

Sironi, M., F.L. Sciacca, C. Matteucci, M. Conni, A. Vecchi, S. Bernasconi, A. Minty, D. Caput, P. Ferrara, F. Colotta, and A. Mantovani. Regulation of endothelial and mesothelial cell function by interleukin-13: selective induction of vascular cell adhesion molecule-1 and amplification of interleukin-6 production. *Blood*, 84:1913–1921, 1994.

Slowik, M.R., L.G. De Luca, W. Fiers, and J.S. Pober. Tumor necrosis factor activates human endothelial cells through the p55 tumor necrosis factor receptor but the p75 receptor contributes to activation at low tumor necrosis factor concentration. *Am. J. Pathol.*, 143:1724–1730, 1993.

Smith, W.B., J.R. Gamble, I. Clarklewis, and M.A. Vadas. Chemotactic desensitization of neutrophils demonstrates interleukin-8 (IL-8)-dependent and IL-8-independent mechanisms of transmigration through cytokine-activated endothelium. *Immunology*, 78:491–497, 1993.

Sporn, M.B., A.B. Roberts, L.M. Wakefield, and B. Crombrugghe. Some recent advances in the chemistry and biology of transforming growth factor-beta. *J. Cell Biol.*, 105:1039–1045, 1987.

Strieter, R.M., S.L. Klunkel, H.J. Showell, D.G. Remick, S.H. Phan, P.A. Ward, and R.M. Marks. Endothelial cell gene expression of a neutrophil chemotactic factor by TNF-alpha, LPS, and IL-1 beta. *Science*, 243: 1467–1469, 1989.

Strieter, R.M., S.L. Klunkel, V.M. Elner, C.L. Martonyi, A.E. Koch, P.J. Polverini, and S.G. Elner. Interleukin-8. A corneal factor that induces neovascularization. *Am. J. Pathol.*, 141:1279–1284, 1992.

Strieter, R.M., P.J. Polverini, S.L. Klunkel, D.A. Arenberg, M.D. Burdick, J. Kasper, J. Dzuiba, J. Van Damme, A. Walz, D. Marriott, S.Y. Chan, R. Roczniak, and A.B. Shanafelt. The functional role of the ELR motif in CXC chemokine-mediated angiogenesis. *J. Biol. Chem.*, 270:27348–27357, 1995.

Suen, Y., M. Chang, S.M. Lee, J.S. Buzby, and M.S. Cairo. Regulation of interleukin-11 protein and mRNA expression in neonatal and adult fibroblasts and endothelial cells. *Blood*, 84:4125–4134, 1994.

Suschek, C., H. Rothe, K. Fehsel, J. Enczmann, and V. Kolb Bachofen. Induction of a macrophage-like nitric oxide synthase in cultured rat aortic endothelial cells. IL-1 beta-mediated induction regulated by tumor necrosis factor-alpha and IFN-gamma. *J. Immunol.*, 151:3283–3291, 1993.

Tachado, S.D., P. Gerold, M.J. McConville, T. Baldwin, D. Quilichi, R.T. Schwartz and L. Schofield. Glycosylphosphatiylinositol toxin of plasmodium induces nitric oxide synthase expression in macrophages and vascular endothelial cells by a protein tryosine kinase-dependent and protein kinase C-dependent signaling pathway. *J. Immunol.*, 156:1897–1907, 1996.

Takeya, M., T. Yoshimura, E.J. Leonard, and K. Takahashi. Detection of monocyte chemoattractant protein-1 in human athersclerotic lesions by an anti-monocyte chemoattractant protein-1 monoclonal antibody. *Hum. Pathol.*, 24:534–539, 1993.

Tartaglia, L.A., D. Pennica, and D.V. Goeddel. Ligand passing: the 75-kDa tumor necrosis factor (TNF) receptor recruits TNF for signaling by the 55-kDa TNF receptor. *J. Biol. Chem.*, 268:18542–18548, 1993.

Thanos, D. and T. Maniatis. NF-kappa B: A lesson in family values. *Cell*, 80:529–532, 1995.

Tijan, R. and T. Maniatis. Transcriptional activation: a complex puzzle with few easy pieces. *Cell*, 77:5–8, 1994.

Valente, A.J., S.R. Fowler, E.A. Sprague, J.L. Kelley, C.A. Suenram, and C.J. Schwartz. Initial characterization of paripheral blood mononuclear cell chemoattractant derived from cultured arterial smooth muscle cells. *Am. J. Pathol.*, 117:409–417, 1984.

Valente, A.J., D.A. Graves, C.E. Vialle-Valentin, R. Delgado, and C.J. Schwartz. Purification of a monocyte chemotactic factor secreted by nonhuman primate vascular cells in culture. *Biochemistry*, 27:4162–4168, 1988.

van der Kar, N.C., L.A. Monnens, M.A. Karmali, and V.W. van Hinsbergh. Tumor necrosis factor and interleukin-1 induce expression of the verocytotoxin receptor globotriaosylceramide on human endothelial cells: implications for the pathogenesis of the hemolytic uremic syndrome. *Blood*, 81:2755–2764, 1992.

van der Kar, N.C., T. Kooistra, M. Vermeer, W. Lesslauer, L.A. Monnens, and V.W. van Hinsbergh. Tumor necrosis factor alpha induces endothelial galactosyl transferase activity and verocytotoxin receptors. Role of specific tumor necrosis factor receptors and protein kinase C. *Blood*, 85:734–743, 1995.

van Hinsbergh, V.W., M. Vermeer, P. Koolwijk, J. Grimbergen, and T. Kooistra. Genistein reduces tumor necrosis factor alpha-induced plasminogen activator inhibitor-1 transcription but not urokinase expression in human endothelial cells. *Blood*, 84:2984–2991, 1994.

Vidal Alles, V., B. Bottazzi, G. Peri, J. Golay, M. Introna, and A. Mantovani. Inducible expression of PTX3, a new member of the pentraxin family, in human mononuclear phagocytes. *Blood*, 84:3483–3493, 1994.

Villarete, L.H. and D.G. Remick. Nitric oxide regulation of IL-8 expression in human endothelial cells. *Biochem. Biophys. Res. Commun.*, 211:671–676, 1995.

Voest, E.E., B.M. Kenyon, M.S. O'Reilly, G. Truitt, R.J. D'Amato, and J. Folkman. Inhibition of angiongenesis *in vivo* by inerleukin-12. *J. Natl. Cancer Inst.*, 87:581–586, 1995.

von Asmuth, E.J., M.A. Dentener, V. Bazil, M.G. Bouma, J.F. Leeuwenberg and W.A. Buurman. Anti-CD14 antibodies reduce responses of cultured human endothelial cells to endotoxin. *Immunology*, 80:78–83, 1993.

von der Ahe, D., C. Nischan, C. Kunz, J. Otte, U. Knies, H. Oderwald, and B. Wasylyk. Ets transcription factor binding site is required for positive and TNF alpha-induced negative promoter regulation. *Nucleic Acid Res.*, 21:5636–5643, 1993.

Vora, M., H. Yssel, J.E. de Vries, and M.A. Karasek. Antigen presentation by human dermal microvascular endothelial cells. Immunoregulatory effect of IFN-gamma and IL-10. *J. Immunol.*, 152:5734–5741, 1994.

Wakabayashi, H., P.G. Cavanaugh, and G.L. Nicolson. Purification and identification of mouse lung microvessel endothelial cell-derived chemoattractant for lung metastasizing murine RAW117 large-cell lymphona cels: identification as mouse monocyte chemotactic protein 1. *Cancer Res.*, 55:4458–4464, 1995.

Wang, J.M., G. Taraboletti, K. Matsushima, J. Van Damme, and A. Mantovani. Induction of haptotactic migration of melanoma cells by neutrophil activating protein/interleukin-8. *Biochem. Biophys. Res. Commun.*, 169:165–170, 1990.

Wang, J.M., A. Sica, G. Peri, S. Walter, I.M. Padura, P. Libby, M. Ceska, I. Lindley, F. Colotta, and A. Mantovani. Expression of monocyte chemotactic protein and interleukin-8 by cytokine-activated human vascular smooth muscle cells. *Arterioscler. Thromb.*, 11:1166–1174, 1991.

Warner, S.J. and P. Libby. Human vascular smooth muscle cells. Target for and source of tumor necrosis factor. *J. Immunol.*, 142:100–109, 1989.

Weber, C., E. Negrescu, W. Erl, A. Pietsch, M. Frankenberger, H.W. Ziegler Heitbrock, W. Siess, and P.C. Weber. Inhibitors of protein tyrosine kinase suppress TNF-stimulated induction of endothelial cell adhesion molecules. *J. Immunol.*, 155:445–451, 1995.

Wen, D., A. Rowland, and R. Derynck. Expression and secretion of gro/MGSA by stimulated human endothelial cells. *EMBO J.*, 8:1761–1766, 1989.

Werner Felmayer, G., E.R. Werner, D. Fuchs, A. Hausen, G. Reibnegger, K. Schmidt, G. Weiss, and H. Wachter. Pteridine biosynthesis in human endothelial cells. Impact on nitric oxide-mediated formation of cyclic GMP. *J. Biol. Chem.*, 268:1842–1846, 1993.

Whelan, J., P. Ghersa, R. Hooft van Huijsduijnen, J. Gray, G. Chandra, F. Talabot, and J.F. DeLamarter. An NF kappa B-like factor is essential but not sufficient for cytokine induction of endothelial leukocyte adhesion molecule 1 (ELAM-1) gene transcription. *Nucleic Acids Res.*, 19:2645–2653, 1991.

Whitley, M.Z. D. Thanos, M.A. Read, T. Maniatis, and T. Collins. A striking similarity in the organization of the E-selectin and beta interferon gene promoters. *Mol. Cell Biol.*, 14:6464–6475, 1994.

Willimann, K., H. Matile, N.A. Weiss, and B.A. Imhof. *In vivo* sequestration of plasmodium falciparum-infected human erythrocytes: a severe combined immunodeficiency mouse model for cerebral malaria. *J. Exp. Med.*, 182:643–653, 1995.

Wojta, J., M. Gallicchio, H. Zoellner, E.L. Filonzi, J.A. Hamilton, and K. Mcgrath. Interleukin-4 stimulates expression of urokinase-type-plasminogen activator in cultured human foreskin microvascular endothelial cells. *Blood*, 81:3285–3292, 1993.

Wolf, F.W., R.M. Marks, V. Sarma, M.G. Byers, R.W. Katz, T.B. Shows, and V.M. Dixit. Characterization of a novel tumor necrosis factor-alpha-induced endothelial primary response gene. *J. Biol. Chem.*, 267:1317–1326, 1992.

Wolf, F.W., V. Sarma, M. Seldin, S. Drake, S.J. Suchard, H. Shao, K.S. O'Shea, and V.M. Dixit. B94, a primary response gene inducible by tumor necrosis factor-alpha, is expressed in developing hematopoietic tissues and the sperm acrosome. *J. Biol. Chem.*, 269:3633–3640, 1994.

Woodroffe, S.B., H.M. Garnett, and J.E. Layton. Cytomegalovirus infection of vascular endothelial cells alters production of GM-CSF and G-CSF. *Immunol. Cell Biol.*, 72:187–190, 1994.

Yan, S.F., I. Tritto, D. Pinsky, H. Liao, J. Huang, G. Fuller, J. Brett, L. May, and D. Stern. Induction of interleukin 6 (IL-6) by hypoxia in vascular cells. Central role of the binding site for nuclear factor-IL-6. *J. Biol. Chem.*, 270:11463–11471, 1995.

Yao, L., V. Bengualid, F.D. Lowy, J.J. Gibbons, V.B. Hatcher, and J.W. Berman. Internalization of staphylococcus aureus by endothelial cells induces cytokine gene expresssion. *Infect. Immun.*, 63:1835–1839, 1995.

Yellin, M.J., J. Brett, D. Baum, A. Matsushima, A. Szabolcs, D. Stern, and L. Chess. Functional interactions of T cells with endothelial cells: the role of CD40L-CD40-mediated signals. *J. Exp. Med.*, 182:1857–1864, 1995.

Yla Herttuala, S., B.A. Lipton, M.E. Rosenfeld, T. Sarkioja, T. Yoshimura, E.J. Leonard, J.L. Witztum, and D. Steinberg. Expression of monocyte chemoattractant protein 1 in macrophage-rich areas of human and rabbit atherosclerotic lesions. *Proc. Natl. Acad. Sci. USA*, 88:5252–5256, 1991.

Yoshizumi, M., M.A. Perella, J.C. Burnett, Jr., and M.E. Lee. Tumor necrosis factor downregulates an endothelial nitric oxide synthase mRNA by shortening its half-life. *Circ. Res.*, 73:205–209, 1993.

Yu, X., S. Dluz, D.T. Graves, L. Zhang, H.N. Antoniades, W. Hollander, S. Prusty, A.J. Valente, C.J. Schwartz, and G.E. Sonenshein. Elevated expression of monocyte chemoattractant protein 1 by vascular smooth muscle cells in hypercholesterolemic primates. *Proc. Natl. Acad. Sci. USA*, 89:6953–6957, 1992.

Zeiher, A.M., B. Fisslthaler, B. Schray Utz, and R. Busse. Nitric oxide modulates th expression of monocyte chemoattractant protein 1 in cultured human endothelial cells. *Circ. Res.*, 76:980–986, 1995.

Zhang, R., W. Min, and W.C. Sessa. Functional analysis of the human endothelial nitric oxide synthase promoter. Sp1 and GATA factors are necessary for basal transcription in endothelial cells. *J. Biol. Chem.*, 270:15320–15326, 1995.

Ziche, M., L. Morbidelli, E. Masini, S. Amerini, H.J. Granger, C.A. Maggi, P. Geppetti, and F. Ledda. Nitric oxide mediates angiogenesis *in vivo* and endothelial cell growth and migration *in vitro* promoted by substance P. *J. Clin. Invest.*, 94:2036–2044, 1994.

Zoja, C., J.M. Wang, S. Bettoni, M. Sironi, D. Renzi, F. Chiaffarino, H.E. Abobud, J. Van Damme, A. Mantovani, G. Remuzzi, and A. Rambaldi. Interleukin-1 beta and tumor necrosis factor-alpha induce gene expression and production of leukocyte chemotactic factors, colony-stimulating factors, and interleukin-6 in human mesangial cells. *Am. J. Pathol.*, 138:991–1003, 1991.

10

THE ROLE OF THROMBIN IN ANGIOGENESIS

Michael E. Maragoudakis

University of Patras Medical School
Dept. of Pharmacology
Patras, Greece

1. INTRODUCTION

Angiogenesis describes collectively a complex of cascade of events leading to the formation of new blood vessels. The process involves endothelial cell activation proliferation and migration, cell rearrangement or programmed cell death as well as production, degradation and realignment of the extracellular matrix.

There is a growing interest in this phenomenon, because it is recognized that many disease states are associated with a derangement of angiogenesis. Practically every medical specialty deals with situations where angiogenesis is involved e.g. tumor growth and metastasis, diabetic retinopathy, inflammation, psoriasis, hemangiomas, wound healing etc.

The concept of angiogenesis-based therapy is widely accepted both for treatment and diagnosis. Suppressors of angiogenesis are of interest in conditions where abnormal proliferation of blood vessels is related to disease progression. On the other hand promoters of angiogenesis have therapeutic applications in conditions resulting from insufficient blood supply, e.g. wound healing, peptic ulcers, coronary artery disease etc. Angiogenesis is also a marker for diagnosis, prognosis and guide to therapy for a variety of cancers. Both vascular density of tumors and assessment of angiogenic factors in the body fluids can be used for this purpose. Angiogenesis is also a prominent event in wound healing and many physiological process such as in the endometrium during ovulation, in the mammary gland during lactation, in hair growth etc.

There are fundamental differences, however, between angiogenesis in the above physiological and repair processes and that occurring in angiogenic diseases. Both the time frame and the regulation of angiogenesis are different. In angiogenic diseases the blood vessel proliferation persists uncontrolled for months or years and contributes to the pathology of these diseases. On the contrary, physiological angiogenesis is self limited with stringent control and returns to the quiescent state after a short time of days or weeks.

Vascular Endothelium, edited by Catravas *et al.*
Plenum Press, New York, 1998

Understanding the overall control mechanism, the factors that activate or suppress angiogenesis and the transduction mechanism involved is of paramount importance both in locating the derangement of angiogenesis in pathological conditions and in developing specific modulators with therapeutic potential.

The list of modulators known to trigger or influence angiogenesis has increased over the past 15 years and it has been proposed by some investigators that the numerous factors may constitute the elements of an as yet unidentified cascade similar to that responsible for blood coagulation. We may be in the same position regarding angiogenesis that we were with blood coagulation before the interrelation of various factors had been defined. Others believe that there is an interplay between redundant promoters and inhibitors of angiogenesis, the balance of which results in the quiescent state of angiogenesis under normal physiological conditions. In every tissue, however, angiogenesis can be activated at a moments notice by an imbalance of the endogenous modulators resulting from over production of a promoter or disappearance of a suppressor.

1.1. Thrombin and Cancer

The relation between blood coagulation and cancer growth and metastasis was first observed by Trousseau in 1872. Many investigators have subsequently verified the frequency of spontaneous coagulation induced in cancerous patients and provided clinical and pathological data in support of the notion that a systemic activation of the blood coagulation cascade is evident in patients with cancer. For example, Ricles and Edwards (1983) have shown that patients with cancer have high circulating fibrinopeptides, an indication of intravascular coagulation and fibrinolysis. Others have shown that tumor cells express the transmembrane protein tissue factor, which when exposed to circulating factor VII activates factor X leading to generation of thrombin and the formation of fibrin (Sloan et al., 1986). Thus tumor cells elicit procoagulant activity directly or through interaction with platelets, leukocytes and endothelial cells.

More recent findings show that thrombin-treated tumor cells have increased ability for pulmonary metastasis (up to 156-fold) and thrombin *in vivo* enhances dramatically the number of colonies formed by injection of B16 melanoma cells in mice. These results suggested the test of anticoagulants in animal tumor models and the successful outcome of these experiments in animals promoted several clinical trials using warfarin in human cancers (Taparelli et al., 1993). The aforementioned findings provide considerable insight for the mechanisms involved in Trousseau's spontaneous coagulation induced by cancer. They can not explain, however, the mechanism by which the generation of thrombin promotes tumor progression and metastasis or the antitumor effect of anti-coagulants.

1.2. Thrombin and Angiogenesis

Thrombin has a multitude of effects both enzymatic and cellular actions on a variety of cell types (platelets, tumor, heart, smooth muscle, endothelial and leukocytes). None of these actions can provide the basis for an explanation of the role of thrombin in tumor growth and metastasis. A plausible explanation may be our finding that thrombin is a potent promoter of angiogenesis (Tsopanoglou et al., 1993). This effect of thrombin is specific and independent of fibrin formation. In view of the fact that angiogenesis has been shown to play a pivotal role not only in tumor growth and metastasis but also in other pathological conditions such as inflammation, wound healing etc. (Folkman, 1985 & 1995) suggests a new role of thrombin in these conditions.

2. METHODS

The angiogenic action of thrombin was studied using three experimental systems:

a. The chick chorioallantoic membrane (CAM) assay which is used by many investigators as a convenient model for studying angiogenesis. In our laboratory we are using collagenous protein biosynthesis as a biochemical index of angiogenesis (Maragoudakis et al., 1988). This method has been validated using a variety of promoters and inhibitors of angiogenesis and comparing the results in parallel experiments with direct counting of blood vessels (Harris-Hooker et al., 1983) or by computed assisted image analysis (Maragoudakis et al., 1995). This method is sensitive, reproducible, convenient and can accommodate the quantitation of angiogenesis in a large number of samples.

b. The Matrigel tube formation assay was used as described previously by Grant et al. (1989) and Haralabopoulos et al. (1994). Briefly, Matrigel, which was prepared from EHS tumor was used to coat 24-well cluster plates (0.25 ml/well). After polymerization of the Matrigel at 37°C for one hour, the test material (e.g. thrombin) was added in 0.5 ml M199 medium per well at the concentration indicated. After that, the cells (40,000 cells/well) were added suspended in HUVEC'S medium. After 18 hours of incubation, the medium was removed and the cells were fixed with Diff quick (Baxter Scientific). Subsequently tube area was quantitated using the Optamax image analysis system as described previously (Haralabopoulos et al., 1994).

c. The Matrigel plug assay was performed as described by Passanitti et al. (1992). 500μl Matrigel at 4°C containing the test material or vehicle was injected subcuntaneously into C57 black female mice. After injection the Matrigel rapidly formed a plug which is left in the animal for 7 days. For each treatment, 500μl of Matrigel was injected into 4 mice and the experiments repeated twice. At the end of the experiment the skin of the mouse was easily pulled back to expose the Matrigel plug, which remained intact. After photographing and noting quantitative differences, the plug was dissected out of the mouse and fixed with 3.7% formaldehyde/phosphate buffered saline and embedded in paraffin. It was then sectioned and stained with Masson Trichrome stain. Histological slices were evaluated for the relative area of the cells that had infiltrated into Matrigel plug using image analysis in a similar manner as the tube assay. This *in vivo* method of angiogenesis has the advantage that no surgery is required for implantation and involves a simple injection of Matrigel which is liquid at 4°C and solidifies rapidly at the body temperature forming a stable Matrigel plug. In addition, the Matrigel does not elicit any immunological response because it is produced by the same strain of mice.

3. RESULTS

3.1. Effect of Thrombin in the Chick Chorioallantoic Membrane System

It was shown in this system that both α- and γ-thrombin cause a dose-dependent promotion of angiogenesis. As with other actions of γ-thrombin, a 10–15 fold-higher concentrations of γ-thrombin than α-thrombin is required to obtain the maximum angiogenic response (about 80% over that in controls) (Tsopanoglou et al., 1993). A maximum of angiogenic response was obtained with 1 IU (8.4 pmoles) of α-thrombin or 130 pmoles of γ-thrombin. As with many actions of thrombin, higher doses of α- and γ-thrombin cause a drop in the angiogenic action probably as a result of desensitization of the receptor.

Alpha-thrombin is a glycosylated trypsin-like serine protease generated from the circulating plasma prothrombin. Upon autocatalytic and factor Xa cleavage the functional two chain molecule of thrombin is generated consisting of 36 and 259 amino acid residues for A and B chain respectively for human thrombin. The two chains are covalently interconnected by disulfide bridge, the B chain having the catalytic site. In addition to the catalytic domain α-thrombin possesses an exocite which is important for binding and cleavage of fibrinogen to fibrin. This "fibrin-(ogen)-recognizing exocite" or "anion binding exocite" is essential for clotting activity.

Gamma-thrombin is the product of proteolytic or autolytic cleavage of α-thrombin. In γ-thrombin the "anion binding exocite" is missing, therefore, it can not form fibrin or participate in the blood clotting cascade (Fenton, 1988). However, thrombin, like α-thrombin is also a potent cell activator. It can activate platelets, promote shape and permeability changes of endothelial cells, cause Ca^{2+} release, prostacyclin biosynthesis, release of plasminogen activator, etc. The only important difference between α- and γ-thrombin is that γ-thrombin has no blood clotting activity.

The fact that both α- and γ-thrombin promote angiogenesis to the same extent lead us to the conclusion that the action of thrombin on angiogenesis is independent of fibrin formation (Tsopanoglou et al., 1993).

The specificity of the effect of thrombin on angiogenesis is shown by the fact that hirudin, which by itself has no effect, completely abolishes the angiogenic action of thrombin, when hirudin is used in combination with thrombin. Hirudin binds both the catalytic and the anion binding exocites, thus inactivating thrombin. Similarly heparin, which binds to thrombin and accelerates its destruction by antithrombin III also cancels out the angiogenic action of thrombin, when it was used in combination with thrombin (Tsopanoglou et al., 1993). P-PACK thrombin, is chemically modified thrombin in which the catalytic site is inactivated but retains the anion binding exocite.

This analog of thrombin has no effect on angiogenesis in the CAM indicating that the anion binding exocite alone is not sufficient to promote angiogenesis. When P-PACK thrombin, however, was combined with either α- and γ-thrombin, it prevented the stimulatory effect of thrombin presumably by interfering with the binding of thrombin via its catalytic site. Apparently P-PACK thrombin binds to the receptor and prevents its activation by thrombin.

These results suggest that the angiogenic action of thrombin is specific and that the catalytic site is essential for this effect.

Recent studies on the thrombin receptor which is a member of the seven transmembrane receptor family, reveals that the receptor has an extracellular amino terminal extension. This extracellular extension possesses the cleavage site by thrombin. Thrombin proteolytically cleaves a fragment of the receptor thus generating a new NH_2-terminus of the receptor. This new receptor peptide acts as a tethered receptor agonist by binding to an as yet unidentified site of the receptor to effect cell activation. This site specific proteolysis of the thrombin receptor provides all the necessary information for receptor activation. A synthetic peptide TRAP, consisting of 14-amino acids representing the NH_2-terminus of the cloned receptor elicits most but not all the effects of thrombin (Garcia, 1992; Garcia et al., 1993). TRAP can mimic the effects of thrombin in cellular events such as secretion, mitogenesis, second messenger generation etc. (Tapparelli et al., 1993). However, not all actions of thrombin can be seen by substituting TRAP for thrombin e.g. tyrosine kinase activation, endothelial cell activation etc. It is of interest that TRAP also promotes angiogenesis in the CAM system in a dose-dependent fashion. However, the maximum obtained with 7 nmoles is about 50% above control and at higher concentrations there is a decline in the angiogenesis promoting effect of TRAP.

These results with TRAP suggests that promotion of angiogenesis by thrombin is a receptor mediated event, which is distinct form the clotting mechanism.

3.2. Effect of Thrombin on the Tube Formation of the Matrigel System *in Vitro*

Within 18 hours human umbilical cord endothelial cells (HUVECS) plated on Matrigel form an anastomozing network of capillary-like tubes (Grant et al., 1989). When thrombin was added there was not a significant effect on the rate and extent of tube formation. However, when the fetal calf serum was reduced from 10% to 2% the presence of thrombin 0.1 and 0.3 IU/ml caused 25% and 55% increase respectively in the area covered by tube like structures. When the concentration of thrombin was further increased there was a decrease in the tube-promoting effect and this is in line with the results obtained in the CAM system and the notion of desensitization of thrombin receptor by high concentrations of thrombin (Haralabopoulos et al., 1997). Similar to the results obtained in the CAM system, P-PACK thrombin was without effect on tube formation. However, when used in combination with thrombin it canceled out the effects of thrombin. These results indicate that the catalytic site of thrombin is important in the promotion of the angiogenic phenotype of HUVECS. The demonstration that thrombin stimulates the proliferation of microvascular endothelial cells could be the first explanation for the angiogenesis promoting effect of thrombin. However, the above experiments in the Matrigel system *in vitro* show that thrombin can induce the angiogenic phenotype in the absence of cell proliferation, since tube formation takes place within 18 hours without cell proliferation in this system.

3.3. Effect of Thrombin on Endothelial Cell Infiltration into Matrigel Plug *in Vivo*

In the Matrigel system *in vivo* in the absence of thrombin or other angiogenic substances the Matrigel plugs were completely clear with no visible blood vessels going into the Matrigel. Plugs containing thrombin (0.3, 1.0 and 3.0 IU/ml Matrigel) appeared pink and a large number vessels could be seen leading into the Matrigel plug. Histological evaluation of the sections by image analysis showed a 15–20 fold increase in the area of cells that have infiltrated into the Matrigel plug containing thrombin (Haralabopoulos et al., 1997).

4. DISCUSSION

These studies *in vitro, in ovo*, and *in vivo* lead to the conclusion that promotion of angiogenesis by thrombin is a receptor mediated event which is distinct from the clotting activity of thrombin. Important questions then arise as to which of the many cellular actions of thrombin are involved in activating the angiogenic cascade and which transduction mechanisms are involved.

Thrombin is not only a key factor in hemostasis, as regulator of blood clotting, but also possesses ubiquitous bioregulatory functions and elicits many cellular actions in various cell types. Particularly in endothelial cells thrombin induces proliferation (Carney, 1992), migration (Pankonin and Teuscher, 1991), cell shape changes (Galdal et al., 1985), disruption of endothelial cell barrier function (DeMichele et al., 1990), synthesis of platelet activating factor (Prescott et al., 1984), formation of PDGF (Shankar et al., 1992), PGI_2 (Jaffe et al., 1987) and thrombomodulin (Bartha et al., 1993) synthesis and cell surface expression of tissue factor (Bartha et al., 1993), synthesis and release of tissue plasminogen activator and its inhibitor (Dichek and Quertermous, 1989) and activation of progelatinase (Zucker et al., 1996).

Many of these effects of thrombin on endothelial cells may be related to activation of angiogenesis e.g. progelatinase activation, PDGF synthesis etc. The relative contribution and the mechanisms involved in the activation of individual steps of the angiogenic cascade remains to be elucidated.

Regarding the question of the transduction mechanisms involved in the activation of angiogenesis by thrombin it seems that the activation of protein kinase C (PKC) plays a major role.

We have shown previously Tsopanoglou et al., (1993) and Tsopanoglou et al., (1994) that activation of PKC by phorbol esters (PMA) or diacylglycerol (DAG) causes a marked stimulation of angiogenesis. On the contrary inhibitors of PKC such as Ro 318220 causes an inhibition of angiogenesis both the basal and that stimulated by PMA and DAG.

It is also known that both α- and γ-thrombin activate phospholipase C leading to inositol triphosphate (IP_3) generation and PKC activation. These events can be seen as early as 5 seconds after the addition of α- and γ-thrombin and requires the active site as shown by the fact that P-PACK thrombin is inactive (Carney et al., 1992). The involvement of G-proteins in the transduction mechanisms of thrombin is evident from many studies. For example pertussis toxin inhibits thrombin-induced activation of phospholipase C and cell proliferation, but it has no similar effects on other mitogenic factors such as EGF, which act via tyrosine kinase activation (Carney et al., 1992).

In view of the above it is reasonable to suggest that PKC activation is one of the early events of thrombin action, which leads to cellular responses leading to activation of the angiogenesis cascade. Indeed the PKC inhibitor Ro 318220 completely abolishes the angiogenic action of thrombin (Tsopanoglou et al., 1994).

We conclude from these studies that thrombin promotes angiogenesis via activation of the cloned receptor. The catalytic site of thrombin is essential of this action and can be mimicked by TRAP, a peptide that represents the NH_2-terminus of the activated thrombin receptor. This promotion of angiogenesis by thrombin is independent of blood clotting activity and represents a new role for thrombin. This finding opens the possibility of developing thrombin mimetic peptides or other related moieties that promote or suppress angiogenesis without effects on the blood coagulation.

This type of agents have therapeutic potential not only in the metastatic spread of cancer but also in other conditions, such as inflammation, wound healing, where angiogenesis is involved (Folkman, 1995) and thrombin has been implicated Tapparelli et al., (1993). The dependence of tumor growth and metastasis on new blood vessel formation is supported by many studies and the involvement of thrombin in cancer is well documented (Silverstein & Nachman, 1992). Also angiogenesis is part of the inflammatory process since thrombin has proinflammatory effects both by activating endothelial cells and monocytes/macrophages. Therefore, local generation of thrombin by macrophage-like cells, which is evident in the synovial fluid of rheumatoid arthritic patients, is likely a contributing factor for inflammatory angiogenesis. Specific, inhibitors of the angiogenic action of thrombin, with no effect on blood coagulation may prove to be novel antitumor and anti-inflammatory agents.

Angiogenesis is also essential in wound healing and thrombin is known to promote incisional wound healing and neovasculization in rats (Carney et al., 1992). Non-thrombogenic analogs of thrombin, therefore, may be valuable agents for treating non-healing wounds, or ulcers in diabetic or older patients.

REFERENCES

Bartha, K., C. Brisson, G. Archipoff, C. de la Salle, F. Lanza, J.P. Cazenave, and A. Beretz. Thrombin regulates tissue factor and thrombomodulin mRNA levels and activities in human saphenous vein endothelial cells by distinct mechanisms. *J. Biol. Chem.*, 268:421429, 1993.

Carney, D.H. Postclotting cellular effects of thrombin mediated by interaction with high-affinity thrombin receptors. In: *Thrombin Structure and Function*, Berliner J.L. ed., Plenum Press, pp. 351–396, 1992.

DeMichele, M.A.A., D.G. Moon, J.W. Fenton, and F.L. Minnear. Thrombin's enzymatic activity increases permeability of endothelial cell monolayers. *J. Appl. Physiol.*, 69:1599–1606, 1990.

Dichek, D. and T. Quertermous. Thrombin regulation of mRNA levels of tissue plasminogen activator and plasminogen activator inhibitor-l in cultured human umbilical vein endothelial cells. *Blood*, 74:222–228, 1989.

Fenton, J.W. Regulation of thrombin generation and functions. *Semin. Thromb. Hemost.*, 14:234–240, 1988.

Folkman, J. Tumor angiogenesis. *Adv. Cancer Res.*, 43:172–203, 1985.

Folkman, J. Angiogenesis in cancer, vascular, rheumatoid and other disease. *Nature Med.*, 1:27–31, 1995.

Galdal, J.G.N. Molecular mechanisms of thrombin-induced human and bovine endothelial cell activation. *J. Lab. Clin. Med.*, 120:513–519, 1992.

Garcia, J.G.N. Molecular mechanisms of thrombin-induced human and bovine endothelial cell activation. *J. Lab. Clin. Med.*, 120:513–519, 1992.

Garcia, J.G.N., C. Patterson, C. Bahler, J. Aschner, C.M. Hart, and D. English. Thrombin receptor activating peptides induce Ca^2+ mobilization, barrier dysfunction, prostaglandin synthesis, and platelet-derived growth factor mRNA expression in cultured endothelium. *J. Cell Physiol.*, 156:541–549, 1993.

Grant, D.S., K.I. Tashiro, B. Segui-Real, Y. Yamada, G.R. Martin, and H.K. Kleinman. Two different laminin domains mediate the differentiation of human endothelial cells into capillary-like structures *in vivo*. *Cell*, 58:933–943, 1989.

Haralabopoulos, G.C., D.S. Grant, H.K. Kleinman, P.I. Lelkes, S.P. Papaioannou, and M.E. Maragoudakis. Inhibitors of basement membrane collagen biosynthesis prevent endothelial cell alignment in Matrigel *in vitro* and angiogenesis *in vivo*. *Lab. Invest.*, 71(4):575–581, 1994.

Haralabopoulos G.C., D.S. Grant, H.K. Kleinman, and M.E. Maragoudakis. Thrombin promotes endothelial cell alignment in Matrigel *in vitro* and angiogenesis *in vivo*. *Am. J. Physiol.*, (submitted for publication).

Harris-Hooker, S.A., C.M. Gajdusec, T.N. Wight, and S.M. Schwartz. Neovascular response induced in cultured aortic endothelial cells. *J. Cell Physiol.*, 114:302–310, 1983.

Jaffe, E.A., J. Grulich, B.B. Weksler, G. Hample, and Watanabe. Correlation between thrombin induced prostacyclin production and inositol triphosphate and cytosolic free calcium levels in human cultured endothelial cells. *J. Biol. Chem.*, 262:8557–8565, 1987.

Maragoudakis, M.E., M. Sarmonika, and M. Panoutsacopoulou. Rate of basement membrane biosynthesis as an index to angiogenesis. *Tissue and Cell*, 20(4):531–539, 1988.

Maragoudakis, M.E., G.C. Haralabopoulos, N.E. Tsopanoglou, and E. Pipili-Synetos. Validation of collagenous protein synthesis as an index for angiogenesis with the use of morphological methods. *Microvascular Res.*, 50:215–222, 1995.

Maragoudakis, M.E., N.E. Tsopanoglou, E. Sakkoula, and E. Pipili-Synetos. On the mechanism of promotion of angiogenesis by thrombin. *FASEB J.*, 9:A587. 1995.

Pankonin, G., and F. Teuscher. Stimulation of endothelial cell migration by thrombin. *Biomed. Biochem. Acta*, 50:1073–1078, 1991.

Passaniti, A., R.M. Taylor, R. Pili, Y. Guo, P.V. Lowrey, J.A. Haney, R.R. Daniel, D.S. Grawd, and G.R. Martin, G.R. A simple quantitative method for assessing angiogenesis and antiangiogenic agents using reconstituted basement membrane heparin, fibroblast growth factor. *Lab. Invest.*, 67:519–528, 1992.

Prescott, S.M., G.A. Zimmerman, and T.M. McIntyre. Human endothelial cells in culture produce platelet-activating factor (1-alkyl-2-acetyl-sn-glycero-3-phosphocholine) when stimulated with thrombin. *Proc. Natl. Acad. Sci. USA*, 81:3534–3538, 1984.

Rides, F.R., and R.L. Edwards. Activation of blood coagulation in cancer: Trousseau's syndrome revisited. *Blood*, 64:14–31, 1983.

Silverstein, R.L. and R.L. Nachman. Cancer and clotting - Trousseau's warning. *N. Engl. J. Med.*, 327:1163–1164, 1992.

Sloan, B.F., J. Rozhin, K. Johnson, H. Taylor, J.D. Crissman, and K.V. Honn. Cathepsin B: Association with plasma membrane in metastatic tumors. *Proc. Natl. Acad. Sci. USA*, 83:2483–2487, 1986.

Tapparelli, C., R. Metternich, C. Ehrhardt, and N.S. Cook. Synthetic low molecular weight thrombin inhibitors: molecular design and pharmacological prolife. *TIPS*, 14:366–376, 1993.

Tapparelli, C., R. Metternich, and N.S. Cook. Structure and function of thrombin receptors. *TIPS*, 14:426428, 1993.

Tsopanoglou, N.E., E. Pipili-Synetos, and M.E. Maragoudakis. Thrombin promotes angiogenesis by a mechanism independent of fibrin formation. *Am. J. Physiol.*, 264 (Cell Physiol. 33):C1302–1307, 1993.

Tsopanoglou, N.E., E. Pipili-Synetos, and M.E. Maragoudakis. Protein kinase C involvement in the regulation of angiogenesis. *J. Vasc. Res.*, 30:202–208, 1993.

Tsopanoglou, N.E., G.C. Haralabopoulos, and M.E. Maragoudakis. Opposing effects on modulation of angiogenesis by protein kinase C and cyclic AMP-mediated pathways. *J. Vasc. Res.*, 31:195–204, 1994.

Zucker S., Conner C., DiMassmo, B.I., Ende, H., Drew, M., Seiki, M and Bahou, W.F. Thrombin induces the activation of progelatinase A in vascular endothelial cells. Physiologic regulation of angiogenesis. *J. Biol. Chem.*, 270:23730–23738, 1996.

V. CEREBRAL CIRCULATION/ ATHEROSCLEROSIS

11

OVERVIEW OF CEREBROVASCULAR DISEASE: ROLE OF THE ENDOTHELIUM

Allan D. Callow

Boston University School of Medicine and
The Whitaker Cardiovascular Institute
Boston City Hospital and
The University Hospital
Boston, Massachusetts

1. INTRODUCTION

The circulation of the brain differs from that of other organs in several respects, most notably by the rigorous application of the universal principle that the volume of blood entering an organ cannot exceed that which exits. Unlike other solid organs, the brain is confined within a non-resilient, rigid, bony box — the skull. Parenchymatous organs such as the spleen, kidney, liver and the abdominal viscera, even the skeletal musculature, have the capacity to accommodate transient changes in blood volume and blood pressure. Not so for the brain. An increase in brain volume causes brain substance to press against the unyielding skull and, in severe instances, to seek egress through various channels called foramina — portals via which nerves exit, and blood vessels enter and also exit the cranium. The largest of these channels is the foramen magnum, the pathway for the spinal cord. Literally, the brain is forced out of the skull by the build-up of pressure within it. This is known as herniation of the brain. It is accompanied by loss of consciousness and, if unchecked, may progress to coma and death. Lesser degrees of intracranial pressure while not fatal, result in compromise of neuronal function.

2. CEREBRAL AUTOREGULATION

The control of blood flow in the brain is based on cerebral autoregulation which means that intracranial blood flow remains constant in the presence of wide variations in systemic *pressure*, at least until physiologic limits are breached. Avoidance of increased intracranial pressure which might be occasioned by increases in systemic blood pressure,

Vascular Endothelium, edited by Catravas *et al.*
Plenum Press, New York, 1998

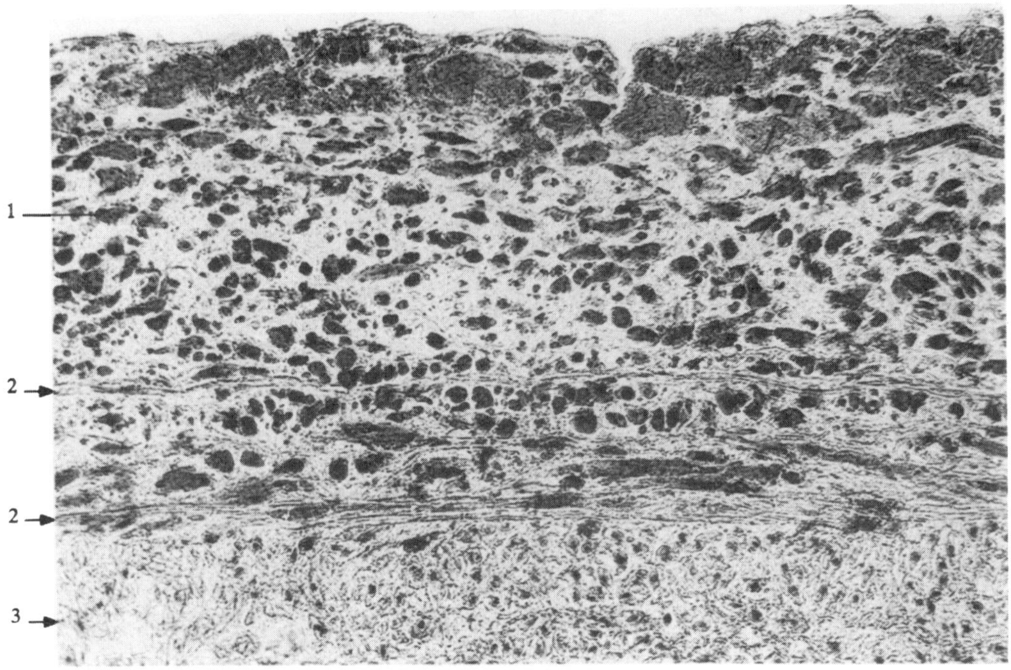

Figure 1. A frozen section micrograph of a lipid streak in the aorta stained with Sudan-hematoxylin showing 1) a thickened intima containing deposits of Sudan-positive lipids, both within and between cells plus myointimal cells with phagocytosed lipid, 2) a middle layer, the media, with fragmented elastic fibers, 3) a loose adventitial layer containing fat cells and nutrient vessels, the vasa vasorum. From: Sandritter's Color Atlas and Textbook of Histopathology, 7th edition 1984, page 76. With Permission.

as with exercise, changes in head position from vertical to horizontal, and during emotional crises, is accomplished by cerebral autoregulation. Intracranial blood flow remains constant despite wide fluctuations in systemic pressure. Thus even with extreme exertion, as in competitive athletic events, blood flow through the brain does not follow systemic blood pressure variations. This is true until physiologic limits are breached. Examples are severe, life-threatening blood loss, drug overdose and cardiac pump failure. The mechanisms of cerebral autoregulation are not well understood despite decades of study. Conventional wisdom credits the endothelial cell with the predominant role.

3. THE BLOOD BRAIN BARRIER

This constitutes the second major difference between blood flow in the brain and blood flow in other organs. Mediated through the endothelium, this barrier provides a degree of protection to the brain that borders on physiological isolation. The *structural* differences making up the barrier are chiefly the tight junctions between the endothelial cells of the capillaries plus a heavy astrocyte backup coverage behind them. *Functional* differences consist of the highly selective nature of the exchange functions of the endothelium and an equally highly selective set of responses to the humoral and neurogenic stimuli operative throughout the systemic circulation. Movement of a wide range of molecules, vasoactive and otherwise, is controlled by specific molecule transporters in and on the endothelial cells. Two examples are: 1) the vasodilation produced by acidosis and hypoxia is much greater in the cerebral vasculature than in other vascular territories; 2) little to no response of cerebral vessels to neural stimuli via their sympathetic systems. The integrity of the blood brain barrier is easily disturbed by hypoxia, ischemia and acute hypertension. Fluid and solutes move out of the cerebral capillary. Brain edema develops and critical autoregulatory mechanisms fail.

The signs and symptoms of **Diseases of the Cerebral Vessels** must be interpreted first as the consequence of specific vascular disease entitles, their primary effects, and secondly as the consequences of the general circulatory impairments they engender. Secondary effects are impairment of blood flow, intravascular thrombosis, and loss of a circulatory bed or a portion thereof. These secondary effects are common to all vascular diseases whether in the brain or in the general circulation.

Atherosclerosis, by several orders of magnitude, is the leading vascular cause of disorders of the brain. Although intracranial atherosclerosis is a serious threat to brain function and survival, symptomatic atherosclerosis is most often due to lesions in the extracranial vessels of which the internal carotid artery is the most frequent locus. The atherosclerotic plaque is characteristically located at the bifurcation of the common carotid artery and extends into the internal carotic branch for several centimeters. Curiously, the distal edge of the plaque rarely extends beyond the cervical portion of the vessel. It is this feature, segmental limitation of disease, which permits surgical removal of the plaque, the procedure known as carotid endarterectomy. Neurologic signs and symptoms are due to embolization from the plaque into the intracranial vessels, most often the middle cerebral artery or its branches. Rarely, and then only in the presence of advanced disease, may signs and symptoms be due to restriction of blood flow. Intracranial atherosclerotic occlusive disease is rarely amenable to surgical treatment.

4. LUXURY PERFUSION

A circulatory aberration, seemingly peculiar to the intracranial circulation has become known as "luxury perfusion." It carries the luxury label based on the perception that more blood per unit of time is delivered to a circumscribed area of the brain than the needs of the area require. This may result from capillary bed dilatation secondary to focal hypoxemia, hypercarbia and change in tissue pH, all attributable to an occluded arterial tributary. On arteriography, this focal vasodilatation appears as an angiographic "blush," or early angiographic appearance of a draining vein, secondary to a localized increase in circulation speed. The usual cause of luxury perfusion is an ischemic stroke leaving an area of localized cerebral infarction. After a major hemispheric stroke, the angiogram will usually reveal secondary hemodynamic effects for several days to weeks following lysis of an embolus. These changes reflect antegrade perfusion to a recently ischemic region of the brain. They reflect loss of normal autoregulation of blood flow in the originally ischemic tissue and the immediate surrounding area. The result is arteriolar and capillary hypervascularity associated with early venous filling.

5. ARTERITIDES AND OTHER ANGIOPATHIES

A rather large number of other conditions, loosely categorized as angiopathies and arteritides, target the cerebral vessels producing neuronal dysfunction and tissue destruc-

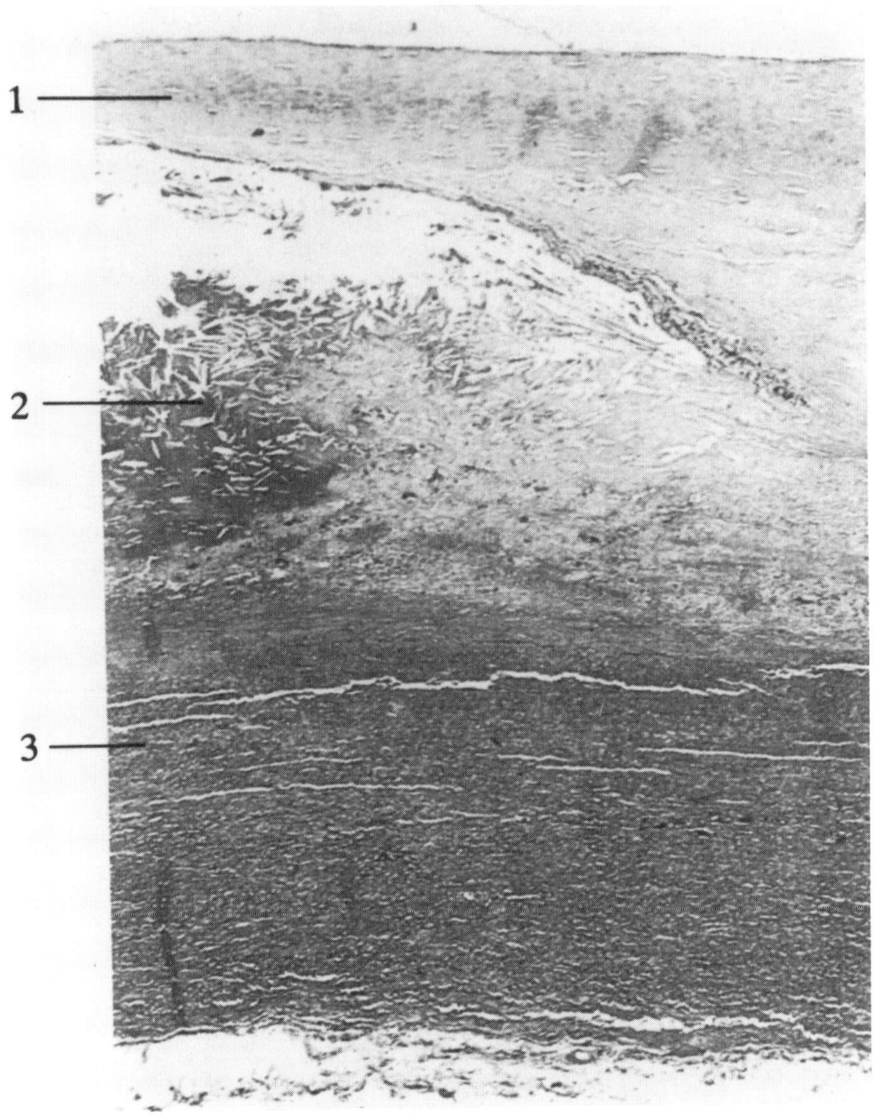

Figure 2. Hematoxylin and eosin stained atheromatous plaque with necrosis; 1) cushion of thickened intima, 2) a paucity of cells and much collagen and disintegration of the collagen and cellular components 3) with loss of tensile strength of the plaque and arterial wall. Original Magnification 40x. From: Sandritter's Color Atlas and Textbook of Histopathology, 7th edition 1984, page 76. With Permission.

tion. Several of these non-atherosclerotic vascular diseases involve the intracranial vessels as part of a systemic vascular disorder. A few affect large and medium-size arteries; others have as their targets smaller arteries and still others select only the arterioles, venules and capillaries. Specific etiology of most of these diseases is unknown, but conventional opinion accepts an autoimmune response as the common denominator. An inflammatory process leading to necrosis of the structural elements of the vessel wall, followed by thrombosis, and terminating in ischemia and infarction, often microscopic in extent, is the sequence of events. For most of these diseases, there is no laboratory or pathognomic

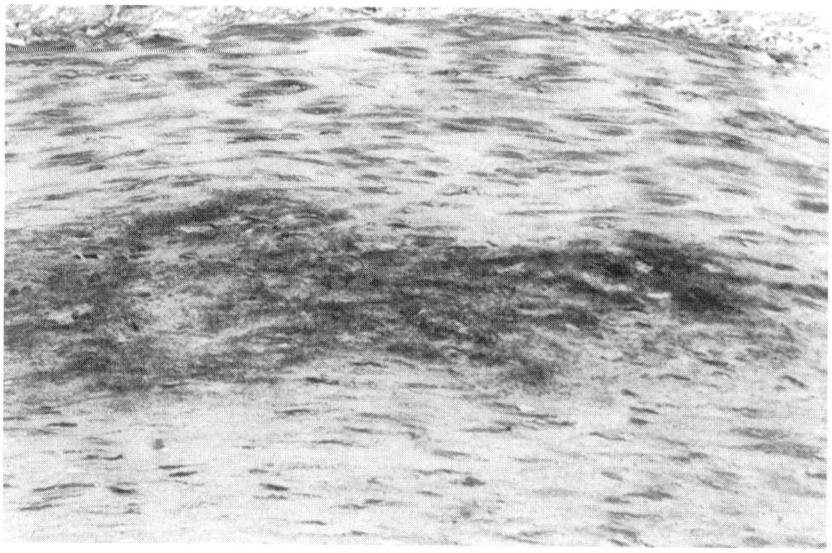

Figure 3. Deposits of calcium within the media. From: Sandritter's Color Atlas and Textbook of Histopathology, 7th edition 1984, page 76. With Permission.

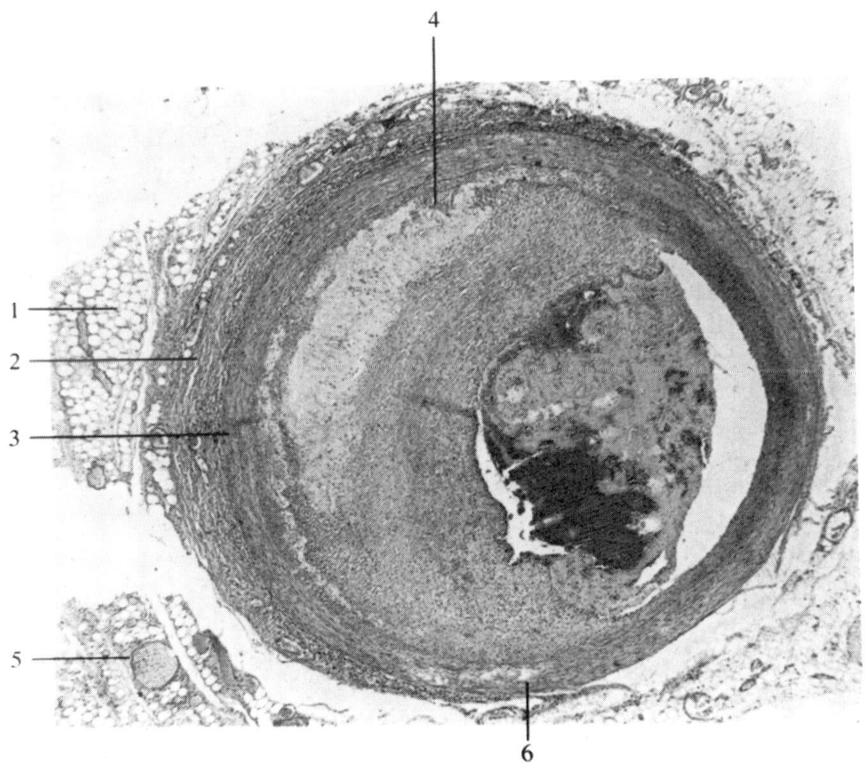

Figure 4. Severe atherosclerotic occlusive disease in a coronary artery with recent thrombus and minimal residual lumen; 1) subepicardial fatty tissue 2) adventitia 3) media 4) thickened intima with a crescent-shaped fibrous tissue collection showing some necrosis which also involves the media (6) a small peripheral nerve is displayed at 5. Original magnification 26x. From: Sandritter's Color Atlas and Textbook of Histopathology, 7th edition 1984, page 78. With Permission.

clinical sign to provide a confident diagnosis. Rather, diagnosis rests on the sum of clinical, radiologic and histologic data. Presumably, circulating antigen-antibody complexes find their way in the wall of the vessel and the complement cascade is activated. Polymorphonuclear leukocytes infiltrate the wall, engulf the complexes, and release proteases which further damage the vessel. Several primary type vasculitides are characterized by granuloma formations. Treatment usually is based on a carefully developed and monitored schedule of anti-inflammatory agents, especially the steroids.

6. TAKAYASU'S DISEASE

A Japanese ophthalmologist described a young Japanese female in 1906 who suffered progressive impairment of vision, attacks of vertigo and syncope, conjunctival injection and alopecia. Because of its gradual encroachment upon the lumen of the brachiocephalic vessels, it became known as "pulseless disease." This form of arteritis has since been seen to involve the entire length of the aorta and its branches. Typically, it is confined to the aortic arch and is now known as obliterative brachiocephalic arteritis for it affects the arch, the subclavian and the carotid arteries.

Intracranial vessels escape. Women are more frequently affected than men. Its onset is usually in the mid to late teens. All three layers of the artery, intima, media and adventitia are involved in an inflammatory process ranging from acute to chronic. The elastic lamina becomes fragmented. Macrophages and giant cells are abundant. Acute lesions display edema of the wall and lymphocyte infiltration. Segments of the artery escape the process. Aneurysm formation may develop because of patchy thinning of the wall. Progressive narrowing is more commonly followed by thrombosis and occlusion. Prodromal symptoms are constitutional: lassitude, weight loss, anorexia, easy fatiguability, anemia and meandering musculoskeletal pain. The cause is unknown but tuberculosis, common in some parts of rural China where large numbers of these cases were found, and cigarette smoking which is wide spread have been implicated. Ophthalmodynamometry may reveal reduced retinal artery pressures. Neoangiogenesis may develop about the optic nerve head. The course of Takayasu's arteritis is unpredictable but slow progression is usual. Gradual loss of vision ending in blindness and repeated episodes of cerebral infarctions mark the eventual outcome. It is probably best managed by very intensive treatment with steroidal anti-inflammatory agents, but, on occasion, endarterectomy and bypass surgery may be needed.

7. GIANT CELL ARTERITIS

This disease affects medium and large size arteries. Granulomas being part of the histologic picture; this is also known as granulomatous arteritis. When seen in conjunction with rheumatoid arthritis, it has been labeled polymyalgia rheumatica. The same histology is seen in the arteritis of the superficial temporal artery, identified as cranial arteritis in the older literature. Giant cell arteritis is the appropriate term for all these inflammatory artery conditions. Nomenclature confusion is increased by the observation that giant cell arteritis may "merge" with systemic vasculitis such as polyarteritis nodosa. The inflammatory process consists of lymphocytes, plasma cells, multinucleated foreign body giant cells, fragmentation of the internal elastic lamina and gradual replacement of the media by con-

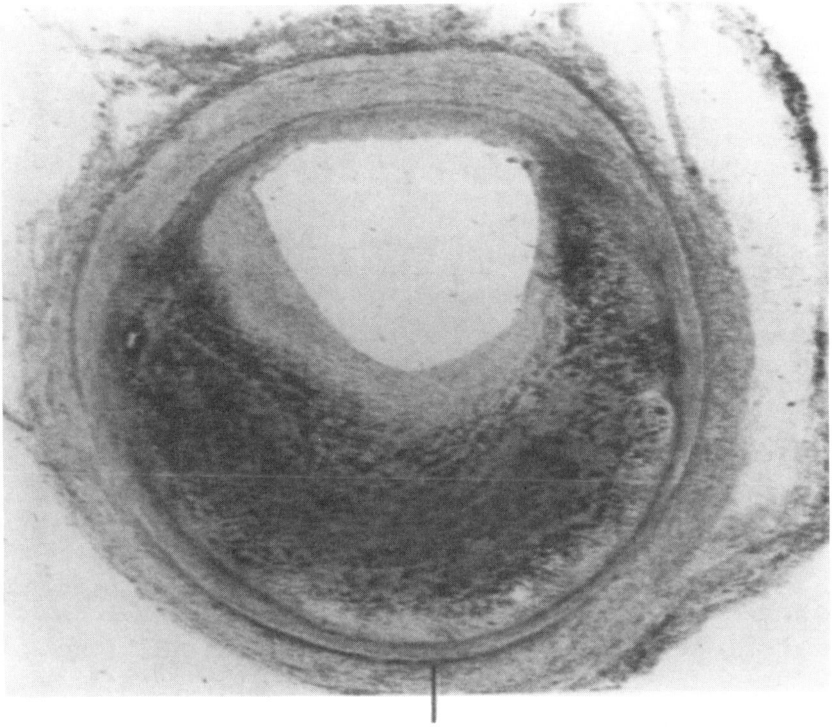

Figure 5. Advanced atheroma in a coronary artery with necrosis of the media (arrow). No thrombus and no plaque fissuring are seen. Orig. mag. 14x. From: Sandritter's Color Atlas and Textbook of Histopathology, 7th edition, 1984, page 78. With Permission.

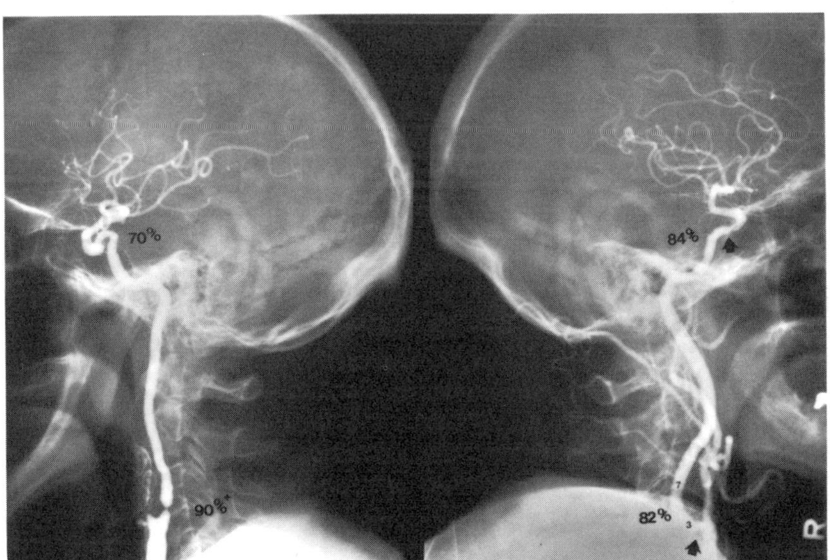

Figure 6. Lateral view arteriograms of LEFT and RIGHT common carotid arteries and portions of the intracranial middle cerebral arteries showing both extra- and intracranial atherosclerotic occlusive disease of both sides. Numbers indicate estimates of degree of lumen reduction.

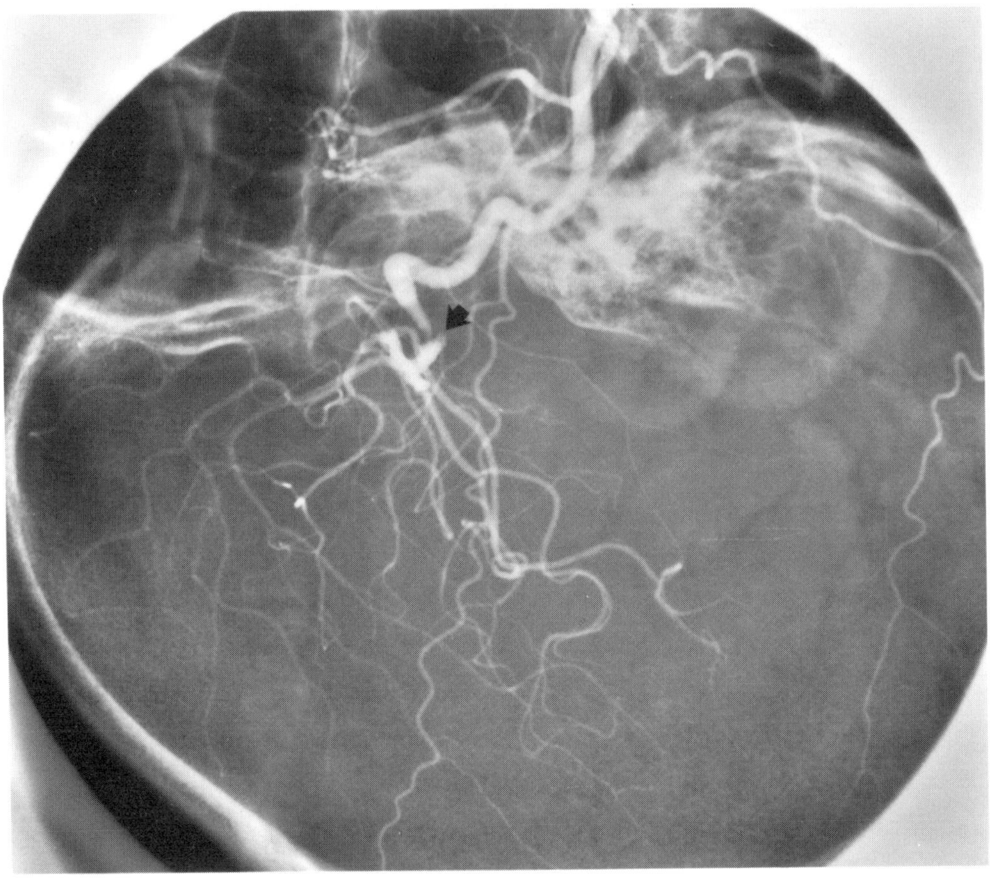

Figure 7. Lateral view arteriogram showing severe intracranial atherosclerotic occlusive disease (arrow) at the siphon of the internal carotid artery distribution. The stenosis is preocclusive.

nective tissue. The two most characteristic and most prominent findings are disruption of the media and elastic lamina and a foreign body giant cell reaction leading to granulomatous degeneration. The lesions are segmental, of a few millimeters to centimeters in length, and with entirely normal artery between them. Diagnosis is by biopsy of an involved segment. Tenderness and redness along the superficial temporal artery demands prompt action. The vessel may be tender to touch, pulsations may be diminished to absent, and the artery may be felt as a cord-like, non-compressible structure. During the active phase, an elevated erythrocyte sedimentation rate is characteristic. Loss of vision of the ipsilateral eye is due to thrombosis of the branches of the ophthalmic artery supplying the optic nerve. Recovery of vision almost never occurs. It is not unusual for cranial arteritis to run its entire course from onset of symptoms to unilateral blindness in less than 24 hours. There may be other signs of reduced external carotid artery blood flow such as pain in the muscles of mastication, alopecia and even partial necrosis of the tongue and scalp. Fortunately, temporal or cranial arteritis is a rare disease affecting not more than 25 persons per 100,000. Treatment consists of prompt and intensive administration of anti-inflammatory steroids.

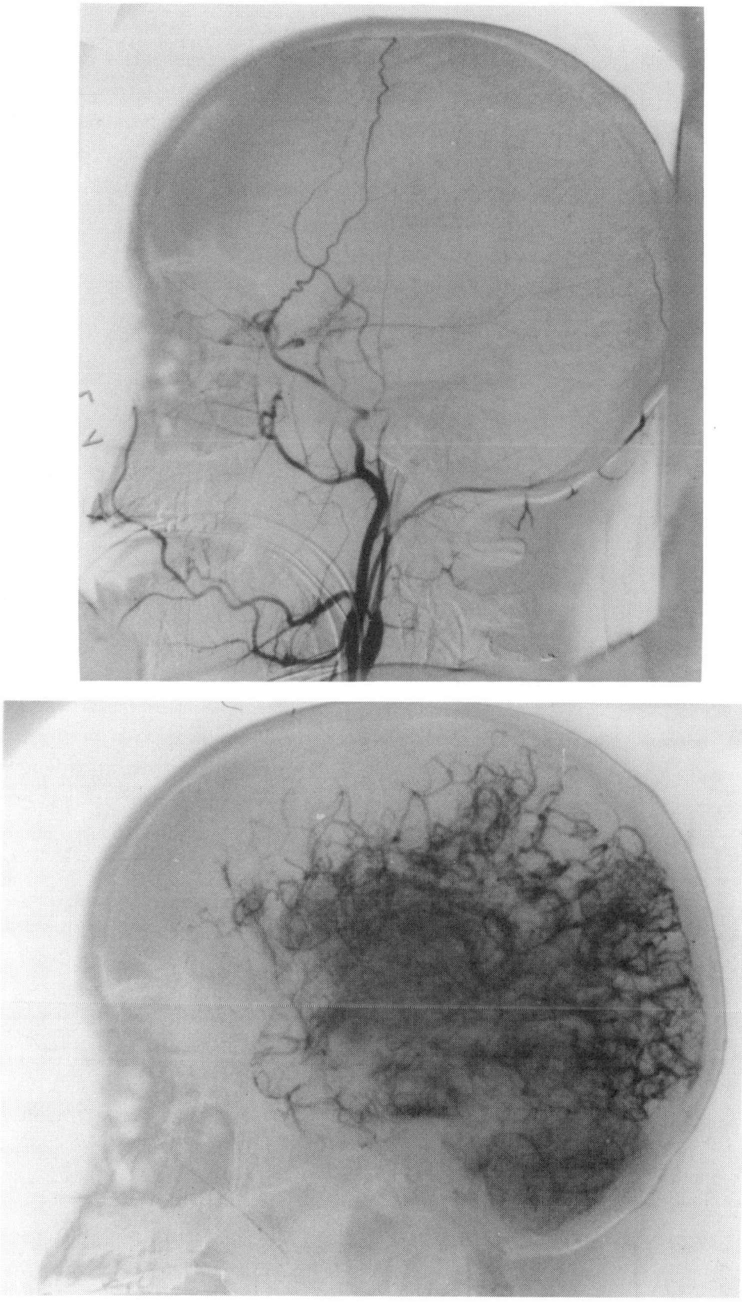

Figure 8. **A.** Arteriogram illustrating an example of Moyamoya Disease: Extracranial injection of cervical carotid artery showing branches of the external carotid artery distributed about the face and scalp but no filling of the intracranial distribution of the internal carotid artery. The same lack of internal carotid filling was present on the opposite side. **B.** Arteriogram illustrating an example of Moyamoya Disease: Lateral view of multiple, small, mostly unnamed intracranial vessels filling via pial collateral vessels from the lest posterior cerebral artery. The patient was a 32 year old woman treated with steroids for several weeks for systemic lupus erythematosus. She developed a right hemiparesis. Vascular involvement begins as a perivascular inflammatory process with small vessel occlusion, stenoses, and ruptures.

8. MOYAMOYA DISEASE

The name derives from the description given to it in 1966 by Suzuki who likened the angiographic appearance of the intracranial vessels to "a puff of cigarette smoke." These very fine unnamed vessels formed a net or cloud-like configuration. They were thought to arise in response to multiple areas of stenosis and occlusion in the distal intracranial branches of the internal carotid arteries resulting in both cerebral infarctions and areas of hemorrhage. The neurologic signs and symptoms are multiple depending on the area of the brain that is ischemic and include cortical blindness, motor aphasia, seizures and death. Histologic findings include concentric lamellar fibrous thickening of the intima, and a relatively normal media and adventitia. The arteriogram looks like a blush of vessels rather than a collection of discrete vessels. Etiology is unknown but evidence exists which permits both a congenital and an acquired origin. It is chronic. It does not affect large vessels, and it has no relationship to the extracranial cervical arteries. Variations of extra-intracranial vascular operations have been of some success in a limited number of highly selected patients.

9. POLYARTERITIS NODOSA

This is a systemic disease with central nervous system involvement in about 10 to 50% of cases. Subarachnoid and intracerebral hemorrhages with hemorrhagic infarction are the manifestation. Multifocal small hemorrhages may occur throughout one or both hemispheres. There is inflammation of the adventitia, polymorphonucleated and eosinophilic cell infiltration of the media, and a reactive cellular proliferation with thickening of the arterial wall and diminution of the lumen. Thrombus may eventually occlude the vessel. Aneurysmal dilatation and rupture due to necrosis of the vessel wall cause the intracranial hemorrhages. Polyarteritis displays a gender preference of men to women by approximately 4:1 with its onset in the 3rd to 5th decades. Myalgia, low grade fever and anemia characterize its onset. Renovasccular hypertension secondary to renal artery stenosis is frequent. Peripheral neuropathy creates a common second set of signs and symptoms. Rarely there may be stenoses and occlusions not only of the intracranial arteries but also of a major artery such as the carotid or basilar with dramatic and even lethal consequences. Remissions are frequent, occur in about half the cases, and may last for years. Diagnosis is established by biopsy of the involved area, as for example, brain, kidney, skeletal muscle, etc. Anti-inflammatory steroids provide at least temporary amelioration but are unable to prevent stenosis and occlusion of the involved vessel.

10. NON-INFECTIOUS GRANULOMATOUS CENTRAL NERVOUS SYSTEM ANGITIS

It is characterized by low grade fever, and elevated sedimentation rate in an age group centering on about the 6th decade. The cause is unknown, but an autoimmune reaction is suspect. The process involves the smaller cerebral arteries and veins with infiltration of the intima by lymphocytes, monocytes and occasional giant cells. The lumen undergoes gradual stenosis and small areas of cerebral infarction occur which are widely distributed throughout both hemispheres. The disease runs a slow, insidious and ultimately

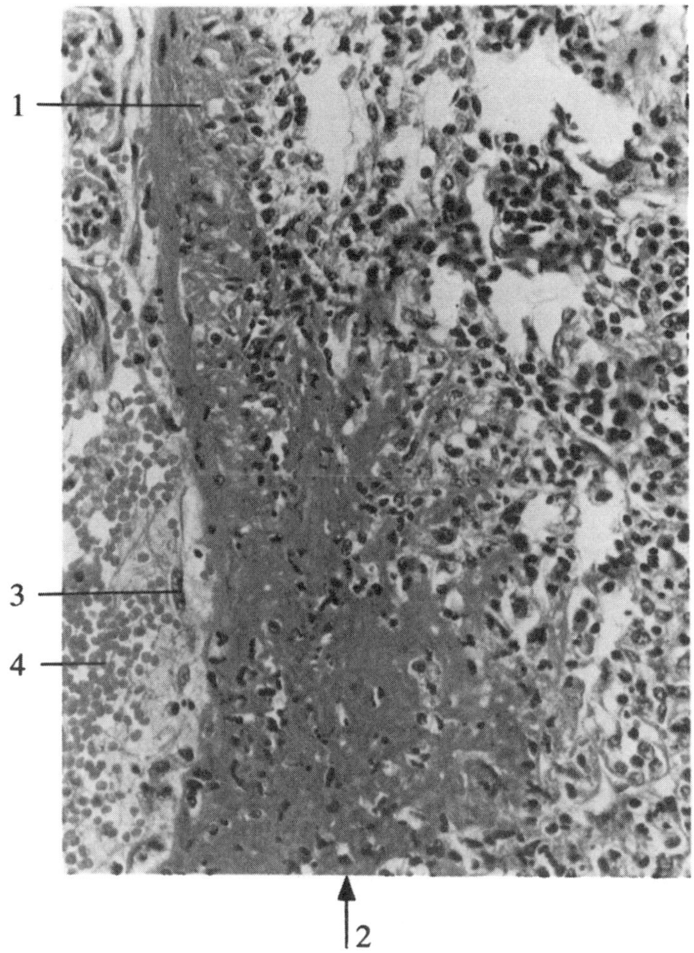

Figure 9. Early stage of fibrinoid necrosis 1) in a lesion of periarteritis nodosa (H & E stain) 2) the intima is lifted off the area of fibrinoid necrosis which is most apparent in segments of the intima and media 3) and endothelial cell, 4) a loose network of fibrin and erythrocytes on the endothelium. From: Sandritter's Color Atlas and Textbook of Histopathology, 7th edition, 1984, page 80. With Permission.

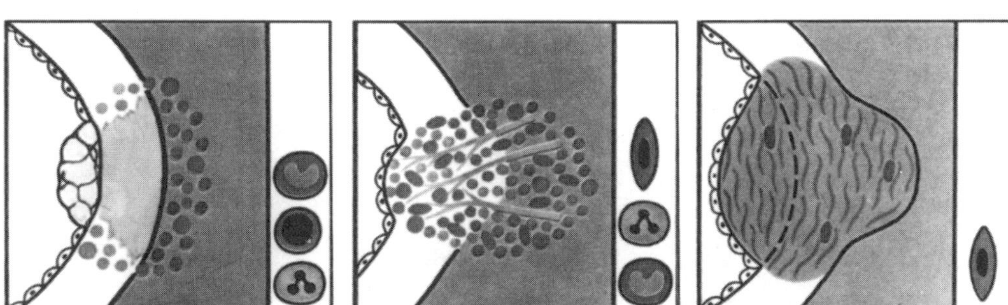

Figure 10. The stages of periarteritis nodosa (polyarteritis nodosa) consist of necrosis of all layers of the wall and especially the internal elastic membrane, cellular granulation tissue formation and fibroblast invasion and proliferation leading to scar formation and reduction or obliteratin of the lumen. From: Sandritter's Color Atlas and Textbook of Histopathology, 7th edition, 1984, page 81. With Permission.

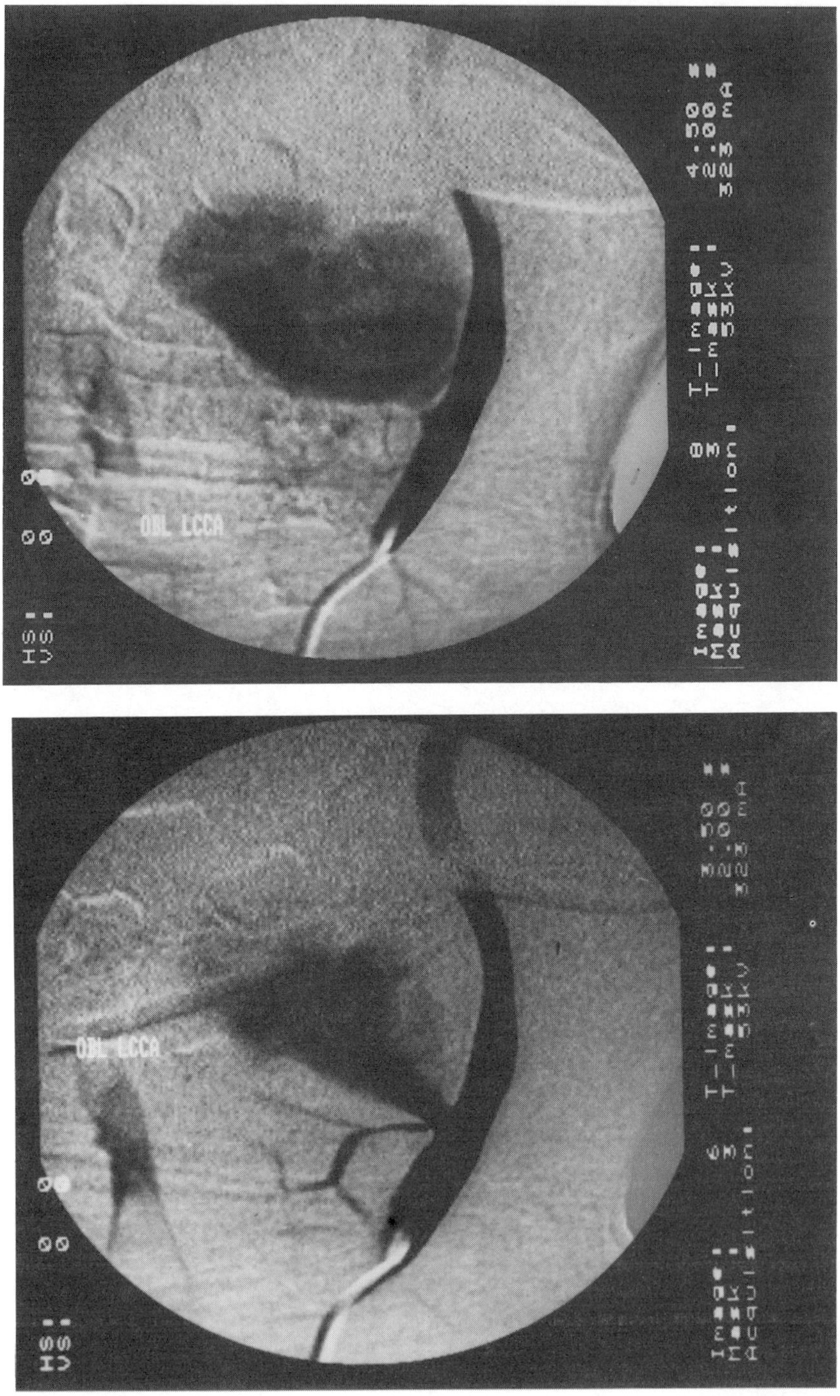

Figure 11. A. Arteriogram of mycotic aneurysm of left common carotid artery: Injection of a radioopaque dye reveals the common carotid artery and a large mass which is the aneurysm. **B.** Arteriogram of mycotic aneurysm of left common carotid artery: Delayed film reveals ejection of the dye from a perforation of the aneurysmal sac. Penetrating trauma to the neck accounts for most aneurysms in the cervical region.

fatal course for up to two years consisting of headache, seizures, stupor, coma and mental deterioration. Arteriography reveals areas of segmental stenosis alternating with areas of dilatation and occlusions of the small cortical arteries. Brain biopsy may be necessary. Steroids, including the anti-metabolites have been recommended.

11. WEGENER'S GRANULOMATOSIS

This is a polyarteritis, but unlike the classical type, it preferentially involves the central nervous system. It is a generalized necrotizing vasculitis and is not easily distinguished on histologic examination from polyarteritis nodosa. Small arteries and veins are primarily affected. Respiratory symptoms such as chronic sinusitis and infections of the broncho-pulmonary tree are common and are believed to be secondary to arteritis in this organ system. Respiratory symptomatology usually precedes involvement of the central nervous system which consists of a polyneuropathy, a consequence of extension of granulomas from the nasal sinuses to the base of the skull or to infarction secondary to venous thrombosis, also a complication of chronic infection. Intracerebral and subarachnoid hemorrhages follow as a consequence of the arteritis. Cyclophosphamide and prednisone have been offered as therapy.

REFERENCES

1. Callow, A.D. and Ernst, C.B. Eds. *Vascular Surgery: Theory and Practice*, Stamford, CT: Appleton & Iaange, 1995.
2. Callow, A.D. *Surgery of the Carotid and Vertebral Arteries for the Prevention of Stroke*, Baltimore, MD: Williams & Wilkins, 1996.
3. Kissel, J.T. Neurologic manifestations of vasculitis. In: *Neurologic Manifestations of System Disease. Neurologic Clinics*, 7:655–673, 1989.
4. Launes, J., Livanainen, M., Erkininjuntti, T. et al. Isolated angiitis of the central nervous system. *Acta. Neurol. Scand.*, 74:108, 1986.
5. Shimizu, K. and Sano, K. Pulseless disease. *J. Neuropathol. Clin. Neurol.*, 1:37–47, 1951.
6. Ishikawa, K. and Yonekawa, Y. Regression of carotid stenosis after corticosteroid therapy in occlusive thromboaortopathy (Takayasu's disease). *Stroke*, 18:677–679, 1987.
7. Shelhamer, J.H., Volkman, D.J., and Parrillo, J.E. Takayasu's arteritis and its therapy. *Ann. Int. Med.*, 103: 121–126, 1985.
8. Yamamoto, S., Ogawa, S., Kitano, T., et al. Complete evaluation of the cardiovascular lesions in 24 patients with Takayasu's aortitis using four-image, intravenous digital subtraction angiography. *Am. Heart J.*, 114:1426–1431, 1987.
9. Derby, B.M. and Cravioto, H.M. Cerebral arteritis. In: *Cerebrovascular Surgery*, Vol. II, Fein, J.M. and Flamm, E.S. eds. New York: Springer Verlag, 1985.
10. Bengtsson, B.A. Incidence of giant cell arteritis. *Acta. Med. Scand.* (Suppl), 658:15–17, 1987.
11. Caselli, R.J., Hunder, G.O., and Whisnant, J.R. Neurologic disease in biopsy-proven giant cell temporal arteritis. *Neurol.*, 38:352–359, 1988.
12. Suzuki, J. and Takaku, A. Cerebrovascular "moyamoya" disease: Disease showing abnormal net-like vessels in the base of the brain. *Arch. Neurol.*, 20:288–289, 1969.
13. Yonekawa, M.D., Handa, H., and Okuno, T. Moyamoya disease: diagnosis, treatment and recent achievement. In: *Stroke: Pathophysiology, Diagnosis and Management*, Vol. 2, Barnett, H., Stein, B., Mohr, J., and Yatsu, E. eds. New York: Churchill Livingstone, 1986.
14. Yamashita, M., Tanaka, K., Matsouo, T., et al. Cerebral dissecting aneurysms in patients with moyamoya disease. *J. Neurosurg.*, 58:120–125, 1983.
15. Havlik, R.J., Fried, I., Chyatte, D., and Modlin, I.M. Encephalo-mental synangiosis in the management of moyamoya disease. *J. Neurosurg.*, 49:679–688, 1978.

16. Olds, M.V., Griebel, R.W., Hoffman, H.J., Craven, M., et al. The surgical treatment of childhood moyamoya disease. *J. Neurosurg.*, 66:675–680, 1987.
17. Matsushima, Y., Fukai, N., Tanaka, K., et al. A new surgical treatment of moyamoya disease in children: a preliminary report. *Surg. Neurol.*, 15:313–320, 1986.

12

GENETIC FACTORS IN ATHEROSCLEROSIS

From Humans to Mice and Back Again

Hooman Allayee,[1] Brad Aouizerat,[1] Richard Davis,[1] Thomas A. Drake,[2] Jennifer Gu[1,2] Aldons J. Lusis,[1] Dietrich Machleder,[1] Shahab Mehdizadeh,[1] Margarete Mehrabian,[1] Jian-Hua Qiao,[1] Karen Rooke,[1] and Carrie Welch[1]

[1]Department of Medicine and
Department of Microbiology and Molecular Genetics
[2]Department of Pathology and Laboratory Medicine
University of California
Los Angeles, California

1. INTRODUCTION

Atherosclerosis, the cause of coronary artery disease (CAD) and stroke, is in reality not one disease but many different diseases, each with a similar endpoint (vascular lesions) but differing in genetic and environmental contributions. In genetic terms, atherosclerosis is very "heterogeneous." Within particular populations, genetic differences appear to be of primary importance in determining who will develop atherosclerosis, and its consequent clinical sequelae, and who will not. Between populations, particularly between those populations consuming different levels of dietary fat, environmental influences predominate.

There are three important reasons for studying the genetic factors contributing to atherosclerosis. The first, and most important, is that knowledge of genetic factors can help pinpoint important processes and pathways involved in the disease. Such knowledge is likely to help in the development of new therapies. The most outstanding example of this is the set of studies of familial hypercholesterolemia (FH) by Michael Brown, Joseph Goldstein and colleagues (Goldstein et al., 1995). Although FH is not a common cause of atherosclerosis (frequency about one in 500 in Western populations), these studies have provided many lessons pertaining to common forms of the disease. In fact, the knowledge of cholesterol homeostasis derived from this work has led, in part, to the development of important therapies, such as statins and bile acid chelators, which are used for treatment of

Vascular Endothelium, edited by Catravas *et al.*
Plenum Press, New York, 1998

Table 1. Naturally occurring variations in mice relevant to atherosclerosis

• Atherosclerotic lesions
• Calcification
• Plasma lipoprotein levels
• Dietary responsiveness
• Arterial lipofuscin
• Type 1 diabetes
• Type 2 diabetes and obesity
• Blood pressure
• Inflammation
• Autoimmune disease
• Bile acid metabolism

many forms of atherosclerosis (Brown and Goldstein, 1991). The second reason for studying genetic factors in atherosclerosis is that such studies promise to provide new, more specific diagnostic tests for the identification of individuals at risk for the disease. With few exceptions, such as analysis of apolipoprotein E (apoE) alleles for the identification of certain hyperlipidemias, the promise of genetic-based tests for CAD has yet to be realized. Undoubtedly, as the genes contributing to risk for the disease are discovered, new tests allowing risk assessment or disease subclassification for optimal therapeutic treatment will be developed. Most likely, such tests will be used in combination with traditional risk assessment measurements, such as blood lipids and blood pressure. Thirdly, studies of the genetic factors contributing to the disease have the potential to reveal new basic information about biological processes. For example, studies of the low density lipoprotein (LDL) receptor in relation to the development of FH led to the identification of the receptor-mediated endocytosis pathway.

A number of Mendelian factors contributing to the development of atherosclerosis, including deficiencies of proteins involved in lipid metabolism, homocysteine metabolism or blood pressure regulation, have been characterized (Lifton, 1996; Keating and Sanguinetti, 1996) (chromosomal locations shown in Figure 1). In these cases, the disorder is clearly associated with the inheritance of a variant allele at a single genetic locus. But such Mendelian disorders explain only a small fraction of disease incidence (less than ten percent). The vast majority of CAD cases result from the interactions of multiple genetic variations, none of which is able to cause disease by itself. As yet, we have little understanding of such "complex" or "multifactorial" forms of atherosclerosis. Many association studies and linkage studies of candidate genes for CAD have been reported, but most such studies have, for the most part, only been able to provide suggestive evidence, not mechanistic details or novel genes (Figure 1). The identification of novel genes for complex forms of the disease remains a challenge for the future.

An alternative to direct human studies of atherosclerosis are genetic studies in animal models, and the mouse is by far the most useful mammal for such studies (Lusis, 1993; Breslow, 1996). Many different inbred strains of mice have been developed which exhibit genetic variations in numerous traits relevant to atherosclerosis (Table 1), and studies in mice provide important advantages for the mapping, cloning, and functional analysis of genes involved in complex traits (Figure 2). In this chapter, we discuss how studies of the mouse animal model can usefully complement human studies directed at understanding genetic factors in atherosclerosis.

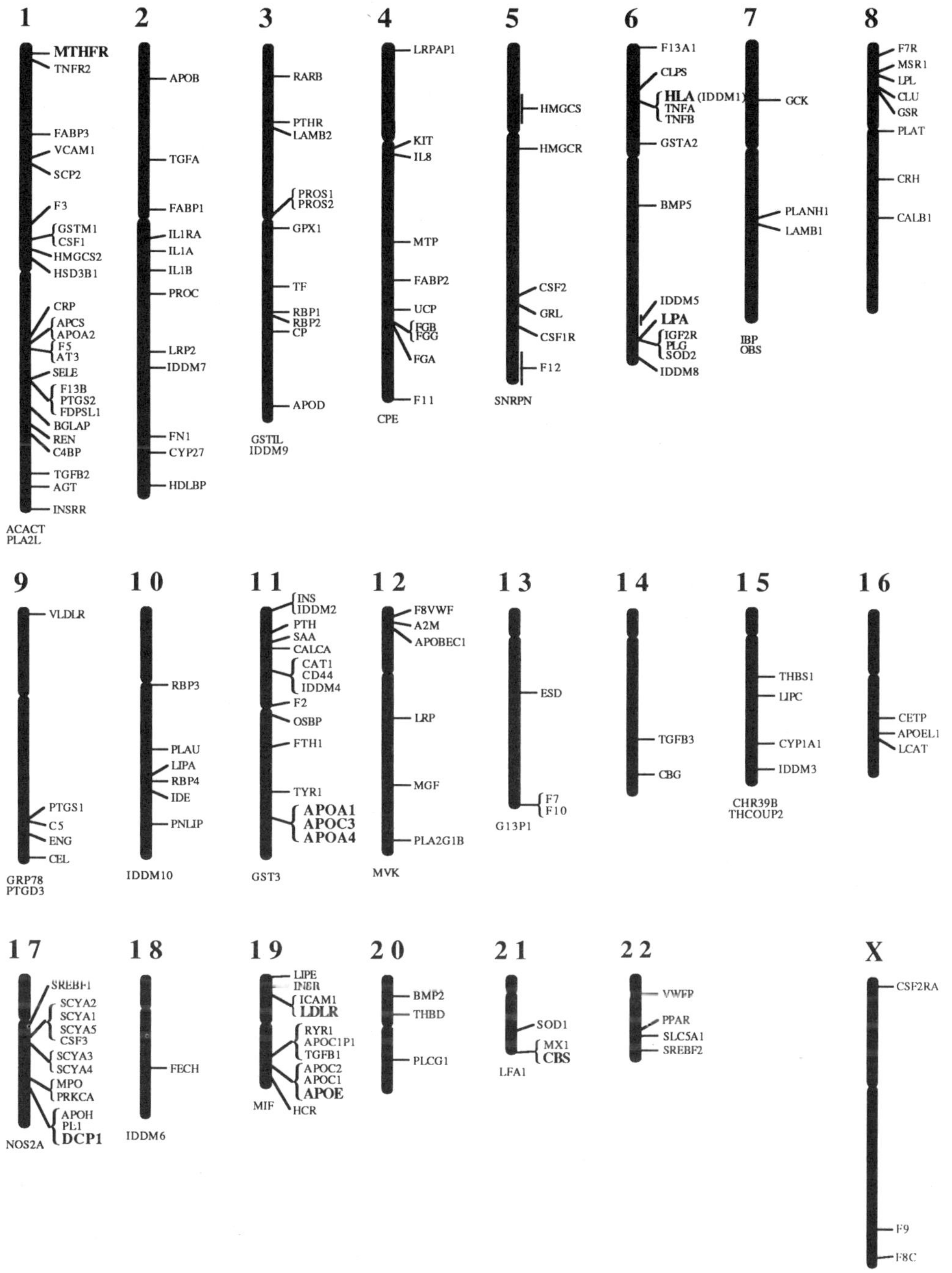

Figure 1. Human chromosomal map of candidate genes for atherosclerosis. Candidate genes have been placed along each chromosome according to their composite locations taken from the Genetic Location Data Base (LDB) and the Human Genome Data Base (Anonymous, 1996; Collins et al. 1992). The relative positions of the genes placed within brackets is not certain. Candidate genes whose regional locations are not known are listed below their corresponding chromosomes. Certain candidate genes that have been clearly shown to be associated with an increased risk in atherosclerosis or a trait relevant to the disease are indicated in bold print. Chromosomes are drawn to scale based on the total length taken from LDB (Anonymous, 1996).

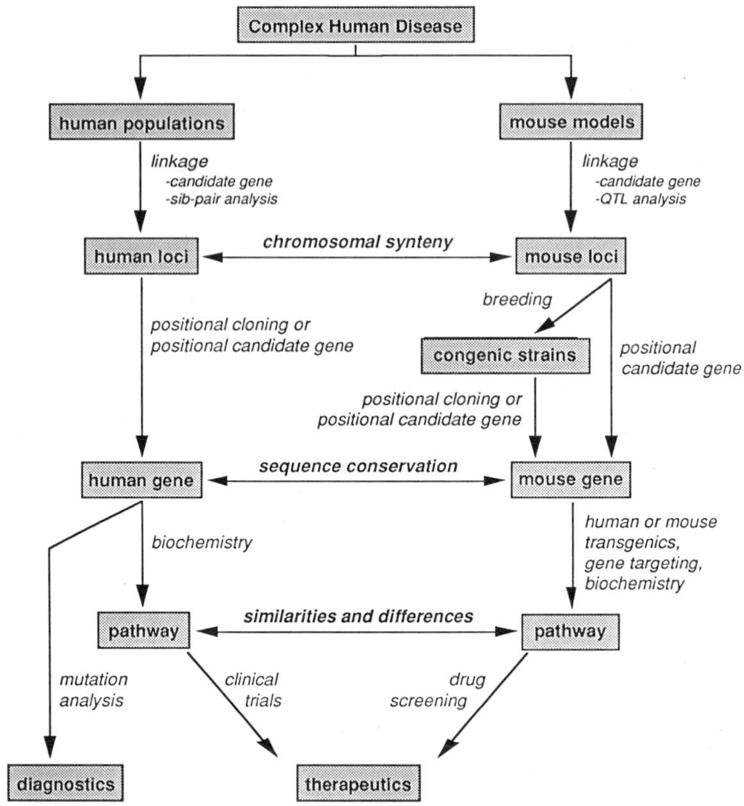

Figure 2. Schematic diagram showing how mouse models can accelerate the identification of genes contributing to complex genetic traits. The left side of the diagram indicates the steps involved in cloning a gene for a complex trait directly in humans, and the right side indicates the steps used in the mouse model. The searches in humans and mice are bridged by conserved chromosomal syntenies (allowing prediction of the chromosomal location of a gene in one species based on its location in the other species) and sequence conservation (allowing the identification of homologous genes by crosshybridization). A "congenic" strain contains a small chromosomal region derived from one strain that has been transferred onto the genetic background of a second strain by breeding. The use of congenic strains allows the isolation of individual genes contributing to complex traits. Adapted from Shih et al., 1995.

2. STUDYING GENETIC FACTORS IN COMPLEX DISEASES SUCH AS ATHEROSCLEROSIS IS DIFFICULT

Figure 2 shows a flow chart for the analysis of complex genetic traits in humans and rodent animal models. There are two approaches for identifying genes contributing to atherosclerosis directly in humans. The first is the candidate gene approach, involving either association or linkage analyses using known genes involved in pathways relevant to atherosclerosis. The second is a search of random parts of the genome to detect linkage to anonymous genes contributing to atherosclerosis or its risk factors. These days, such searches usually involve whole genome scans using polymorphic markers. Once a locus contributing to an atherosclerosis-related trait is identified, genes in the region are examined for possible involvement. In the case of Mendelian disorders, such a "positional cloning" strategy is straightforward (although laborious) given enough informative meioses.

But for complex diseases, fine mapping (to within a centiMorgan (cM) or less) of the gene requires extremely large numbers of individuals.

A fundamental problem with human studies is that genetic heterogeneity and environmental influences make it very difficult to detect the effects of moderate genetic influences, such as those likely to be involved in most forms of atherosclerosis. In addition, it can be difficult to extend such studies to the molecular level. In the case of Mendelian disease, mutations with obvious consequences for gene expression (e.g. splice site or termination mutations) are usually involved, whereas with common genetic variations, the genetic changes are likely to be subtle and difficult to distinguish from silent polymorphisms. Given such difficulties, it is probably not surprising that many cardiovascular researchers view such genetic studies with considerable skepticism. "I have resolved never to touch an association study; I want clean results" is one complaint we have heard. Are such studies without merit? We do not think so, since they provide directions for further research and thinking. The problems inherent to studies of common genetic variations of complex triats are very large and they will not be solved overnight or by purely biochemical approaches.

3. HUMAN CANDIDATE GENE STUDIES HAVE BEEN QUALIFIED SUCCESSES

Epidemiological studies have revealed a number of risk factors for atherosclerosis, some such lipoprotein levels, diabetes, and blood pressure having a clear genetic component. Further, biochemical studies have identified a number of pathways and genes involved in some of these risk factors (e.g. the renin-angiotensin pathway for control of blood pressure). At the same time, cell biology and pathology studies, in humans and animal models, have revealed some of the likely pathways contributing to the cellular and molecular vascular interactions in the disease process. Together, the genes identified in such studies can be considered "candidate genes" for atherosclerosis (Lusis, 1987); that is, variations in these genes could conceivably contribute to the genetic component of atherosclerosis. Figure 1 shows the chromosomal locations of some of these candidate genes (the list is rapidly expanding, and what constitutes a valid candidate gene is not always clear). Many of these genes have now been examined in population association or in family linkage studies for an effect on cardiovascular disease or, more often, a relevant risk factor. The studies have produced a number of striking findings. For example, common variations of the apoE gene (APOE, Chr. 19) explain about 10% of the genetic variance in plasma cholesterol levels in Western populations (as well as being associated with type III hyperlipidemia, late onset Alzheimer's disease and longevity). Another example is the apo(a) gene (LPA, Chr. 6) which determines the levels of the atherogenic Lp(a) lipoprotein. Other candidate genes have also proved to be the cause of a number of rare Mendelian disorders influencing atherosclerosis, such as depressed high density lipoprotein (HDL) levels (apoAI deficiency; APOAI, Chr. 11) and very high homocysteine levels (cystathione β-synthetase; CBS; Chr. 21). But, frequently the association between the candidate gene and the disorder have not been entirely definitive. For example, a number of studies have found a significant population association between an insertion-deletion polymorphism of the angiotensinogen converting enzyme gene (DCP1, Chr. 17), but several other studies have failed to reveal an association (Lifton, 1996). Many other conclusions of candidate gene studies have flat-out failed the test of replication, in most cases probably because the criteria for significance were insufficiently stringent. Often, there has been a failure to compensate for the overestimation of significance that arises when multiple comparisons are performed.

4. MOST GENES FOR ATHEROSCLEROSIS ARE STILL UNKNOWN

Our present list of candidate genes probably includes only a small fraction of the genes that contribute to common forms of atherosclerosis. There are about 100,000 genes in a mammalian genome, but we understand the functions of only a small fraction of these. Studies in the mouse, discussed below, have shown that even for a relatively well-understood process such as lipoprotein metabolism, most genetic variations are *not* due to the familiar lipid transport genes (apolipoproteins. lipoprotein receptors, etc.).

It seems likely that many genetic variations involved in atherosclerosis may be regulatory in nature. For example, studies of FH have revealed the importance of the LDL receptor in determining levels of LDL in the circulation. Yet, structural gene mutations of the LDL receptor explain only a small fraction of LDL level variations. Could genes which control the expression of the LDL receptor be responsible for much of the interindividual variation in LDL levels in the human population? Studies from Goldstein, Brown, and colleagues have resulted in the identification of two important transcription factors, sterol regulatory element binding proteins -1 and -2 (SREBF-1, Chr. 17 and SREBF-2, Chr. 22) which control LDL receptor transcription in response to intracellular cholesterol levels (Wang et al., 1994; Hua et al., 1995). These are now very attractive candidates for the regulation of plasma cholesterol levels. But, of course, the expression of SREBPs is determined by other genes. For example, a protease is required to cleave the membrane bound SREBP proteins, releasing the mature SREBP amino terminal fragment and allowing it to migrate to the nucleus to activate gene transcription. A variety of additional proteins are likely to be involved in the SREBP pathway, including factors that "sense" intracellular cholesterol levels. Finally, the mature SREBP peptide almost certainly interacts with other nuclear factors. Thus, LDL receptor realization is likely to be determined by a network of regulatory proteins, and genetic variations in the genes encoding these proteins may contribute to plasma cholesterol levels.

Can we realistically expect to untangle rather subtle genetic variations affecting the many different networks which contribute to atherosclerotic susceptibility in the human population? Although it is clearly feasible to map genes contributing to complex genetic traits (Lander and Schork, 1994), it has not yet been feasible to determine the locations of these genes with sufficient accuracy to allow positional cloning (Kruglyak and Lander, 1995). In contrast to Mendelian traits, it is not possible to unambiguously trace the inheritance of a susceptibility gene for a complex trait through families, thus, the mapping must rely on statistics rather than clear-cut chromosomal crossovers. Researchers in the field are optimistic about the possibility that linkage disequilibrium will solve the problem in some cases, but several issues are still unclear.

Only a small number of whole genome scanning studies of complex traits have been performed thus far, the most advanced being studies of type 1 diabetes. These scans involved typing of microsatellite markers at intervals of about 10 cM throughout the 3500 cM human genome followed by sib-pair analysis (Davies et al. 1994). These studies resulted in the identification of a number of putative loci contributing to the disease, with the MHC complex on chromosome 6 explaining about one-third of the genetic contribution (Figure 1). The chromosomal regions identified are far too large (several cM) to allow a positional cloning strategy at present, although they are useful in directing candidate gene searches. Efforts are underway to reduce the size of the regions by typing additional individuals and by making use of linkage disequilibrium (Copeman et al. 1995).

5. THE MOUSE IS THE MOST USEFUL MAMMALIAN MODEL FOR GENETIC STUDIES

There are several important differences between the dissection of complex traits in mice as compared to humans (Figure 2). First, mapping is much more straightforward in the mouse; crosses can be planned, large numbers of progeny can be produced, and environment and age can be controlled. Because of conserved "chromosomal synteny," the corresponding region of the human genome can be tested for genetic factors influencing the trait. Second, and perhaps most important, it is easier to go from a locus to a gene in the mouse. As in humans, candidate genes can be tested. But, in contrast to human studies, if none of the candidates are found to underlie the trait, it is still possible to isolate the chromosomal region on a common genetic background in mice by constructing a "congenic strain." This, in turn, makes possible biochemical and genetic studies, including positional cloning. Once the mouse gene has been identified, the human homolog can be easily obtained by DNA-hybridization, since most gene coding sequences are highly conserved between species. Finally, the mouse offers important advantages for functional studies. Over the last ten years, transgenic mice and knockout mice have become essential tools for examining the role of specific genes in lipoprotein metabolism, hypertension, diabetes and obesity, coagulation, vascular cell interactions and inflammation.

The use of rodents for the identification of genetic factors in atherosclerosis is only possible, of course, if strains of mice (or rats) differ in traits relevant to atherosclerosis. As shown in Table 1, rodents differ in many of the traits important in atherosclerosis. It should be mentioned that rats offer some advantages as compared to mice for analysis of certain traits. For example, insulin resistance and hypertension are more easily studied in larger animals and have been more intensively investigated in rats. Furthermore, transgenic rats can be created routinely, although knockout rats have not been produced as yet. Genetic studies are difficult in larger mammals, such as pigs and rabbits, because of the longer generation times and the much greater maintenance expenses.

6. MAPPING COMPLEX TRAITS IN THE MOUSE

As previously discussed, inbred strains of mice differ in numerous traits relevant to heart disease (Table 1). Some of these trait differences are due to major genes which can be dissected in a straight forward manner, but the vast majority are complex (like the differences between people). Thus, when two inbred strains of mice differing in a complex trait of interest (for example, cholesterol levels) are bred, the backcross or F2 mice will exhibit a continuum of values for the trait (from low to high cholesterol levels with no discrete groupings). This results from the fact that the progeny animals will exhibit many different combinations of the parental chromosomes. Nevertheless, because the mice can all be maintained under similar environmental conditions, and because they all differ for the same *set* of genetic factors, the chromosomal regions responsible can be defined much more easily than could those for the comparable human traits. Since the locations of the chromosomal regions harboring the genes of interest are unknown, genetic markers (such as highly polymorphic simple repeat sequences) must be typed along all parts of the genome (usually at intervals of about 10 cM in the 1500 cM mouse genome). Figure 3 shows the locations of chromosomal regions harboring genes for various traits related to cardiovascular disease.

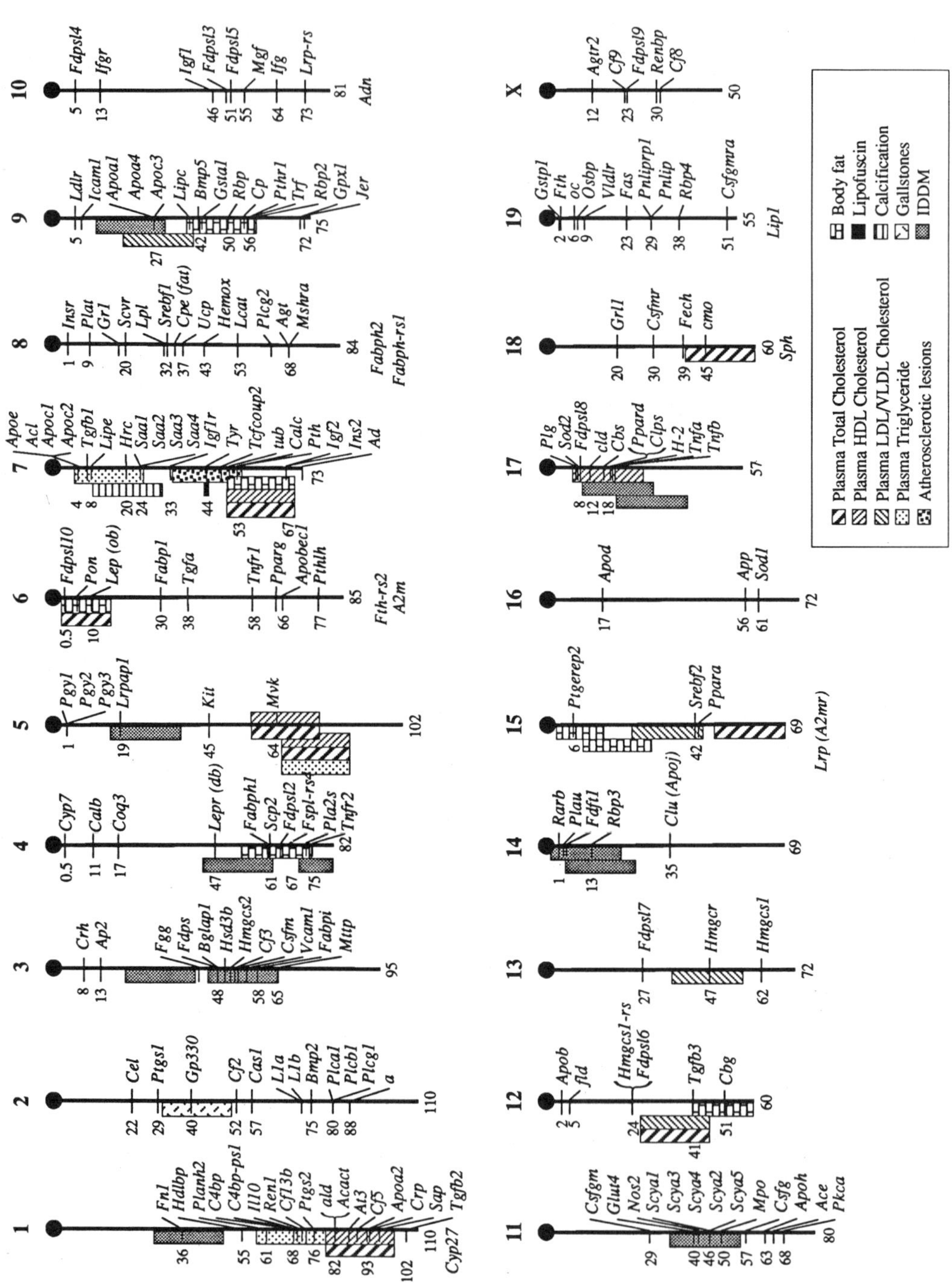
Plasma Total Cholesterol
Plasma HDL Cholesterol
Plasma LDL/VLDL Cholesterol
Plasma Triglyceride
Atherosclerotic lesions
Body fat
Lipofuscin
Calcification
Gallstones
IDDM

Figure 3. Mouse chromosomal map of quantitative trait loci (QTLs) and candidate genes for atherosclerosis-related traits. QTLs, represented by shaded rectangles, are indicated as 20 cM regions centered around the locations of peak LOD scores. With the exceptions of the loci for gallstones and insulin-dependent diabetes mellitus (IDDM), all of the QTLs are supported by LOD scores > 3. (A LOD score is a statistical measure of whether a gene contributing to a given trait is linked to a particular genetic marker. A LOD score > 3 is generally considered to be strong evidence of linkage. However, the precise threshold depends on various factors including the nature of the experimental cross, the number of markers typed, and the number of loci contributing to the trait). The candidate genes are indicated to the right of the chromosomes; recombination distances (cM) relative to the centromere are shown to the left. Brackets indicate uncertain gene order. Genes assigned to chromosomes but not regionally mapped are listed below the corresponding chromosomes. Data from a variety of mouse crosses were combined with respect to anchor loci to form this composite map. The order of loci between anchors can only be inferred on the basis of distances and, thus, the order of loci and QTL positions on this map should be considered provisional. Chromosomes are drawn to scale, based on total recombinational lengths taken from the Chromosome Committee composite maps of the mouse (Silver et al., 1994). The QTLs are described in Qiao et al. 1993; Purcell-Huynh et al., 1995; Shih et al., 1995; Warden et al., 1995; Ivandic et al., 1996. Gene symbols and locations are given in Anonymous, 1995 and Silver et al., 1994.

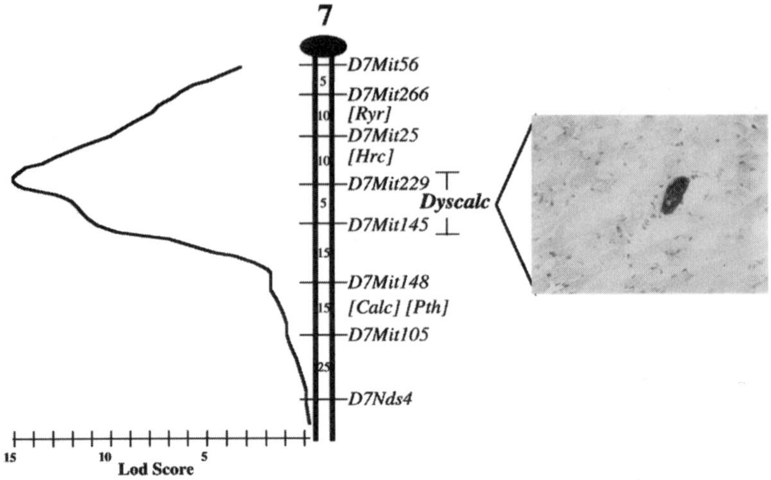

Figure 4. Chromosomal localization of a locus contributing to myocyte calcification *(Dyscalc)* on mouse Chr. 7 using a (C57BL/6J × C3H/HeJ)F2 intercross. The chromosome is drawn with the centromere at the top. Microsatellite marker and gene symbols are given to the right of the chromosome, and the genetic distance (in cM) between each marker or gene is indicated between the bars of the chromosome. A plot of the Lod scores for linkage, calculated at two cM intervals along the chromosome, is shown on the left. The peak Lod score indicates the most likely position of the gene underlying the trait; the 95% confidence interval for the position of *Dyscalc* is indicated on the right. Candidate genes for *Dyscalc,* previously mapped to the same region of Chr. 7, are indicated in brackets.

As an example of mapping complex traits, let us consider the *Dyscalc* gene locus on Chr. 7 causing myocyte calcification. Most advanced atherosclerotic lesions in humans exhibit significant calcification, and some evidence suggests that this may contribute to clinical events. Qiao and collaborators (1994) showed that inbred strains of mice exhibit various forms of calcification (including calcification of coronary arteries, aorta, and myocytes) and that calcification exhibits a clear genetic component. Because calcification is difficult to quantitate in humans, is highly age-dependent, and is at least partly dependent upon lesion development, evidence for a genetic component in arterial calcification in humans is still not available. Genetic crosses revealed that the patterns of calcification in mice are complex, being under the influence of multiple genes as well as environmental factors. Ivandic and colleagues (1996) recently mapped a major gene in mice for a form of myocyte calcification called dystrophic cardiac calcinosis (*Dyscalc*, Chr. 7). The gene was identified in a genetic cross between inbred strains C3H (susceptible) and C57BL/6 (resistant). The Fl heterozygotes were all resistant, indicative of a recessive pattern of inheritance (Figure 4). Each of the F2 animals was typed for about 150 genetic markers spanning the entire mouse genome, and the segregation patterns of the markers were compared with the segregation pattern of the calcinosis. A genetic marker on mouse Chromosome 7 (*D7Mit229*) was found to be strikingly associated with the calcinosis. Thus, mice which had inherited both alleles of the marker from strain C57BL/6 and mice which were heterozygous for the marker tended to be resistant (22/143 animals with calcinosis) whereas those which inherited both alleles from strain C3H tended to be susceptible (26/36 animals with calcinosis). When the data were analyzed using a statistical program which determines the significance of the data at intervals between markers, the plot shown in Figure 5 was obtained. The vertical-axis corresponds to the locations of various markers on Chr. 7, with distances shown in cM, and the horizontal-axis corresponds to a "lod

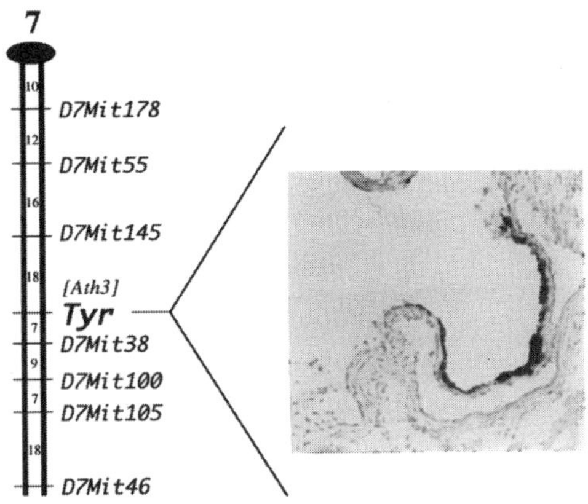

Figure 5. Chromosomal localization of a gene contributing to cardiac lipofuscin on mouse Chr. 7 using a (B6.C-H25^c × C)F2 intercross. The chromosome is drawn with the centromere at the top. Microsatellite marker and genes symbols are given to the right of the chromosome, and the genetic distance (in cM) between each marker or gene is indicated between the bars of the chromosome. The tyrosinase gene *(Tyr)* was identified as the underlying genetic determinant of the trait; mutations of *Tyr* which result in the production of an inactive protein prevent the deposition of lipofuscin granules in cardiac tissues of albino mice. Lipofuscin deposition (black) in the left ventricular side of the aortic valves in a susceptible strain is shown. A locus contributing to atherosclerotic lesion formation, *Ath3* (indicated by a shaded bar), was previously mapped to the same region of Chr. 7.

score" (the logarithm base ten of the odds ratio). A lod score is a measure of the likelihood that the data are explained by a gene in the region rather than by chance, and the peak lod score corresponds to the most likely location of the gene.

7. THE POWER OF A "POSITIONAL CANDIDATE GENE" APPROACH

Once the chromosomal regions controlling the trait of interest have been mapped, the list of possible candidate genes for the trait has been greatly reduced. Thus, biochemical and molecular studies can be focused on those genes which coincide with the identified chromosomal region. This approach, called the "positional candidate gene approach," is proving useful in studies with humans as well as mice (Collins, 1995).

The approach is nicely illustrated by the studies of Qiao, Welch, and colleagues (1993) on arterial lipofuscin deposition. Lipofuscin is a yellowish-brown pigment consisting of oxidized lipids and protein, which frequently accumulates in macrophages in atherosclerotic lesions in humans. Its potential role in atherosclerosis has been unknown. Mehrabian et al., (1991) identified lipofuscin in heart valves and arteries in mice and showed that inbred strains of mice differ in the deposition of lipofuscin. Qiao et al., (1993) analyzed a genetic cross between a strain which develops lipofuscin (B6.C-H25^c) and a strain which is resistant (BALB/c), and observed that pigment deposition segregated as a recessive trait mapping to mouse Chromosome 7 (Figure 5). The region identified contained the gene for tyrosinase (*Tyr*. Chr. 7), which catalyzes the first step in melanin synthesis. The tyrosinase gene at first seemed like a rather poor candidate, since lipofuscin is clearly distinct from melanin and tyrosinase is not known to be expressed in macrophages or the artery wall. But it was a rather straightforward candidate gene to examine, since many mutations of tyrosinase can be identified by virtue of their easily recognized consequence (albinism). When Qiao et al., (1993) studied a spontaneous point mutation of the tyrosinase gene on a susceptible genetic background, they observed an absence of lipofuscin, demonstrating a direct link between tyrosinase gene expression and lipofuscin. The nature of the link is as yet unclear, but it may be related to the pro-oxidant effects of tyrosinase. This study illustrates how defining chromosomal regions contributing to traits can help identify an underlying gene. In the absence of the knowledge of the chromosomal location of the lipofuscin gene, many more attractive candidate genes than tyrosinase would have been considered, and the testing of each candidate gene for a possible role would not have been feasible.

8. INDIVIDUAL LOCI FOR COMPLEX TRAITS CAN BE ISOLATED USING CONGENIC STRAINS

Although most of the 100,000 or so mammalian genes have now been partially sequenced, only a small fraction have been functionally characterized. Thus, the chances of finding the gene underlying a complex trait from a list of candidate genes in the region of interest is small, even for relatively well-characterized processes such as lipoprotein metabolism. In the mouse, individual loci for a trait can be isolated, that is, separated from the other genes influencing the trait, on common genetic backgrounds (Silver, 1995). Once the individual locus has been isolated, the trait becomes Mendelian in nature, since only a single genetic factor, differing between the congenic strain and background strain, now contributes to the trait. It should be noted that congenic strains could be difficult to construct if the trait results from interactions among several genes. In addition, positional cloning using the congenic strain would only be feasible in those instances where the individual locus had a large enough impact on the trait to produce clearly distinguishable classes among the backcross or F2 mice.

One of the first examples of the use of congenic strains to isolate a locus for a complex trait relevant to atherosclerosis was for multigenic obesity. During the late 1980's, we observed that progeny of genetic crosses between the common laboratory strain C57BL/6 and stocks of the interfertile species *Mus spretus* were frequently obese, despite the fact that neither parental strain was obese. Thus, when (C57BL/6 × *M. spretus*)F1 females were crossed to C57BL/6 mice, the backcross progeny ranged from less than a few percent body fat to over 50% body fat. In collaborative studies with Dr. Jan Fisler, UCLA, we mapped four genes contributing to the multigenic obesity, designated *Mob1*, *Mob2*, *Mob3* and *Mob4* on Chr. 7, 6, 12, and 15, respectively ("*Mob*" comes from *m*ultigenic *ob*esity) (Warden et al., 1995). Each of the loci had distinctive effects on obesity and related traits such as plasma insulin and lipoprotein levels. Particularly interesting was the finding that a locus controlling hepatic lipase activity coincided with the Chr. 7 (*Mob1*) obesity locus, providing a possible link between obesity and plasma lipoprotein levels. The loci also exhibited distinct nonadditive interactions, explaining how two non-obese parental strains could produce progeny which are extremely obese (Figure 6). Fortuitously, the Chr. 7 locus (*Mob1*) resides near a minor histocompatibility locus (*H-1*) for which congenic strains had been constructed a number of years ago. One congenic strain consisted of a C57BL/10 genetic background and chromosome 7 donor regions derived from strain 129. Although the C57BL/10 strain closely resembles the C57BL/6 strain used in our cross, the donor strain was very different from the *M. spretus* mice, which diverged from common laboratory strains (*M. musculus*) several million years ago. Nevertheless, we reasoned that the genetic variations influencing body fat in the C57BL/6 × *M. spretus* cross may be relatively common, since many inbred strains of labora-

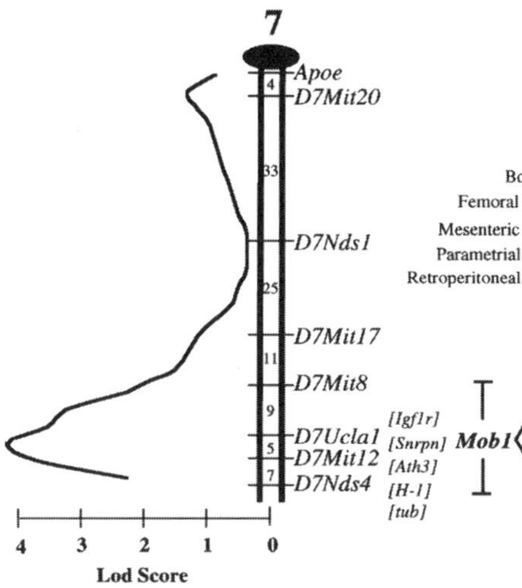

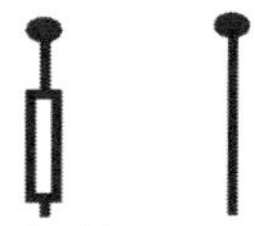

	B10.129	B10	P value
Body fat (%)	9.20±1.40	14.10±0.60	.01
Femoral fat pad (g)	0.14±0.03	0.27±0.02	.006
Mesenteric fat pad (g)	0.25±0.02	0.24±0.04	NS
Parametrial fat pad (g)	0.26±0.07	0.53±0.04	.005
Retroperitoneal fat pad (g)	0.02±0.01	0.10±0.01	.0006

Figure 6. Chromosomal localization of a locus contributing to multigenic obesity *(Mob 1)* on mouse Chr. 7 using a (C57BL/6 × *M. spretus)*F1 X C57BL/6 backcross. The chromosome is drawn with the centromere at the top. Microsatellite marker and gene symbols are given to the right of the chromosome, and the genetic distance (in cM) between each marker or gene is indicated between the bars of the chromosome. A plot of the Lod scores for linkage, calculated at two centiMorgan intervals along the chromosome, is shown on the left. The peak Lod score indicates the most likely position of the gene underlying the trait; the 95% confidence interval for the position of *Mob 1* is indicated on the right. Candidate genes for *Mob 1,* previously mapped to the same region of Chr. 7, are indicated in brackets; the relative positions of these genes were taken from Anonymous (1995). Two of the backcross progeny demonstrate the range of body fatness observed in the cross (lower right). Congenic strains whose donor strain regions overlap *Mob 1* have markedly different body fat profiles, confirming that a gene contributing to obesity is located in this region (top right).

tory mice vary in body fat and other obesity-related traits. Therefore, we examined the congenic strain for differences in body fat as compared to the background strain. The results showed highly significant effects of the Chr. 7 locus on body weight, retroperitoneal fat pad weight, percent body water (inversely proportional to body fat) and plasma cholesterol (Figure 6). For example, the congenic strain exhibited about 40% less body fat than the background C57BL/10 strain. These results strongly supported the linkage analysis. Because this congenic strain differs from the background strain only at the Chr. 7 locus, it should now be feasible to map the *Mob1* gene with much more precision than in the C57BL/6 × *M. spretus* cross where multiple loci influencing obesity were simultaneously segregating. The *Mob1* locus resides near a Mendelian mutation resulting in obesity, called tubby (*tub*), which has recently been identified by positional cloning (Kleyn et al., 1996; Noben-Trauth et al., 1996). It will be of interest to examine whether the *Mob1* is a subtle variation of the tubby gene.

9. CONSERVED "CHROMOSOMAL SYNTENY" CAN BE USED TO PREDICT THE LOCATIONS OF GENES CONTRIBUTING TO COMPLEX TRAITS

Because linkage groups are conserved between mouse and human chromosomes (sometimes called conserved "chromosomal synteny"), knowledge of the location of a

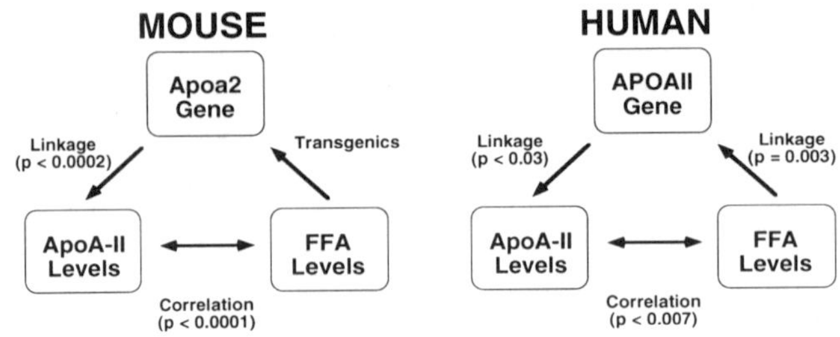

Figure 7. Variations of the mouse and human apoAII genes have similar consequences on the complex traits of plasma apoAII levels and plasma free fatty acid (FFA) levels. The mouse apoAII gene *(Apoa2,* Chr. 1) has exhibited linkage to plasma apoAII levels in several genetic crosses, including a (C57BL/6 *x M spretus)*F1 × C57BL/6 backcross ($p < 0.0002$). In the mouse, apoAII levels are strongly correlated with plasma FFA levels ($p < 0.0001$) and FFA levels are altered in a transgenic mouse overexpressing apoAII. Similarly, the human apoAII gene (APOA2, Chr. 1) exhibited linkage with plasma apoAII levels ($p<0.03$) and plasma FFA levels ($p<0.0003$), and plasma apoAII levels were correlated with plasma FFA levels ($p<0.007$) (Warden et al., 1993a).

chromosomal region contributing to a complex trait in mice can in some cases allow one to predict where a particular locus is likely to reside in the human genome, even without knowledge of the underlying gene. For example, in collaborative studies with Jan Fisler, UCLA, and Claude Bouchard, Laval University, we have made use of mouse multigenic obesity loci to predict the locations of human obesity loci (in preparation). The human and mouse genetic loci for complex traits will, of course, not always correspond. Genetic factors represent differences in expression patterns of genes between individuals, and which genes vary in the population will often depend on chance rather than selection. For example, the genes contributing to type 1 diabetes (insulin dependent diabetes mellitus) in humans appear to be predominantly different from those influencing type 1 diabetes in a mouse model (Figures 1 and 3, IDDM in humans and *Idd* in mice).

Our studies of the mouse apoAII locus (*Apoa2*, Chr. 1) provide an example of how a gene can similarly influence complex traits in different species (Figure 7). A number of genetic crosses in mice revealed that the distal region of mouse Chr. 1, containing the apoAII gene, influenced plasma apoAII protein levels, HDL size, plasma HDL-cholesterol levels and plasma free fatty acid (FFA) levels (Doolittle et al., 1990; Mehrabian et al., 1993; Purcell-Huynh, et al., 1995). Sequencing of the apoAII gene in multiple strains of mice revealed differences influencing both apoAII synthesis and plasma apoAII half-life (Figure 8). Because of the results in mice, we examined the human apoAII gene locus (APOA2, Chr. 1) for possible linkage to the same traits. The apoAII gene was followed using an informative simple sequence repeat in a total of 306 individuals in 25 families. Robust sib-pair linkage analysis was performed using the SAGE linkage programs, and the results revealed close congruence with the results in mice (Warden, et al. 1993a). Thus, in humans as in mice, the apoAII locus was linked to levels of both plasma apoAII and plasma FFA, and the latter two were highly correlated (Figure 7). In mice, the apoAII gene exhibited linkage to plasma HDL-cholesterol levels, and although we did not observe significant linkage to HDL-cholesterol levels in humans, significant linkage was observed with levels of plasma apoAI in humans (Bu et al., 1993). This represents one of the first examples of how the identification of a genetic factor contributing to a complex trait in mice can guide human studies. This is likely to be an important strategy in the future.

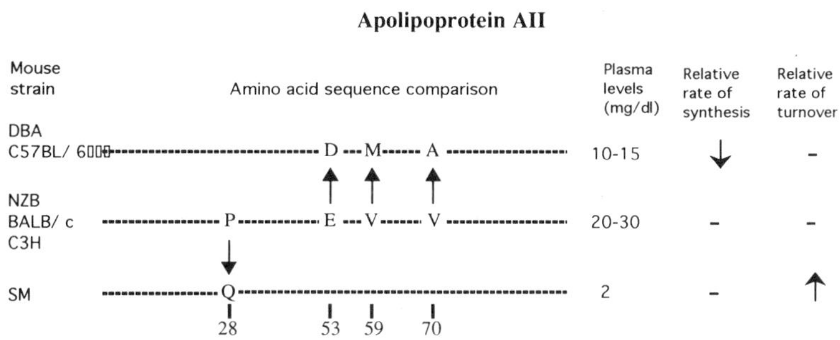

Figure 8. Sequence variations of mouse apoAII influencing expression. Mouse apoAII levels in plasma exhibit considerable variation between inbred strains of mice, ranging from about 2 mg/dl (strain SM) to about 30 mg/dl (strain C3H). This variation is not due to transcriptional differences, since all the strains examined have similar mRNA levels. Strains DBA and C57BL/6 have about two-fold decreased plasma apoAII levels due to substitutions in the mRNA which decrease translational efficiency (Doolittle et al., 1990). Strain SM has dramatically reduced plasma apoAII levels due to an amino acid substitution increasing turnover in the circulation (Purcell-Huynh et al., 1995).

10. "MODIFIER GENES" CAN BE IDENTIFIED THROUGH GENETIC CROSSES OF ANIMALS WITH DEFINED MUTATIONS INFLUENCING ATHEROSCLEROSIS OR RELATED TRAITS

A particularly important future application of gene targeted and transgenic mice is likely to be the identification of "modifier" genes which influence the phenotype resulting from mutations of specific genes. In humans, the existence of a modifier gene that abolishes the hypercholesterolemia associated with LDL receptor null mutations has been reported by Hobbs and coworkers (Hobbs et al., 1989). These investigators studied a large kindred with FH and, through the use of genetic markers, were able to identify individuals who had mutant LDL receptors but relatively normal LDL levels. The authors postulated the existence of a dominant hypercholesterolemia suppressor gene, but it did not correspond to any of the usual suspects (apolipoproteins and lipid metabolizing enzymes), and the family did not contain enough informative meioses to allow a whole genome scanning approach for mapping of the gene. Modifier genes, such as this cholesterol lowering gene, will be very difficult to track down in human families because of the limited number of individuals that can be studied within a kindred (assuming that only affected individuals are informative). It seems likely, however, that studies with genetically modified animals will be useful in identifying modifier genes. For example, to identify modifier genes for FH, LDL-receptor-deficient mice could be bred onto various genetic backgrounds by crossing them with different inbred strains of mice, and the resulting animals could then be tested for hypercholesterolemia. If differences were observed, the modifier genes could first be mapped and subsequently cloned by the same strategy used for identification of genes for other complex genetic traits. The best example of the use of animals to identify modifier genes is in the area of cancer research. The adenomatous polyposis coli (APC) gene was first identified in human genetic studies. A defective copy of the gene leads to the development of hundreds or thousands of polyps in the colon, which, unless they are removed, progress to become carcinomas. A mouse with the APC mutation also exhibited intestinal polyps, and it was noted that the number of polyps was greatly influenced by the genetic background, suggesting the existence of modifier genes. To map the modifier genes, backcrosses between the APC mutant and various other strains were constructed and the progeny analyzed for polyp number as well as genetic markers spanning the genome. When the results were statistically analyzed, a region on distal Chr. 4 was found to be strongly associated with polyp number (Dietrich et al., 1993). Several genetically engineered mice are likely to be useful in searches of modifier genes for atherosclerosis in the mouse. Of particular interest are several stocks which develop advanced atherosclerotic lesions, including apoE knockout mice, LDL receptor knockout mice, and human apoB transgenic mice (discussed in Shih et al., 1995).

11. POSITIONAL CLONING IS EASIER IN MICE

The mouse offers important advantages for positional cloning of novel genes. In particular, large genetic crosses can be constructed, facilitating fine mapping. Thus, with a cross of one or two thousand mice, linkage analysis in mice can narrow the region of interest to a few hundred kilobases or less. Also, interspecific crosses can be constructed, greatly simplifying the identification of polymorphic markers for linkage purposes. For example, in collaborative studies with Mark Doolittle and colleagues at the Wadsworth Veteran's Administration, UCLA, we are attempting to clone the gene for "combined lipase deficiency" (*cld*, Chr. 17), a mutation causing deficiency of both hepatic lipase and lipoprotein lipase in mice (Paterniti et al., 1983; reviewed in Reue and Doolittle, 1996). Thus far, the *cld* mutation has been mapped with respect to two deletions within the *t* locus, t^{h20} and t^{w18} (Figure 9) (Artzt et al., 1991). We are constructing a detailed linkage map of the region using interspecifc crosses with about one thousand backcross and F2 mice (K. Rooke, Y.-R. Xia, A.J. Lusis, unpublished). The mapped markers will then be used to screen yeast artificial chromosome (YAC) libraries, or other libraries carrying large DNA fragments such as bacterial artificial chromosomes (BACs), to construct a "contig" (a series of overlapping clones) of the region. Since the *cld* mutation is recessive, the clones can be screened for the presence of the *cld* gene by complementation in transgenic mice, or in fibroblast cell lines derived from *cld* mice (discussed in Reue and Doolittle, 1996).

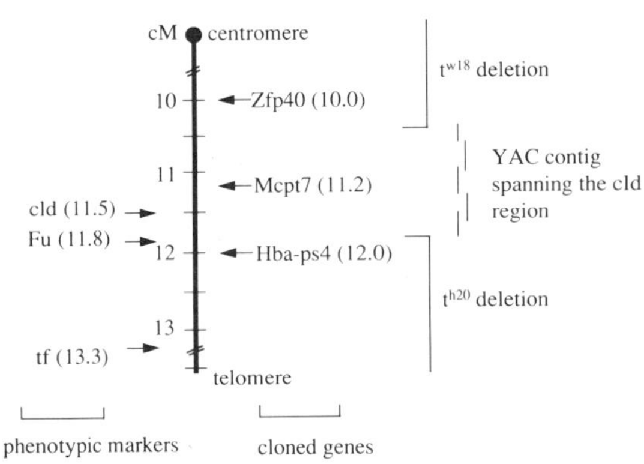

Figure 9. Region of Chr. 17 containing the *cld* mutation. The composite chromosome is drawn with the centromere at the top. The cloned and phenotypic markers are ordered as they would appear on a wild-type chromosome; the markers would be inverted on the *t* complex. The markers are as follows: *Zfp4O*, a zinc finger protein; *Mcpt7*, a mast cell protease; *Hba-ps4*, a pseudogene of hemoglobin A; *cld*, combined lipase deficiency; *Fu*, a dominant mutation that affects skeletal development; *tf*, a recessive mutation that affects hair pattern. The t^{w18} deletion removes *Zfp4O* but not *cld*; the t^{h20} deletion removes *Fu, Hba-ps4* and *tf* but not *cld*. Thus, t^{w18} and t^{h20} are shown to flank the *cld* gene. It is important to note that although the *cld* mutation has been tentatively placed at 11.5 cM from the centromere the actual localization could differ by 1–2 cM on either side. A tiling path of YAC clones will be constructed as indicated to ensure that the YAC contig contains the *cld* gene.

12. THE MOUSE IS A USEFUL TEST TUBE FOR EXAMINING GENE FUNCTIONS IN VIVO

The mouse offers advantages for elucidation of gene functions and pathways, not only because of the feasibility of physiologic studies unethical in humans but also because of the ability to perform planned genetic modifications (Breslow, 1994; Smithies and Maeda, 1995). The ability to generate transgenic and knockout mice has already had a major impact on research in cardiovascular disease, and the use of such genetically modified animals is rapidly expanding. Once a gene contributing to a complex trait has been identified, transgenic or gene targeted mice can help define the molecular mechanisms involved. The apoAII gene provides a good example. As discussed above, differences in apoAII expression influence several aspects of lipoprotein metabolism in both mice and humans, but the most provocative result involved the effect of apoA2 expression on atherosclerotic lesions. Thus, genetic studies involving a congenic strain suggested that increased apoAII expression resulted in elevated HDL-cholesterol levels accompanied by *increased* susceptibility to atherosclerosis (Mehrabian et al., 1993). Subsequently apoAII transgenic mice were constructed and found to exhibit spontaneous atherosclerosis despite having dramatically elevated HDL-cholesterol levels (Warden et al., 1993), confirming the conclusions from genetic studies. This was very different from the results with apoAI transgenic mice, where increased apoAI expression increased HDL-cholesterol but reduced susceptibility to atherosclerosis. Subsequent studies have shown that HDL from apoAII transgenics and nontransgenic litter mates is about equally effective at promoting cholesterol efflux from macrophages, but that the transgenic HDL has diminished ability to protect against LDL oxidation. This led in turn to the finding that the transgenic HDL is deficient in the enzyme paraoxonase, which protects against LDL oxidation in tissue culture (L. Castellani, B. Van Lenten, M. Navab, A. Lusis, unpublished). Recently, we have created paraoxonase knockout mice to test the hypothesis that the enzyme is responsible, in part, for the antiatherogenic properties of HDL (D. Shih, J. Gu, Y.-R. Xia, A.J. Lusis, unpublished). These studies illustrate the utility of a genetic approach. Whereas apoAII was previously considered a passive passenger on HDL, it is now clear that it can dramatically influence the functional characteristics of HDL. Indeed, some human epidemiological studies have suggested that HDL particles containing apoAII may be less protective against atherosclerosis than particles containing apoAI alone (reviewed in Hedrick and Lusis, 1994).

13. CONCLUSIONS

Using studies from our laboratory, we have attempted to illustrate how the mouse model can usefully contribute to understanding complex forms of atherosclerosis. We have focused on the advantages of the mouse for gene discovery, but, in addition, the mouse will likely be particularly useful for examining genetic interactions between multiple loci. One early example of this comes from analysis of the obesity loci *Mob1*, *Mob2*, *Mob3*, and *Mob4*. The loci exhibited clear, nonadditive interactions, and each locus exerted characteristic effects on body fat, lipoprotein levels, and insulin levels. Also, the loci influenced various fat depots in a highly specific manner, reinforcing the notion that genetic factors determine not only total body fat but the distribution of the fat (Warden et al., 1995). Since new techniques and analytical methods are likely to dramatically transform studies of complex diseases in the years to come, it is difficult to predict future strategies. But, clearly, the mouse will play a major role in efforts to solve the puzzle of atherosclerosis and to develop ways of dealing with the disease.

ACKNOWLEDGMENTS

Our research was supported by NIH grants HL28481, HL42488, and HL30568 and by the Laubisch Fund, UCLA.

REFERENCES

Anonymous. Mouse Genome Database. Mouse Genome Informatics Project, The Jackson Laboratory, Bar Harbor, Maine. World Wide Web.(URL:http//www. informatics.jax.org), 1995.

Anonymous. GDB(TM) Human Genome Data Base. Johns Hopkins University, Baltimore, Maryland. World Wide Web. (URL:http://gdbwww.gdb.org/db5. 6lbrowser/docs/topq/html), 1996.

Artzt, K., Barlow, D., Dove, W.F., Lindahl, J., Klein, J., Lyon, M.F., and Silver. L.M. Maps on mouse chromosome 17: First report. *Mamm. Genome*, 1:5–29, 1991.

Breslow, J.L. Insights into lipoprotein metabolism from studies in transgenic mice. *Ann. Rev. Physiol.*, 56:797–810, 1994.

Breslow, J.L. Mouse models of atherosclerosis. *Science*, 272:685–688, 1996.

Bu, X., Warden, C.H., Xia, Y-R., DeMeester, C., Puppione, D.L., Teruya, S., Lokensgard, B., Daneshmand, S., Brown, J., Gray, R., Rotter, J.I., and Lusis, A.J. Linkage analysis of the genetic determinants of high density lipoprotein concentrations and composition: evidence for involvement of the apolipoprotein A-II and cholesteryl ester transfer protein loci. *Hum. Genet.*, 93:639–648, 1994.

Collins, A., Lawrence, S., and Shields, D.C. Algorithms for a local database. *Ann. Hum. Genet.*, 56:223–232. (URL:http://cedar.genetics.soton.ac.uk/public .html/gene.html, 1992.

Collins, F. Positional cloning moves from perditional to traditional. *Nat. Genet.*, 9:347–350, 1995.

Copeman, J.B., Cucca, F., Hearne, H.M., Cornell, R.J., Reed, P.W., Ronningen, K.S., Undlien, D.F., Nistico, L., Buzzetti, R., Tosi, R., Pociot, F., Nerup, J., Cornelis, F., Barnett, A.H., Bain, S.C., and Todd, J.A. Linkage disequilibrium mapping of a type 1 diabetes susceptibility gene (IDDM7) to chromosome 2q31–q33. *Nat. Genet.*, 9:80–85, 1995.

Davies, J.L., Kawaguchi, Y., Bennett, S.T., Copeman, J.B., Cordell, H.J., Pritchard, L.F., Reed, P.W., Gough, S.C.L., Jenkins, S.C., Palmer, S.M., Balfour, K.M., Rowe, B.R., Farrall, M., Barnett, AH., Bain, S.C., and Todd J.A. A genome-wide search for human type 1 diabetes susceptibility genes. *Nature*, 371:130–136, 1994.

Dietrich, W.F., Lander, E.S., Smith, J.S., Moser, A.R., Gould, K.A., Luongo, C., Borenstein, N., and Dove, W. Genetic identification of *Mom*-1, major modifier locus affecting *Min*-induced intestinal neoplasia in the mouse. *Cell*, 75:631–639, 1993.

Doolittle, M.H., LeBoeuf, R.C., Warden, C.H., Bee, L.M., and Lusis, A.J. A polymorphism affecting apolipoprotein A-II translational efficiency determines high density lipoprotein size and composition. *J. Biol. Chem.*, 265: 16380–16388, 1990.

Fisler, J.S., Warden, C.H., Pace, M.J., and Lusis, A.J. BSB: A new mouse model of multigenic obesity. *Obesity Res.*, 1:271–280, 1993.

Goldstein, J.L. and Brown, M.S. The low density lipoprotein receptor: A key for unlocking a multifactorial disease. In: *Etiology of Human Disease at the DNA Level*. Lindsten, J. and Petterson, U., eds. pp. 129–140, Raven Press, N.Y., 1991.

Goldstein, J.L., Hobbs, H.H., and Brown, M.S. Familial hypercholesterolemia. In: *The Metabolic and Molecular Bases of Inherited Disease*. Scriver, C.R., Beaudet, AL., Sly, W.S., and Valle, D., editors. pp. 1981–2030, McGraw-Hill, Inc., New York, 1995.

Hedrick, C.C. and Lusis, A.J. Apolipoprotein A-II: A protein in search of a function. *Can. J. Cardiol.*, 19:453–459, 1994.

Hobbs, H.H., Leitersdorf, E., Leffert, C.C., Cryer, D.R., Brown, M.S., and Goldstein, J.L. Evidence for a dominant gene that suppresses hypercholesterolemia in a family with defective low density lipoprotein receptors. *J. Clin. Invest.*, 84:656–664, 1989.

Hua, X., Wu, J., Goldstein, J.L., Brown, M.S., and Hobbs, H.H. Structure of human gene encoding sterol regulatory element binding protein-1 (SREBF1) and localization of SREBF1 and SREBF2 to chromosomes 17q11.2 and 22q13. *Genomics*, 25:667–673, 1995.

Keating, M.T. and Sanguinetti, M.C. Molecular genetic insights into cardiovascular disease. *Science*, 272:681–685, 1996.

Kleyn, P.W., Fan, W., Kovats, S.G., Lee, J.J., Pvledo, J.C., Wu, Y., Berkemeier, L.R., Misumi, D.J., Holmgren, L., Charlat, O., Woolf, E.A., Tayber, O., Brody, T., Shu, P., Hawkins, F., Kennedy, B., Baldini, L., Ebling, C.,

Alperin, G.D., Deeds I., Lakey, N.D., Colpepper, J., Chen, H., Glücksmann-Kuis, M.A., Carlson, G.A., Duyk, G.M., and Moore, K.J. Identification and characterization of the mouse obesity gene *tubby*: A member of a novel gene family. *Cell*, 85:281–290, 1996.

Kruglyak, L. and Lander, E.S. High resolution genetic mapping of complex traits. *Am. J. Hum. Genet.*, 56:1212–1223, 1995.

Ivandic, B.T., Qiao, J-H., Machleder, D., Liao, F., Drake, T.A., and Lusis, A.J. A locus on chromosome 7 determines myocardial cell necrosis and calcification (dystrophic cardiac calcinosis) in mice. *Proc. Nat. Acad. Sci. USA*, 93:5483–5488, 1996.

Lander, E.S. and Schork, N.J. Genetic dissection of complex traits. *Science*, 265:2037–2048, 1994.

Lifton, R.P. Molecular genetics of human blood pressure variation. *Science*, 272:676–680, 1996.

Lusis, A.J. Genetic factors in atherosclerosis: the candidate gene approach. *J. Lipid Res.*, 29:397–429, 1988.

Mehrabian, M., Demer, L.L., and Lusis, A.J. Differential accumulation of the intimal monocyte-macrophages relative to lipoproteins and lipofuscin correspond to hemodynamic forces on cardiac valves in mice. *Arterioscl. Thromb.*, 11:947–957, 1991.

Mehrabian, M., Qiao, J.-H., Hyman, R., Ruddle, D., Laughton, C., and Lusis, A.J. Influence of the apoA-II gene locus on HDL levels and fatty streak development in mice. *Arteriosel. Thromb.*, 13:1–10, 1993.

Noben-Trauth, K., Naggert, J.K., North, M.A., and Nishina, P.M. A candidate gene for the mouse mutation *tubby*. *Nature*, 380:534–538, 1996.

Paterniti, J.R., Brown, W.V., Ginsberg, H.N., and Artzt, K. Combined lipase deficiency (cld): A lethal mutation on chromosome 17 of the mouse. *Science*, 221:167–169, 1983.

Purcell-Huynh, D.A., Weinreb, A., Castellani, L.W., Mehrabian, M., Doolittle, M.H., and Lusis, A.J. Genetic factors in lipoprotein metabolism: Analysis of a genetic cross between mouse strains NZB/BINJ and SM/J using a complete linkage map approach. *J. Clin. Invest.*, 96:1845–1858, 1995.

Qiao, J.-H., Welch, C.L., Xie, P.-Z., Fishbein, M.C., and Lusis, A.J. Involvement of the tyrosinase gene in deposition of cardiac lipofuscin in mice: Association with aortic fatty streak development. *J. Clin. Invest.*, 92:2386–2393, 1993.

Qiao, J-H., Xie, P-X., Fishbein, M.C., Kreuzer, J., Drake, T., Demer, L.L., and Lusis, A.J. Pathology of atheromatous lesions in inbred and genetically engineered mice: Genetic determination of arterial calcification. *Arteriosel. Thromb.*, 14:1480–1497, 1994.

Reue, K. and Doolittle, M.H. Naturally occurring mutations in mice affecting lipid transport and metabolism. *J. Lipid Res.*, 37:1387–1405, 1996.

Shih, D.M., Gu, L., Hama, S., Xia, Y-R., Navab, M., Fogelman, A.M., and Lusis, A.J. Genetic-dietary regulation of serum paraoxanase expression and its role in atherogenesis in a mouse model. *J. Clin. Invest.*, 96:1630–1639, 1996.

Shih, D.M., Welch, C., and Lusis, A.J. New insights into atherosclerosis from studies with mouse models. *Molec. Med. Today*, 1:364–372, 1995.

Silver, L.M., Nadeau, J.H., and Goodfellow, P.N., eds. *Mouse chromosome committee reports. Mamm. Genome*, 5 (Suppl.).

Silver, L.M. *Mouse Genetics: Concepts and Applications*. Oxford University Press, New York.

Smithies, O. and Maeda, N. Gene targeting approaches to complex genetic diseases: Atherosclerosis and essential hypertension. *Proc. Natl. Acad. Sci. USA*, 92:5266–5272, 1995.

Wang, X., Sato, R., Brown, M.S., and Goldstein, J.L. SREBP-1, a membrane-bound transcription factor released by sterol-regulated proteolysis. *Cell*, 77:53–62, 1994.

Warden, C.H., Daluiski, A., Bu, X., Purcell-Huynh, DA., DeMeester, C., Shieh, B.-H., Puppoine, D.L., Gray, R.M., Reaven, G.M., Chen, Y.-D. I., Rotter J.I., and Lusis, A.J. The apolipoprotein AII gene determines plasma apolipoprotein AII levels and free fatty acid levels in both humans and mice. *Proc. Nat. Acad. Sci. USA*, 90:10886–10890, 1993a.

Warden, C.H., Fisler, J.S., Shoemaker, S.M., Wen, P-Z., Svenson, K.L., Pace, M.J., and Lusis, A.J. Identification of four chromosomal loci determining obesity in a multifactorial mouse model. *J. Clin. Invest.*, 95:1545–1552, 1995.

Warden, C.H., Hedrick, C.C., Qiao, J.-H., Castellani, L.W., and Lusis, A.J. Atherosclerosis in transgenic mice overexpressing apolipoprotein A-II. *Science*, 261:469–472, 1993b.

13

ROLE OF APOLIPOPROTEIN E IN ALZHEIMER'S DISEASE

Vassilis I. Zannis,[1] Eleni E. Zanni,[1] Savvas C. Makrides,[1] Dimitris Kardassis,[2] and Sergei Aleshkov[1]

[1]Section of Molecular Genetics
Center for Advanced Biomedical Research
Departments of Medicine and Biochemistry
Boston University Medical Center, Boston, Massachusetts
[2]Department of Biochemistry
University of Crete School of Medicine
Heraklion, Crete, Greece

1. GENERAL BACKGROUND ON ALZHEIMER'S DISEASE

Alzheimer's Disease (AD) is a devastating disease which affects the elderly population and is associated with loss of memory (1–3). The prevalence of AD in people over 65 years of age is approximately 5–10%, and for people over 85 is approximately 50% (1). It is estimated that the current annual cost for care of AD patients is $100 billion. Furthermore, it is projected that the number of people affected by AD will increase 4-fold by the year 2030. Clearly, AD is a major public health problem.

The following four proteins have been implicated in the pathogenesis of AD:

- A membrane protein which maps on chromosome 21 and was designated amyloid precursor protein (APP). This precursor protein gives rise to a pathogenic 39- to 42-amino acid long amyloid peptide $\beta(A\beta)$. Mutations in APP, which will be discussed further, are associated with early onset of familial AD (Figure 1A–C) (4–7). Another mutation in APP is associated with hereditary cerebral hemorrhage with amyloidosis of Dutch type (8). Finally, development of AD is accelerated Down syndrome (9).
- A membrane protein which maps on chromosome 14. This protein was initially named (S-182) and subsequently presenilin 1. Mutations in presenilin-1 are associated with early onset familial AD (Figure 2) (10).
- Another membrane protein closely related to presenilin 2 which maps on chromosome 1. This protein was initially designated STM2 and subsequently presenilin 2. Mutations in presenilin-2 are also associated with early onset of familial AD (11).

Vascular Endothelium, edited by Catravas *et al.*
Plenum Press, New York, 1998

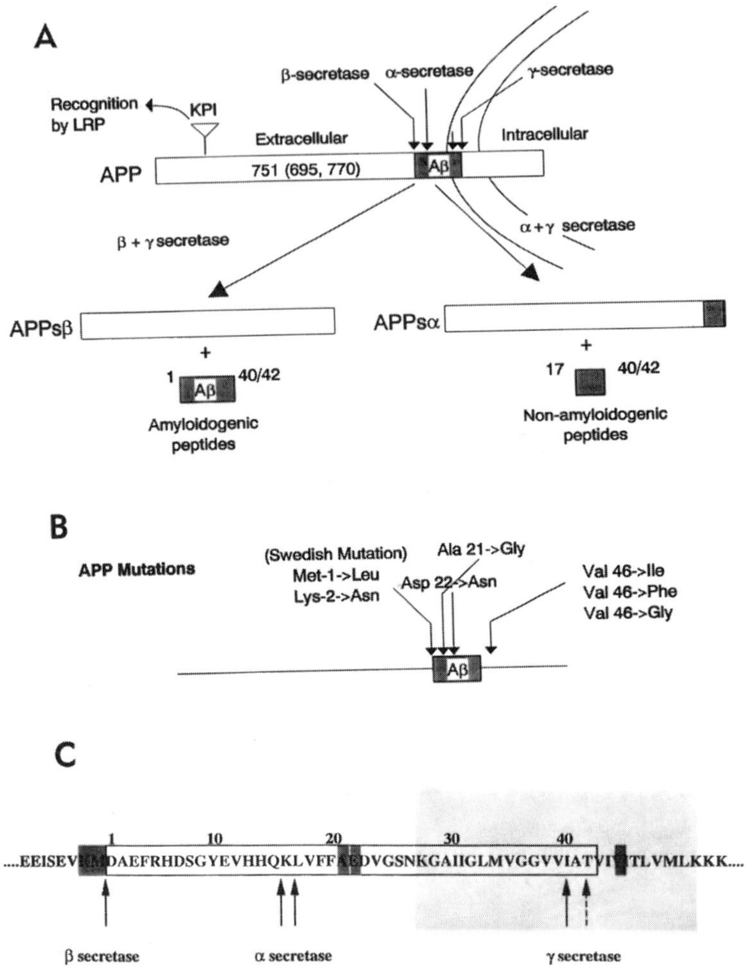

Figure 1. Panel A: Schematic representation of the structure of APP indicating important APP domains, sites of proteolysis and products of different proteolytic pathways. **Panel B**: Locations of the various published APP mutations that are associated with AD or with hereditary cerebral hemorrhage amyloidosis (Dutch type). **Panel C**: Major sites of cleavage of APP by the α, β and γ secretases. Shaded areas indicate the site mutations shown in Panel B. APP derived from the 751 amino acid form is a serine protease inhibitor (24–28) and inhibits the coagulation factor XIa and IXa (29,30). These functions are localized in the KPI domain (24–30). KPI mediates binding, endocytosis and degradation of APP by the LDL receptor related protein (LRP) (31).

A common feature of all three forms of early onset AD is the increased production of the Aβ 1–42 form. In addition, in some cases of APP mutations, there is an increase in the total production of Aβ (12,13).

- The apolipoprotein E which maps on chromosome 19 (14).

2. BRAIN LESIONS ASSOCIATED WITH AD

The lesions encountered in the brain of AD patients encompass degenerations of numerous functional and anatomical classes of neurons (15). Two types of brain lesions are considered to be the hallmark of AD histopathology:

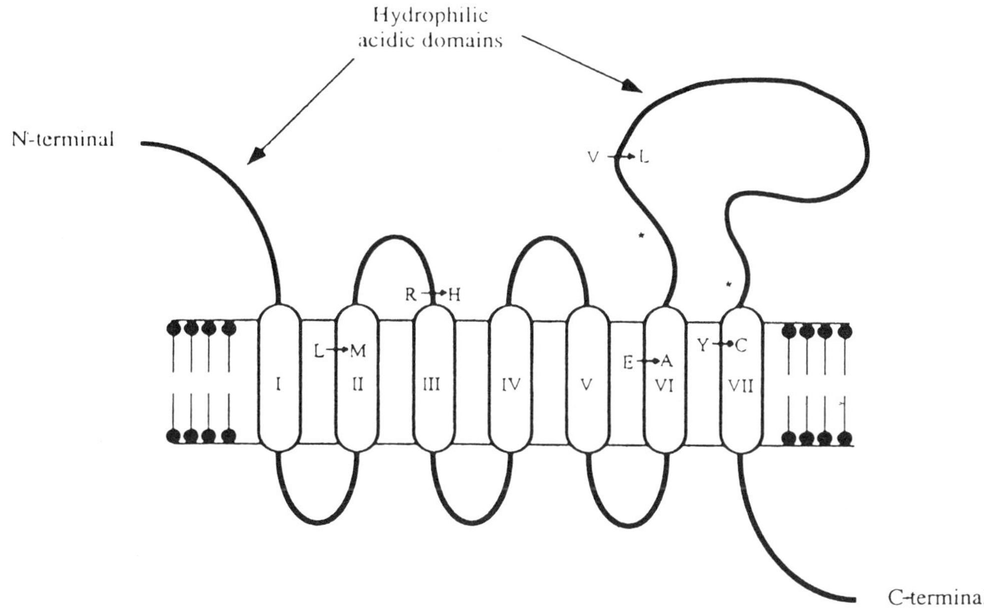

Figure 2. A model showing the putative structural domains of presenilin 1 and sites of mutations that are associated with early onset of AD (from Ref. 10 with permission).

- Abnormal fibers (called paired helical filaments; PHF) in neuronal cell bodies leading to the formation of neurofibrillary tangles (16,17). The filaments have 20 nM width, 80 nM periodicity. This fibrous material is recognized by antibodies to the microtubule-associated protein tau (18,19).
- Amyloid plaques in the brain and amyloid deposits in the walls of blood vessels. The core of the plaque contains deposits of extracellular fibers 6–9 nm in diameter (20) and is surrounded by dystrophic neurites, activated microglia and fibrillary astrocytes (15).

  The formation of PHF in perikarya and neurites may be associated with formation of the amyloid plaques or other insults which cause the injury in the brain of AD patients.

3. AMYLOID PRECURSOR PROTEIN (APP), AMYLOID PEPTIDE β(Aβ) AND THEIR ROLE IN THE PATHOGENESIS OF AD

Sequencing of a 4 KDa peptide found in the amyloid filaments and amyloid plaques of AD patients identified a peptide that was designated amyloid peptide β(Aβ) (21,22). Isolation of cDNAs encoding Aβ showed that it represents a 39- to 42-amino acid residue long proteolytic fragment derived from an ubiquitous larger precursor designated amyloid protein precursor (APP) (23). The APP is a transmembrane protein of unknown function. The APP gene can generate transcripts encoding proteins of 695 to 751 or 770 amino acids. The 751 residue form contains a region homologous to the Kunitz family of serine protease inhibitors (KPI) (24–28).

The APP derived from the 751 amino acid form is also an inhibitor of the coagulation factor IXa and XIa (29,30). Recently, it was shown that the KPI domain of APP mediates the binding, endocytosis and degradation of APP by the LDL receptor-related protein LRP (31). Inhibition of synthesis of APP is also associated with inhibition of cell growth (32), APP may play a role in wound repair (33) and cell adhesion (34).

4. APP PROCESSING AND FORMATION OF Aβ

Following synthesis, the APP undergoes constitutive cleavage resulting in the release of a 110 KDa soluble APP fragment (APPs) into the medium and the retention of an approximately 10 KDa fragment in the membrane (35). Secreted APPs can be detected in many tissues(35–38). A major secreted form of APP, designated APPsα, is released by action of a protease designated α-secretase which cleaves APPα between amino acids 16 and 17 of the Aβ domain (Figure 1C) (35, 39). The α secretase activity is higher in brain cells than in cells derived from other tissues. Another soluble form of APP designated APPsβ, is generated by cleavage of APP by another protease designated β-secretase which cleaves prior to amino acid 1 of Aβ (Figure 1A) (13, 38). It was shown recently that APPsα form has a neuroprotective role and is 100-fold more efficient in protecting hippocampal neurons against excitotoxicity (39). The Aβ form found in brain amyloid plaques consists of 39- to 42 amino acids and contains 28 amino acids of the extracellular domain of Aβ and 11 to 14 amino acids of the transmembrane domain. This fragment involves cleavage of APP by β secretases as well as by an additional protease designated γ secretase (13).

The subcellular localization of these proteases is not clear. The pathway of cleavage which leads to the amyloidogenic peptide(s) may involve the internalization of a surface full-length APP which has been demonstrated to occur in cell cultures (40) and its proteolytic degradation in some cellular compartment such as lysosomes (41). In addition, it is possible that small amounts of APP are released in a physiological context or as a result of membrane damage, and is subsequently cleaved at or near the γ-secretase site by one or more different proteases (42–45). Soluble Aβ forms have been detected in conditioned media of APP-transfected cells (45,46) and in CSF (46,47) and consist mainly of Aβ 1–40 as well as small amounts of Aβ (1–42) and other minor forms (13). The presence of Aβ peptides of different length demonstrates the existence of several alternative pathways of APP-processing. A variety of products that result from such alternative APP processing pathways have been identified (13). Seubert et al. described a secreted form of APP in mixed brain cell cultures as well as in CSF which is cleaved precisely at the amino terminus of Aβ (38). These findings showed that APP can be processed in a potentially amyloidogenic manner by a proteolytic activity which is considerably higher in cultured brain cells than in cells derived from other tissues and corresponds to the β secretase (38). In addition, other investigators identified a population of β-amyloid-related peptides 3 and 4 kDa in size in the conditioned media of a variety of APP-transfected and non transfected cells (48). Most interestingly, the highest levels of the 4 kDa peptide which is identical or closely related to the full length Aβ are secreted by human astrocytes, while the N-terminal truncated 3-kDa form, probably resulting from the action of a-secretase, is mainly secreted by neurons (48). The findings also stress the importance of astrocytes and suggest that this cell type may be a potentially major source of amyloidogenic Aβ in the brain (48).

5. THE ROLE OF PROTEASES AND PROTEASE INHIBITORS AND INFLAMMATORY PROCESSES IN AMYLOIDOGENESIS

The relevance of astrocytes to the amyloidogenic process becomes even more evident when one considers that these cells not only represent a probable source of Aβ in the brain, but also of another major constituent of the amyloid plaques. In senile plaques, the serine protease inhibitor α1-antichymotrypsin (ACT) is found tightly associated with Aβ (49,50). In the AD brain, the amounts of ACT mRNA and protein are significantly increased compared to controls (49). ACT is an acute phase protein which is strongly induced in the liver by IL-1, and which is also highly expressed in reactive astrocytes (51,52). ACT synthesis by reactive astrocytes might be triggered by cytokine upregulation as a result of injury, inflammation or neurodegeneration. The presence of complement components (53) and activated microglia (54) in or around Aβ deposits also strongly suggests the occurrence of inflammatory processes within the senile plaque. Reactive astrocytosis can be observed as a result of diverse neurologic injuries in many neurologic disorders. The phenomenon is complex and, depending on the activating mechanism, includes the upregulation of a variety of molecules such as cytokines, antigen presenting or adhesion molecules, cytoskeletal proteins and enzymes (55). Reactive astrocytes have also been shown to express high levels of APP in response to injury (56). It is interesting that reactive astrocytes express in parallel elevated levels of both proteases and protease inhibitors (57). High expression of ACT in reactive astrocytes is not only seen in AD brains, but also in brains of Huntington's disease patients (46), or in mouse brains after mechanical injury (57). In addition, proteases similar to cathepsin G, an enzyme strongly inhibited by ACT, have been demonstrated in monkey brain astrocytes (58), and a strong cathepsin G-like immunoreactivity has been observed in reactive astrocytes in mechanically injured mouse brains (57). The production of both proteases and protease inhibitors might allow astrocytes to participate in remodeling the extracellular matrix at sites of neural injury and to promote neurite extension. On the other hand, these molecules could be involved in certain pathologic conditions. Increased production of amyloid Aβ by astrocytes might well be a consequence of the disturbance of the sensitive balance between proteases and protease inhibitors.

6. THE ROLE OF Aβ IN AMYLOIDOGENESIS

The extracellular accumulation of Aβ generates diffuse plaques in many brain regions. It also causes local activation of microglial and astrocytes and deposition of Aβ-associated proteins (12). One such protein is apoE which as discussed below can form stable complexes with Aβ. The deposition of Aβ and amyloid-associated proteins imposes progressive neurotoxic and neurotrophic effects on neurites (59). This injury, in turn, leads to secondary degenerative and regenerative changes in intrinsic (cortical) and projection (subcortical) neurons and alterations of cytoskeletal proteins including formation of tau containing PHF in neurites and somata. The end result is neuronal loss, neurotransmitter deficit and other biochemical and cellular changes that are associated with dementia and clinically diagnosed AD.

It was demonstrated that increased production of Aβ, particularly the Aβ 1–42 form, can mediate its aggregation in the brain, leading to the formation of fibrillar deposits. On the other hand, it was shown that aggregated Aβ can be toxic to cultured neurons, astrocytes, microglia and endothelial cells (60–65). Furthermore, transgenic mice, overexpressing a mutant form of APP, overproduce Aβ and develop AD-like pathomorphological changes in the brain (66). These findings led to the hypothesis that Aβ plays an important role in the pathological cascade that leads to AD (15). Thus, the problem of binding, inactivation and clearance of Aβ in the brain emerges as one of the possible rational approaches for the treatment of AD.

7. INVOLVEMENT OF APOLIPOPROTEIN E IN ALZHEIMER'S DISEASE BACKGROUND ON APOE

Apolipoprotein E is an important protein in the lipoprotein transport system (67). It is a ubiquitous 34.2 KDa protein of known sequence (68–72), and it resides in VLDL, IDL, HDL and chylomicron remnants. The three-dimensional structure of the 22 KDa amino terminal domain of apoE has been elucidated and shown to consist of four antiparallel helices (73–76). Computer modeling suggests that the carboxy terminal region may also form similar antiparallel helical structures (Figure 3) (77).

ApoE is synthesized by the liver and the majority of the peripheral tissues (78) including astrocytes (79–87), smooth muscle cells and epithelial cells (88,89). Newly synthesized apoE is modified by O-glycosylation at Thr-194 (90,91), and is secreted either as

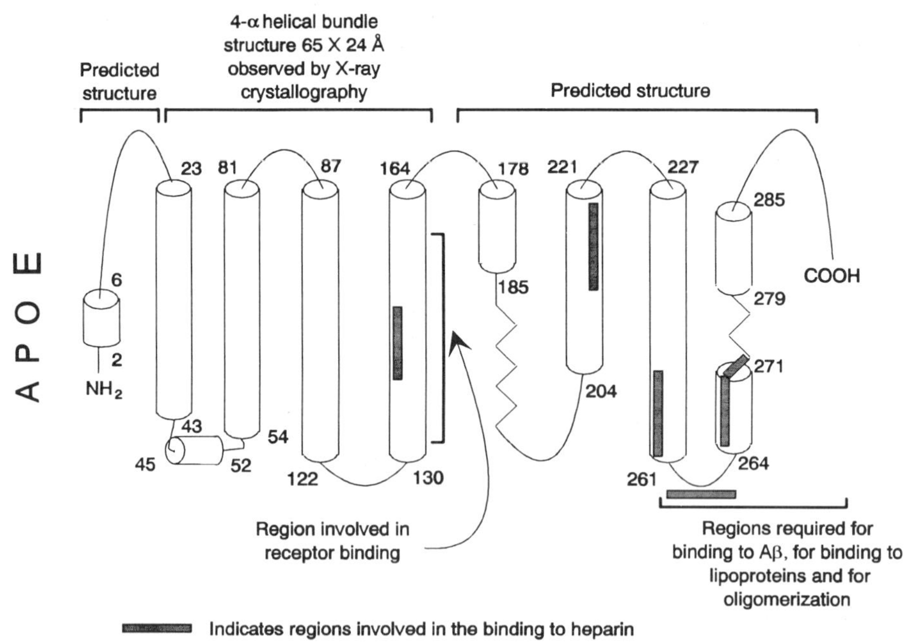

Figure 3. Diagram showing the secondary structure of apoE. Helical sections are denoted as cylinders, β structures as \/\/\/ and turns as U. Sections of random coil are show as irregular curves. The 4 ahelical bundle structure between residues 23 and 164 was determined by x-ray crystallography (73–76). The predicted structure was calculated by Nolte and Atkinson (77) (reproduced from Ref. 7 with permission). The figure also shows domains of apoE involved in receptor binding (102–105), heparin binding (106, 107) lipid and lipoprotein binding (109–112) and for binding to Aβ (148).

a lipid-free apoprotein or as a lipoprotein particle (92,93). ApoE is the ligand which promotes the recognition and catabolism of apoE-containing lipoproteins such as chylomicron remnants, β VLDL, hypertriglyceridemic VLDL, IDL and HDL with apoE by the LDL receptor (94–96), the E/a2M macroglobulin receptor (95,97,98) and possibly the VLDL receptor (99–101). The receptor-binding domain of apoE is located in the vicinity of residues 130–160 (102–105). The protein also contains four heparin-binding domains between residues 142–147, 211–218, 243–273 (106,107) which may help anchor the protein on heparan sulfate proteoglycans prior to its uptake by the cell receptors (108). The carboxyterminal domain of apoE is required for lipid and lipoprotein binding (109–112). Truncated apoE forms fail to associate with lipoproteins (112).

There is a common genetic polymorphism of apoE in humans which is the result of the three common apoE alleles, ϵ4, ϵ3, and ϵ2 at a single genetic locus (113–115). These three alleles give rise to three homozygous (E4/4, E3/3, E2/2) and three heterozygous (E4/3, E3/2, E4/2) phenotypes (113–115). The average frequency of the apoE alleles in Caucasians is ϵ4=0.15, ϵ3=0.77 and ϵ2=0.08. This corresponds to a calculated phenotype frequency of E4/4=3%, E3/3=60%, E2/2=1%, E4/2=22%, E3/2=12%, and E4/2=2% (Figure 4A,B) (116). The three common apoE variants result from amino acid substitutions at residues 112 and/or 158. The E4 isoprotein contains Arg-112, Arg-158, the E3 contains

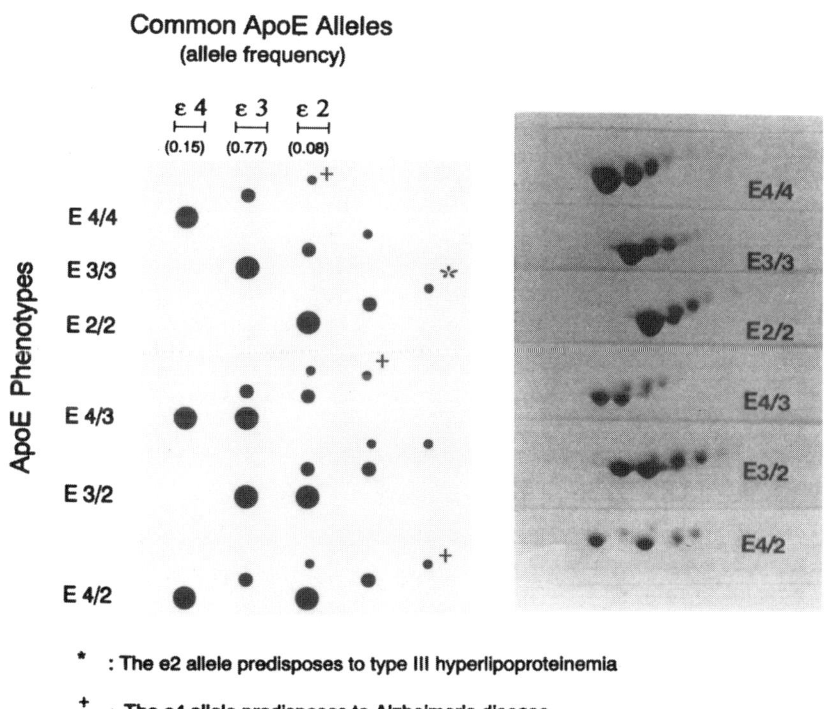

Figure 4. Panel A: Schematic representation of the common and rare apoE alleles and apoE phenotypes. Large and small spheres represent the unmodified and the O-glycosylated sialyated forms of apoE, respectively. The symbols * and + indicate phenotype associated with type III hyperlipoproteinemia and Alzheimer's disease, respectively. **Panel B**: Two-dimensional PAGE patterns of the six apoE phenotypes observed in humans arranged according to their isoelectric points.

Cys-112, Arg-158, and the E2 contains Cys-112, Cys-158 (113–115,117,118). The apoE alleles affect the plasma cholesterol levels in the general population (116). The ϵ4 allele is associated with higher and the ϵ2 allele with lower total and LDL cholesterol levels as compared to the ϵ3 allele (116).

8. THE ROLE OF APOE IN CARDIOVASCULAR DISEASE

Mutations in apoE which prevent its binding to the LDL receptor and possibly other receptors are associated with type III hyperlipoproteinemia (119–126). Depending on genetic and environmental factors, the Arg-158--Cys mutation may result in autosomal recessive form type III hyperlipoproteinemia (67,119,120). Other rare apoE mutations between amino acids 136 to 152 located within the receptor binding domain of apoE are associated with dominant forms of type III hyperlipoproteinemia (121–126). The importance of apoE in cholesterol homeostasis was established unequivocally by studies of human patients and animal models with apoE deficiency (127–138). The human patients had elevated cholesterol levels (530 mg/dl) and developed premature atherosclerosis (127–129). Approximately 90% of the plasma cholesterol was distributed in the VLDL and IDL region (127,128). A similar profile was obtained with homozygous apoE-deficient mice and apoE transgenic mice expressing mutant forms of apoE (130–138).

The data obtained from the mutant patients and the apoE-deficient and apoE transgenic mice establish that apoE is required for the clearance of cholesterol ester-rich lipoprotein remnants which float in the VLDL and IDL region (130). The deficient mice developed coronary and pulmonary atherosclerotic lesions within 10 weeks on chow diet which was further exacerbated when the animals were placed on western-type diets (130,132,133).

The importance of apoE in the clearance of the remnants of lipoprotein metabolism was also established by studies using transgenic mice. In this case, overexpression of the rat apoE gene in mice decreased the VLDL and LDL levels and conferred resistance to diet-induced hypercholesterolemia (139,140). The role of apoE in the clearance of apoE-containing lipoprotein remnants by cell receptors is shown schematically in Figure 5.

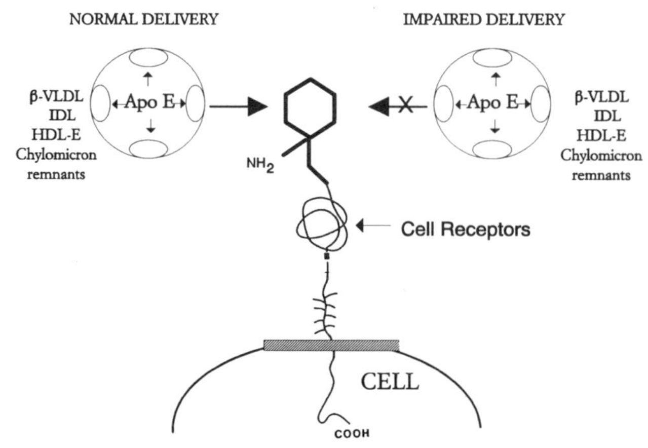

Figure 5. Schematic representation showing cholesterol delivery to cells via the interaction of lipoproteins containing apoE on their surface with cell receptors. X indicates mutations in apoE which prevent the binding of apoE-containing lipoproteins (lipoprotein remnants) to cell receptors. Such mutations are associated with high levels of plasma cholesterol, type III hyperlipidemia, and premature atherosclerosis.

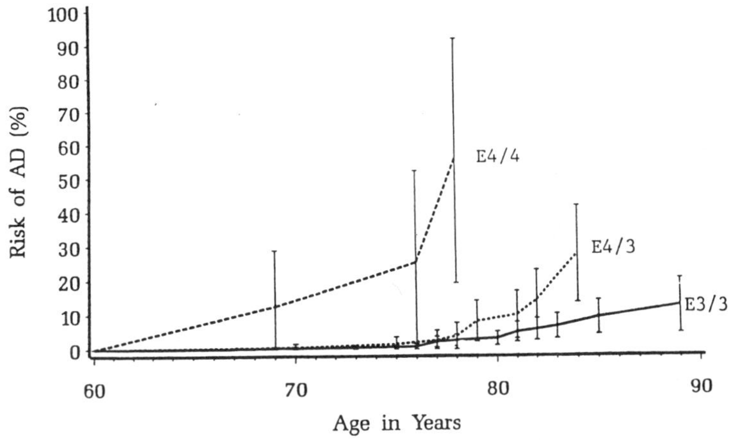

Figure 6. Incidence of Alzheimer's disease in 1030 participants of the Framingham study with different apoE phenotypes. (From Ref. 1 with permission).

9. THE ROLE OF APOE IN ALZHEIMER'S DISEASE

Several observations have linked apoE to AD: (a) AD maps on chromosome 19, and it is linked to apoE (141). (b) ApoE is associated with the senile plaque and conophilic angiopathy of Alzheimer's disease (14,142,143). (c) The ε4 allele is a risk factor for familial late onset Alzheimer's disease (14,144–146). The frequency of the ε4 allele is approximately three times higher in patients with AD than in non-AD subjects (0.42 vs. 0.15), and there is a dosage effect of the ε4 allele on lifetime risk for AD and the age of onset of the disease (Figure 6) (144,146). (d) ApoE can form a stable complex with amyloid peptide β (Aβ) in vitro (147). Differences have been described between the apoE3 and apoE4 in the formation of the complex (148–151). This interaction may be related to the formation of amyloid plaques in the brain. ApoE3, but not apoE4, forms a complex with the microtubule associated proteins tau and Map2C (152–154). This interaction may be related to the formation of neurofibrillary tangles.

10. PUTATIVE MECHANISMS LEADING TO THE PATHOGENESIS OF ALZHEIMER'S DISEASE

The following four hypotheses have been advanced to explain the role of apoE isoforms in the pathogenesis of AD.

- Interaction of apoE with Aβ: The first hypothesis implies extracellular interaction of apoE with Aβ, which may promote the formation of amyloid fibrils that exist in the core of the amyloid plaque in the brain. It has been shown that apoE4 binds more efficiently than apoE3 to Aβ and may favor the formation of the fibrils (148). Initial observations indicated that apoE from cerebrospinal fluid binds with high affinity to Aβ (143,155). *In vitro* experiments have shown that apoE can form a complex with Aβ 1–28 peptide which is stable in 5% SDS at 100EC for 5 min (148). Formation of stable complexes of apoE4 with Aβ is fast whereas formation of complexes of apoE3 with Aβ is slow (requires hours) (148). Interest-

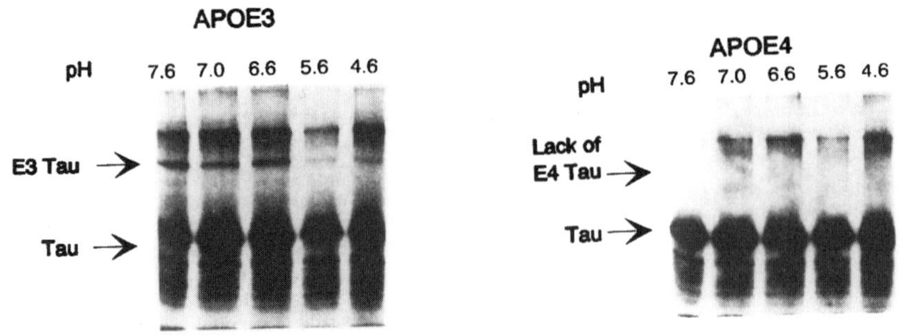

Figure 7. pH-dependent binding of tau to apoE and lack of binding to apoE4. (From Ref. 152 with permission).

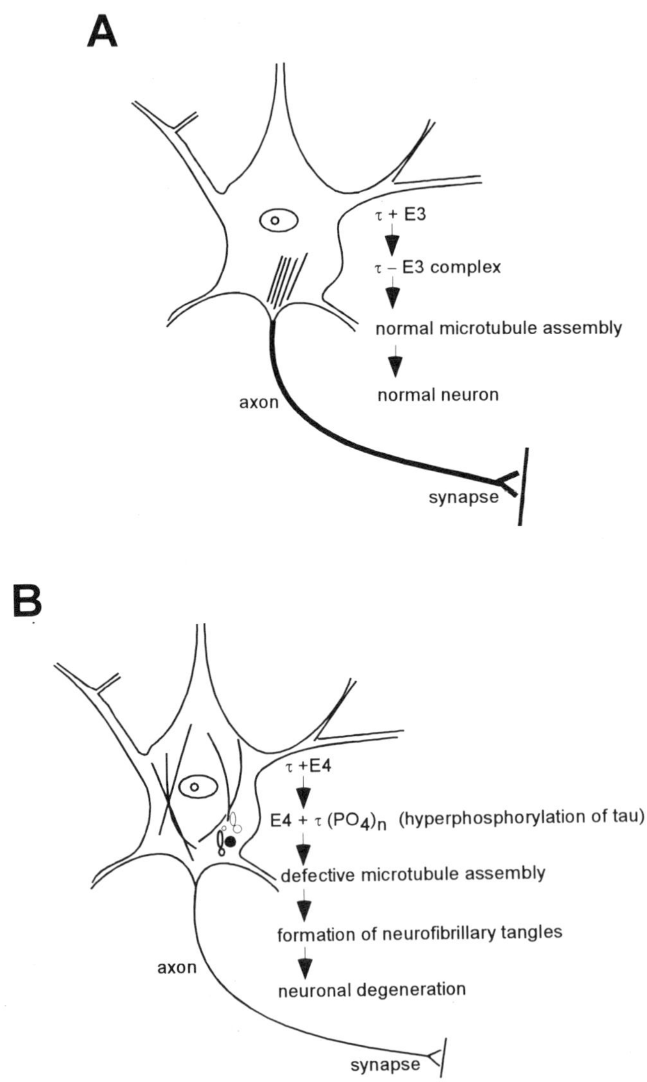

Figure 8. Schematic representation showing how the lack of binding of apoE4 to J may prevent microtubule assembly, promote the formation of pair helical filaments and cause neuronal degeneration.

ingly, two recent studies showed that newly synthesized apoE3 secreted by HEK293 cell interacts stronger with Aβ than apoE4 over the large pH range 4 to 9 (150). However, purification of apoE attenuated the differences in binding of the two apoE isoforms to Aβ (151). Electron microscopy studies showed that a combination of apoE4 with Aβ resulted in the formation of monofibrils 7 nm in diameter. The matrix of monofibrils formed with apoE4 was much denser than that with apoE3 (156).

- Defective interaction of apoE with proteins tau and Map2C: The second hypothesis implies lack of interaction of apoE4 with the microtubule associated protein tau and Map2C (152,154,156). It has been shown that apoE3 binds to tau and Map2C whereas apoE4 does not bind (Figure 7) (156). It has been suggested that binding of tau to apoE3 protects it from phosphorylation in the microtubule-binding domain, while apoE4 does not bind to tau and Map2C (152,156). Consequently, tau is hyperphosphorylated and can no longer bind efficiently and stabilize microtubules, thus leading to the formation of neurofibrillary tangles inside neurons. These events may cause neuronal dysfunction or death (Figure 8A,B) (156).
- Neuronal repair and remodeling: The third hypothesis implies a role of apoE in neuronal repair and remodeling (157). Although the ability of neurons in the CNS to regenerate is minimal, it is well established that specific areas of the brain are capable of synaptic growth to compensate for the loss of synapses caused by neuronal lesions (158,159). Growth is correlated with an increased synthesis of apoE (84,160). In contrast, the ability of the AD brain to compensate for neuronal loss appears to be compromised (161,162). Other studies have shown that apoE3 in the presence of βVLDL increased neurite outgrowth and decreased branching in cultures of dorsal root ganglion neurons, whereas apoE4 decreased neurite outgrowth and branching (163). In addition, apoE4 in the presence of βVLDL can be taken up by neuro-2a cells uptake of apoE4 affects neurite extension as well as tubulin structure (Figures 9 and 10) (157,164). Similarly, neurite extension is affected differently in neuro-2a cells permanently transfected with either apoE3 or apoE4 (165). The requirement for βVLDL indicates that apoE is taken up by neurons via the LDL, LRP, or VLDL receptors (94–101). Upon entry into the cell, it is possible that apoE3 may interact and stabilize the cytoskeleton, and thus support neurite exten-

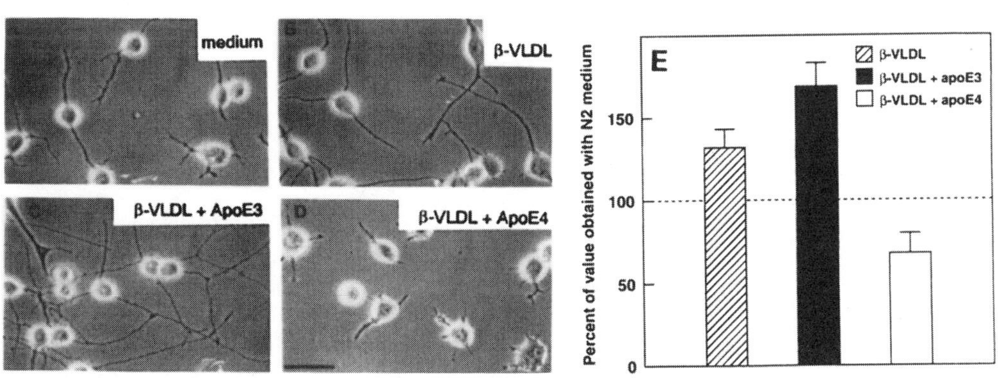

Figure 9. **Panels A–D**: Effect of βVLDL apoE3 and apoE4 bound to βVLDL on neurite outgrowth from neuro 2a cells. Scale bar 25 :M. **Panel E**: Effect on βVLDL and apoE3 and apoE4 bound to βVLDL on neurite extension. Neurite extension in medium alone was taken as 100%. (From Ref 164, with permission).

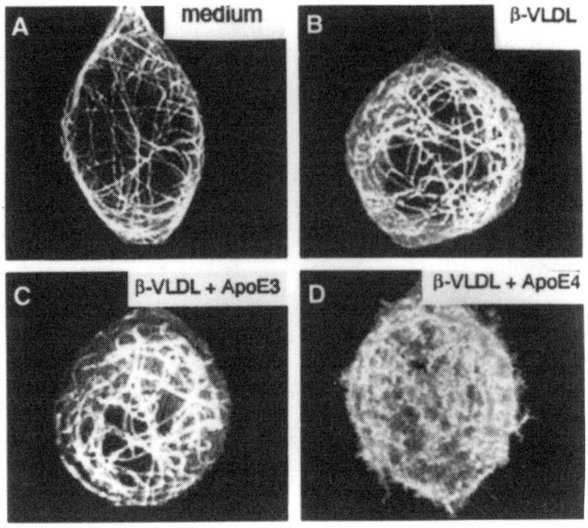

Figure 10. Effect of βVLDL on apoE3 and apoE4 bound to βVLDL on tubulin polymerization and microtubule formation. (From Ref. 164, with permission).

sion. Differences in lipid transport or receptor binding among the apoE isoforms may affect these processes (Figure 11A,B). These interesting findings suggest a potential association of apoE with intracellular cytoplasmic ligands, such as tau and Map2C (152,154,156). The limitations of these in vitro experiments are that βVLDL type particles have not been shown to exist in the brain fluids and that

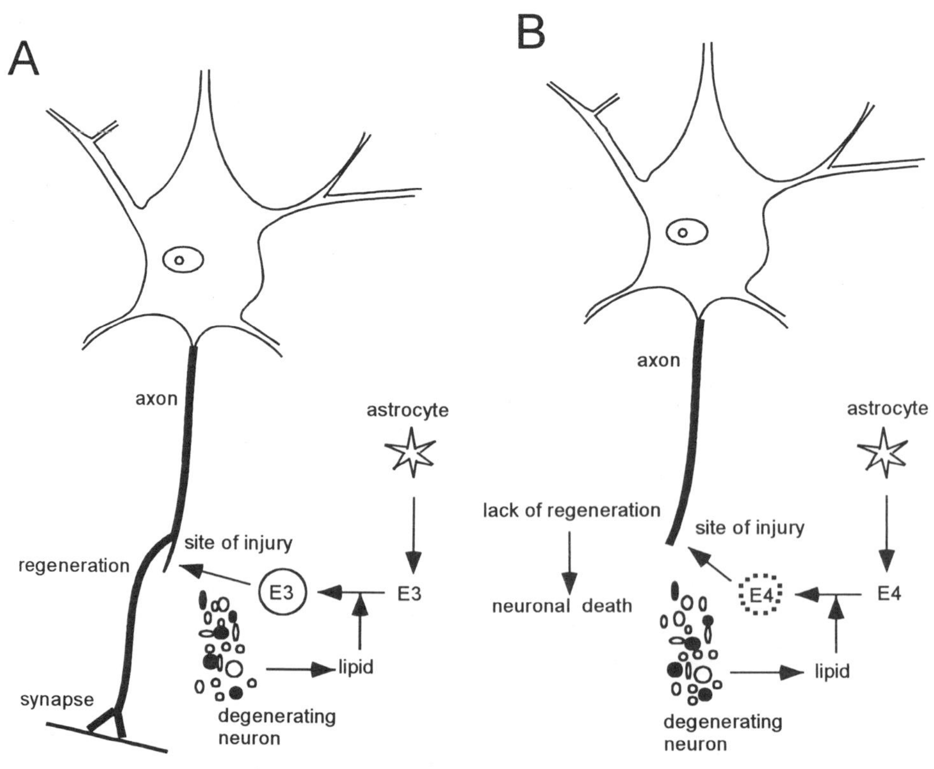

Figure 11. Schematic representation showing how putative differences in lipid transport in subjects with apoE3 (Panel A) or apoE4 (Panel B) phenotype may affect neuronal remodeling.

apoE taken up via the LDL receptor is normally targeted to lysosomes for degradation. Other processes however, such as receptor-mediated retroendocytosis and transcytosis may facilitate the entry of apoE into the cytoplasmic compartment of neurons and subsequent binding to intracellular protein ligands (166). In this regard electron microscopy studies showed the presence of apoE in the cytoplasm of cortical neurons confined in the cell body and the proximal dendrites (167).

- Cholinergic deficits: The fourth hypothesis on the etiology of AD implicates the ε4 allele in the dysfunction of the lipid-transport system, which may affect phospholipid homeostasis and cholinergic function. The perturbation of lipid metabolism in turn may affect the efficiency of synaptogenesis and neuronal cell remodeling during compensatory reinnervation following neuronal cell loss (168,169).

The AD brain has been shown to undergo functional and histological changes. These include a reduction of choline acetyltransferase (ChAT) activity in the hippocampus (162,168,169), frontal cortex (170), and the nucleus basalis of Meynert (171), a population of basal forebrain neurons shown to be particularly affected in AD (172); reduction in nicotinic receptor binding sites (168); loss of proliferating dendrites (161); and a decrease in the length of dendritic spines (173). The severity of the cholinergic deficit in AD patients has been shown to be correlated with the ε4 allele copy number. Thus, ChAT activity in the frontal cortex (170) and the hippocampus (169) of AD patients was shown to be inversely proportional to the ε4 allele copy number. Similarly, the reduction in nicotinic receptor binding sites was ε4 allele-dependent (Figure 12) (168). In addition, the concentration of apoE protein was shown to be reduced in the hippocampus of AD patients carrying the apoε4/ε4 genotype compared to normal controls or AD patients with the apoε3/ε3 genotype (174).

Taken together, these data indicate that the impaired cholinergic system of the AD brain is influenced by the ε4 genotype. The mechanism for this interaction is unclear. The major apolipoproteins in human cerebrospinal fluid are apoE and apoAI, whereas apoB appears to be absent (175). Astrocytes (79–87) within the neuronal lesion synthesize and release apoE which is involved in the mobilization of cholesterol needed for membrane

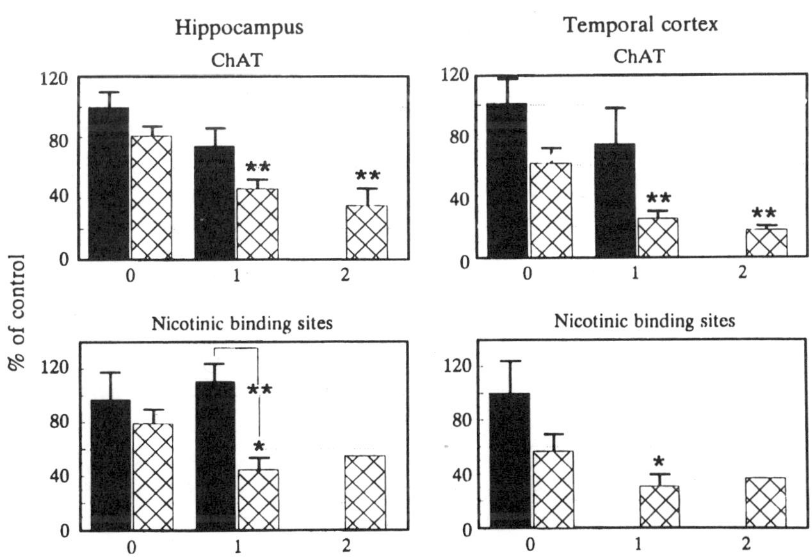

Figure 12. Changes in ChAT activity and nicotinic binding sites in the hippocampus and the temporal cortex of subjects with different apoE genotypes. (From Ref. 168 with permission.)

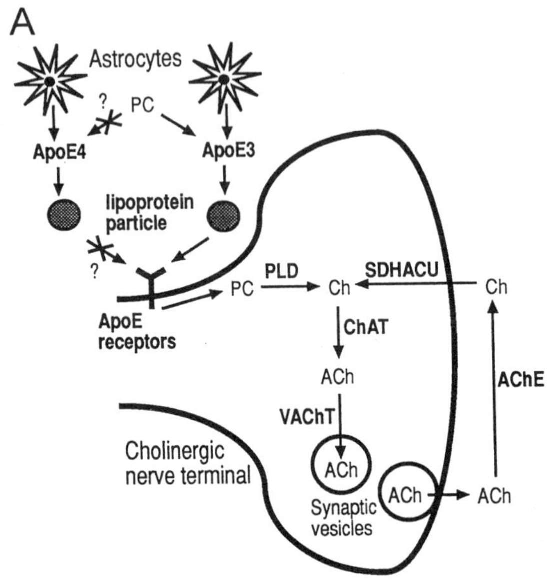

Figure 13. Panel A. Hypothetical role of apoE isoforms in the development of the cholinergic deficit in AD. Abbreviations: phosphatidylcholine (PC); phospholipase D (PLD); choline (Ch); acetylcholine (ACh); choline acetyltransferase (ChAT); vesicular ACh transporter (VAChT); acetylcholinesterase (AChE); sodium-dependent high-affinity choline uptake (SDHACU) system. The diagram shows only the nerve terminal and the most relevant molecules involved. **Panels B,C**. Schematic representation showing neuronal degeneration in AD by direct interaction of Aβ or of amyloid fibers with the cell membranes. This may be prevented by a neuroprotective form of apoE such as apoE3 which binds and removes Aβ.

formation. It has been proposed (168,169) that defects in apoE-dependent lipid transport caused by the apoE4 phenotype may adversely affect cholesterol and phospholipid mobilization from damaged neuronal cells in the brain of AD patients (176,177). This may ultimately cause loss of synaptic integrity and a poor reinnervation process necessary to compensate for neuronal loss.

Alternative by putative defects in phospholipid transport may result in cholinergic deficit in cholinergic neurons. In this class of neurons the membrane phosphatidylcholine (PC) is hydrolyzed possibly by a specialized phospholipase D to generate choline (Ch) which is used for acetylcholine (ACh) synthesis in a reaction catalyzed by choline acetyltransferase (ChAT). The newly formed ACh is transported into secretory vesicles by vesicular ACh transporter (VAChT), and then released into the synapse. Acetylcholinesterase hydrolyzes the intrasynaptic ACh to free Ch, some of which is taken up into the presynaptic terminal by the sodium-dependent high-affinity choline uptake system. The membrane PC used in this process has to be replenished. Some of it is synthesized from free Ch derived from the circulation or from de novo synthesis, and some is taken up as a component of apoE-containing lipoprotein particles which are secreted by the astrocytes. ApoE4 may be less efficient in taking up PC than apoE3 or apoE2. It is possible that the apoE4-containing particles may be transported less well into the neurons than the apoE3- or apoE2-containing particles by apoE-recognizing lipoprotein receptors (VLDL, LDL, or LRP receptors) (94–101). Thus in the individuals carrying the apoE4 allele, the PC loss which occurs in cholinergic neurons, would not be adequately replenished, leading to abnormalities of membrane composition and function, and perhaps loss of viability (Figure 13A).

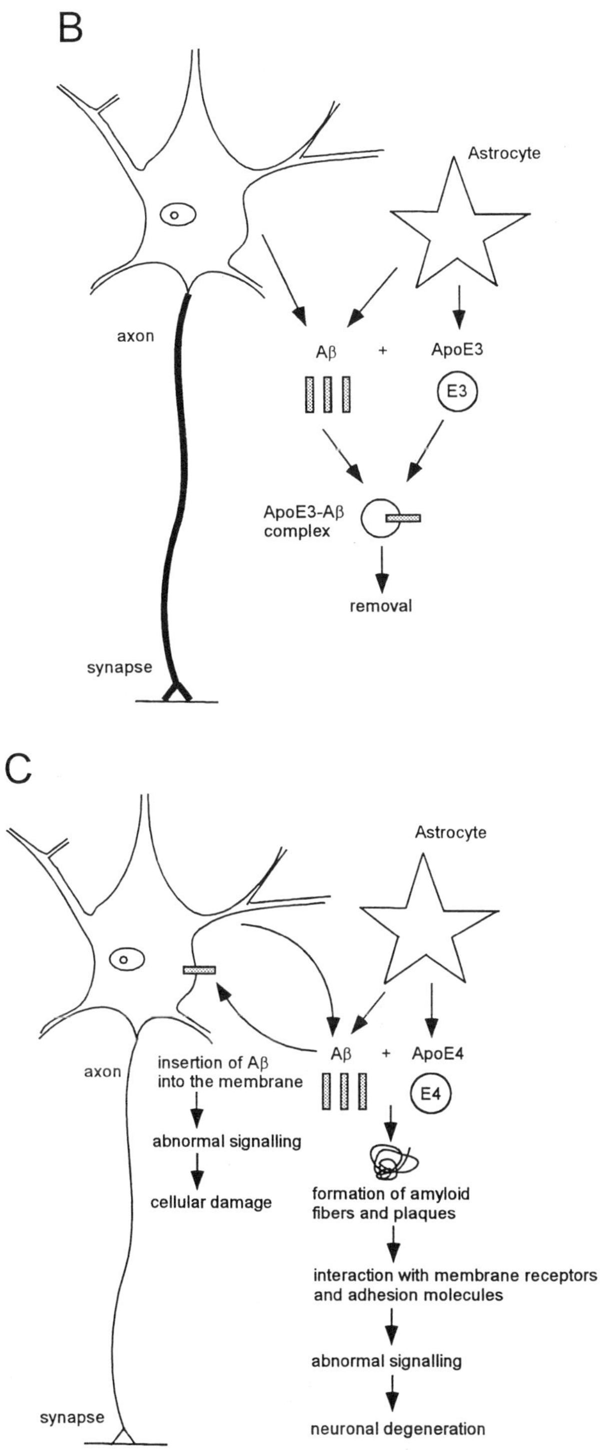

Figure 13. (*Continued*)

Recent studies have shown dendritic remodeling of subcortical neurons of AD patients which may represent a response to neurodegenerative changes (178). Patients with one or two ϵ4 alleles showed more severe neuronal degeneration than those lacking the ϵ4 allele. Furthermore the carriers of the ϵ4 allele displayed significantly reduced ϵ4 copy number dependent dendritic remodeling as compared to the non-ϵ4 subjects. The findings suggest that the ε4 allele affects neuronal remodeling *in vivo* and this may explain the cholinergic deficit and other histopathological changes observed in AD (178).

The mechanism by which apoE isoforms affect dendritic remodeling and neuronal cell viability is not known. Recent studies have shown that Aβ belongs to a class of a-helical peptides which have asymmetric distribution of hydrophobic residues along the axis of the a-helix. This property enables these peptides to adopt a tilted configuration upon insertion into a membrane or a lipid core (179). Recent studies have shown that Aβ acts similarly to the simian immunodeficiency virus (SIV) as a fusogenic peptide and promotes fusion or leakage of small unilamellar lipid vesicle (SUV). It has been proposed that the fusogenic effect of this peptide is related to its unique ability to insert into the membranes at an angle and promote their rearrangement (179). Furthermore, the fusogenic effects of Aβ peptides could be inhibited by apoE2 and apoE3 but not by apoE4 (180).

Other studies have shown that Aβ causes a pH-dependent disruption of membranes containing acidic phospholipids. It was proposed that such disruption of plasma or endosomal membranes could account in principle for the neurotoxic effects of Aβ (181). These studies imply direct perturbation of the cell membrane by Aβ which may account for its neurotoxicity in AD (Figure 13B,C). ApoE2 and apoE3 protected to a greater extent neuronal B12 cells from Aβ as well as hydrogen peroxide cytotoxicity as compared to apoE4 (182). These findings suggest that a protective effect of apoE2 and apoE3 and are consistent with the binding data of Figure 16A and C described below. Such protection may in part be mediated by the strong association of apoE2 and apoE3 with Aβ which may prevent its polymerization and the formation of paired helical filaments.

11. INCREASED REACTIVITY OF APOE2 MONOMERS AND DIMERS TOWARDS Aβ

Population studies have shown convincingly that subjects with either E3/2 or E3/3 phenotype have a decreased risk for development of familial late onset of Alzheimer's disease as compared to subjects with apoE4/4 or E3/4 phenotype (14,146,183). Although there are limited data, it appears that the ϵ2 allele may confer even better protection against Alzheimer's disease as compared to the ϵ3 allele (184). Given the paucity of biochemical data pertinent to the interaction of apoE2 towards Aβ, we studied the reactivity of the naturally occurring apoE2, apoE3 and apoE4 forms towards Aβ. Considering that the reactivity of apoE towards Aβ may be sensitive to the conformational stability of the protein, the experiments were performed with apoE newly expressed by cultures of BHK cells or primary astrocytes electroporated with the recombinant Semliki Forest Virus (SFV) apoE RNAs. The culture medium containing nascent apoE was collected 24–48 hours post-electroporation analyzed for apoE synthesis and utilized for binding studies to Aβ (Figure 14A,B).

This analysis revealed the following important features of the protein that may be relevant to its reactivity towards Aβ. The nascent apoE exists mostly (80–90%) in the modified (sialylated) form as compared to the plasma apoE which consist of approximately 80% of the unmodified form (Figure 15). Binding studies of apoE to Aβ under physiological conditions (pH 7.4, 37°) showed that the intensity of apoE–Aβ complex

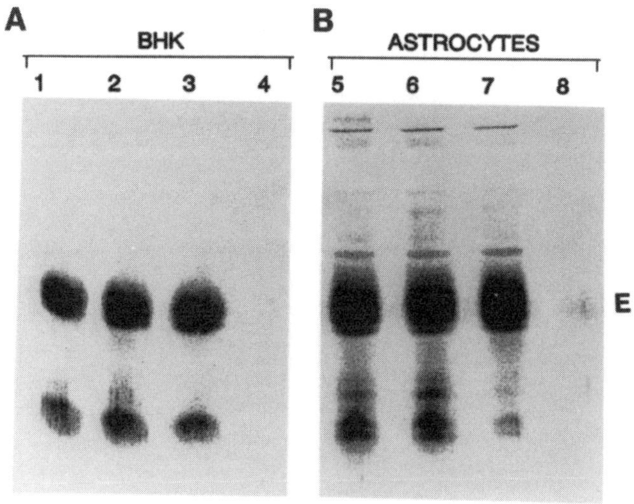

Figure 14. Expression of apoE2, E3 and E4 by recombinant pSFVapoE virus vector in BHK cells and primary astrocytes obtained from apoE deficient mice. SDS-PAGE and immunoblotting of apoE secreted into the media of cells 24 hrs following infection with recombinant apoE containing SF virus. Lanes 4&8: 50:l media of untransfected cells. Lanes 1,2,3: 100:l media from control BHK cells infected with SF virus expressing apoE2, apoE3 and apoE4 respectively. Lanes 5,6,7: 100:l media obtained from primary astrocyte cultures derived of apoE-deficient mice 24 hr following infection with virus expressing apoE2, apoE3 and apoE4 respectively. The blot was treated with antibodies against human apoE. Note: ApoE is heterogeneous indicating sialylation. Lower band of Mr represents a degradation product of apoE.

formed with the different isoforms follows the order apoE2>apoE3>>apoE4 (Figure 16A–C). ApoE2 and apoE3 but not apoE4 form dimers in the culture medium which can be detected by electrophoretic techniques.

The binding of Aβ to apoE2 dimers is greater than that of apoE3 dimers (185). The reactivity of VLDL and E4 or apoE obtained via the baculovirus expression system was different from that observed with the apoE present in the conditioned media (compare Figures 16A,B, 17A,B, 18A,B).

Several factors may account for these differences. As reported previously and demonstrated in the present study, apoE produced by mammalian cells exists in different forms, including monomers, dimers, unmodified, modified, lipid-free, lipid-bound etc. (92,186–188). These apoE forms may have different conformations or different properties and therefore different reactivities towards Aβ. Several observations suggest that the conformation of apoE may be very important in promoting apoE-Aβ interactions. These include the difference in the binding to Aβ of the newly secreted, and presumably native, apoE2, apoE3 and apoE4 monomers as compared to the plasma or commercially available apoE preparations, the changes in the reactivity of apoE following purification or changes in pH and temperature (148,150,151,185) and the increased reactivity of the apoE2 dimers with Aβ as compared to the apoE3 dimers. ApoE2 has two cysteine residues at positions 112 and 158 whereas apoE3 has only the 158 cysteine. X-ray crystallography of the 22 KDa amino terminal portion of the human apoE which contains several 22-residue repeats (189) has shown that it forms a four-helix bundle (73–76). Computer modeling also predicted that similar antiparallel structures exist in the carboxy terminal domain of apoE (77). X-ray crystallography also showed that substitution of Arg for Cys-158 in apoE2 or substitution of Arg for Cys-112 in apoE4 are associated with conformational changes in

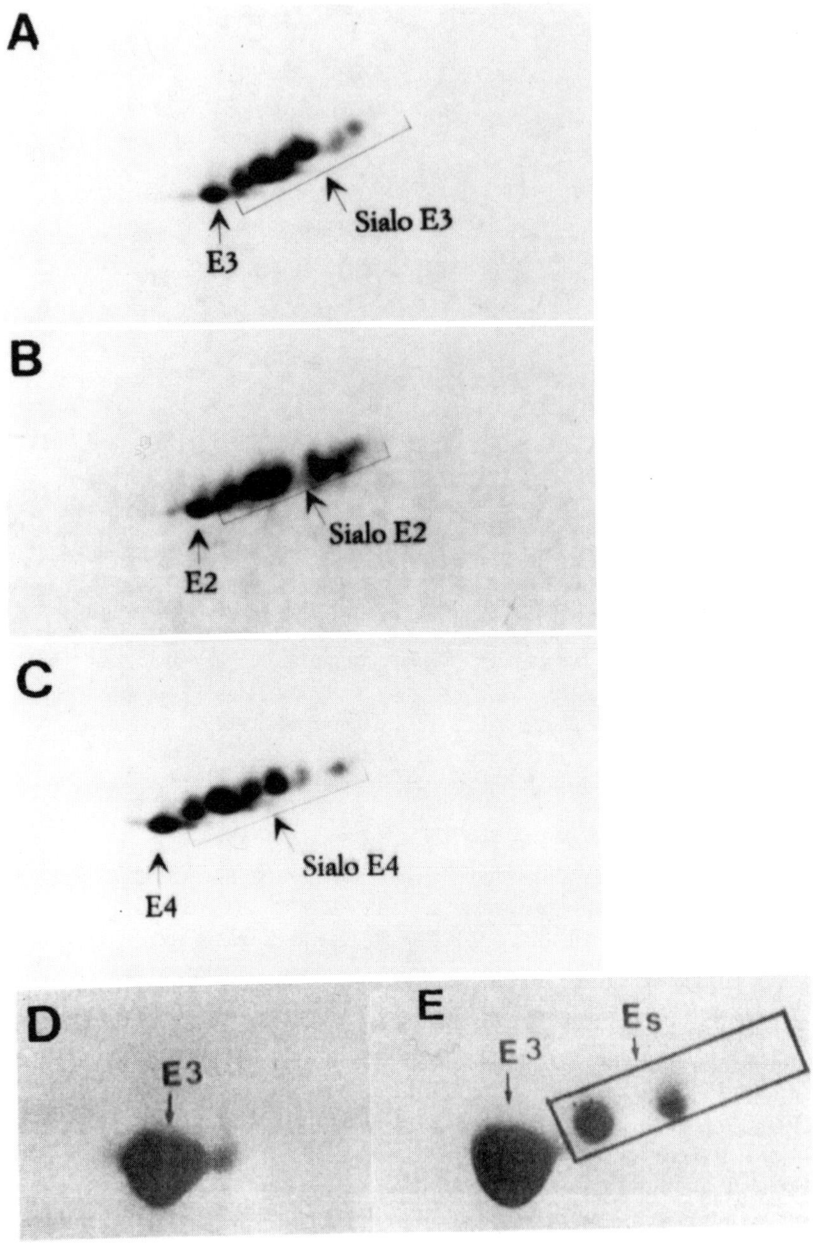

Figure 15. Panels A–C: Two-dimensional gel electrophoresis and autoradiography of apoE secreted by BHK cells expressing apoE2, apoE3 and apoE4. For this analysis ^{35}S-labeled apoE secreted by the different cell clones were labeled with ^{35}S-methionine. The radiolabeled apoE was precipitated with anti-apoE antibodies and analyzed by 2D gel electrophoresis and autoradiography. The gel shows the autoradiogram obtained from this analysis. The relative isoelectric point differences of apoE2, apoE3 and apoE4 was established by comparison with marker plasma apoE which was included in the sample. **Panel D**: two dimensional patterns of plasma apoE. **Panel E**: two-dimensional pattern of plasma apoE3 following treatment with neuraminidase which removes the modified apoE forms.

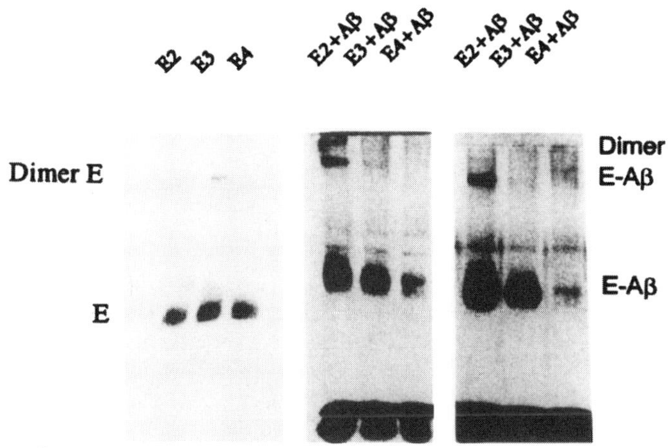

C

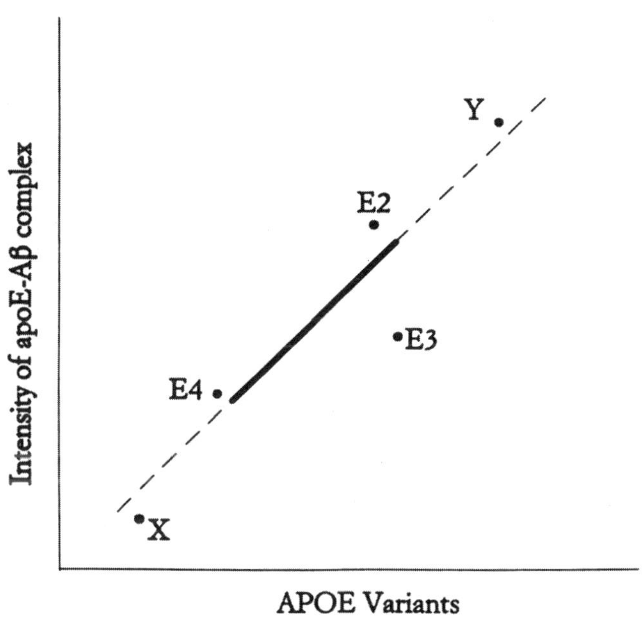

Figure 16. Western blotting of apoE2, apoE3 and apoE4 secreted by BHK-21 cells following interaction with Aβ. **Panel A**: membrane treated with antibodies against apoE. This analysis standardized the quantities of apoE utilized in the binding experiments of Panel B. **Panel B**: the membrane was treated with monoclonal antibodies against Aβ. M indicates the position of the protein MW standards (BioRad prestained). 1=myosin Mr=202 KDa, 2=β-galactosidase Mr=133 KDa, 3=BSA Mr=71 KDa, 4=carbonic anhydrase Mr 41800 KDa, 5=soybean trypsin inhibitor, Mr=30.6 6=lysozyme Mr=17.8 KDa, 7=aprotinin Mr=6.9 KDa. The position of apoE, a dimeric form of apoE, Aβ and complexes of Aβ with the monomeric and dimeric form of apoE are shown. The interaction of apoE with Aβ was performed at 37°C and pH 7.4 for 2 hours using 0.6 μg to 0.7 μg apoE and 15 μg Aβ per assay. **Panel C**: Correlation of the intensity of E-Aβ complex formed between natural apoE isoforms apoE2, apoE3, and apoE4 and the incidence of AD.

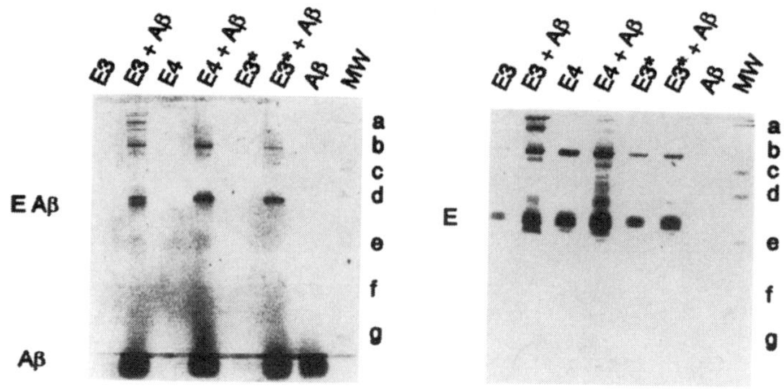

Figure 17. Western blotting of apoE3 and apoE4 produced by the baculovirus expression system following incubation in the presence and absence of Aβ. **Panel A**: filter treated with mouse monoclonal antibody against Aβ. **Panel B**: Blot treated with antibodies against apoE. Lane 1: apoE3; Lane 2: apoE3 + Aβ; Lane 3: apoE4; Lane 4: apoE4 + Aβ; Lane 5: renatured apoE3; Lane 6: renatured apoE3 + Aβ; Lane 7: Aβ; Lane 8: molecular weight standard (New England Biolabs), as follows. a=β galactosidase Mr=175 KDa, b=paramyosin Mr=83 KDa, c=glutamic dehydrogenase Mr=62 KDa, d=aldolase Mr=47.5 KDa, e=triose phosphate isomerase Mr=32.5 KDa, f=lactoglobulin Mr=25 KDa, g=lysozyme Mr=16.5 KDa. Binding assays were performed with 0.6 μg of apoE and 15 μg per assay.

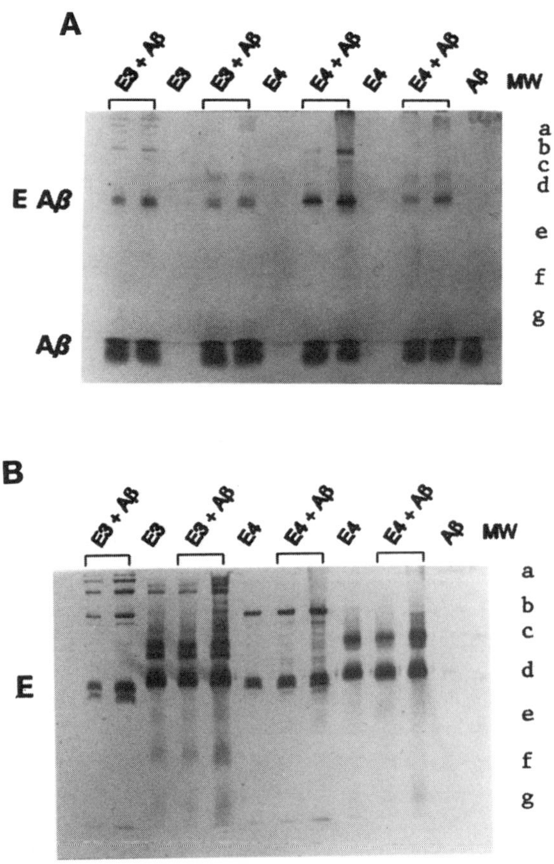

Figure 18. Binding of VLDL E3 and E4 to Aβ. **Panels C,D**: Western blotting of recombinant and VLDL apoE incubated in the presence and absence of Aβ. **Panel A**: Blot treated with antibodies against Aβ. **Panel B**: Blot treated with antibodies against apoE. Lane 1: 3:1 apoE3 + Ab; Lane 2: 9:1 apoE3 + Ab; Lane 3: 3:1 VLDL E3; Lane 4: 3:1 VLDL E3 + Aβ; Lane 5: 9:1 VLDL E3 + Aβ; Lane 6: 3:1 apoE4; Lane 7: 3:1 apoE4 + Aβ; Lane 8: 9:1 apoE4 + Aβ; Lane 9: 3:1 VLDL E4; Lane 10: 3:1 VLDL E4 + Aβ; Lane 11: 9:1 VLDL E4 + Aβ; Lane 12: Aβ alone. The position of the molecular mass standards (which are barely visible) is indicated. They are as in Figure 17.

the 3D structure of apoE and these changes may account for the different reactivities of apoE isoforms towards Aβ (74,76). Alterations in apoE conformation may also occur by binding to different discoidal or spherical lipoprotein particles as compared to the conformation of apoE in solution. In the discoidal particles apoE helices spread out to cover through their hydrophobic surfaces the periphery of the discoidal particle (190,191) whereas in the spherical particles the helices must bend to accommodate their hydrophobic surfaces on the surface of the spherical particle. These differences in conformation may also result in differential reactivities of apoE isoforms towards Aβ. Finally, the length of the Aβ peptide may also be consequential in the apoE-Aβ interaction.

12. IMPLICATIONS OF APOE-Aβ INTERACTIONS IN THE PATHOGENESIS OF AD

The binding data obtained with nascent apoE secreted by mammalian cells in this and in two previous studies (150,151) and the epidemiological data show that strong binding of nascent apoE to Aβ correlates inversely with the risk of developing Alzheimer's disease (10,144–146). ApoE4, which interacts weakly with Aβ, predisposes to AD relative to apoE2 and apoE3. Thus it is possible that apoE2 and apoE3 secreted by astrocytes in the brain in vivo associate more efficiently with Aβ than apoE4. Such association may prevent the polymerization of the Aβ and the formation of amyloid fibers and amyloid plaques in the brain (Figure 16C). Although the concentration of Aβ utilized in the present studies is several orders of magnitude higher than its physiological concentration in the brain milieu (12), it is still possible that under certain conditions the local concentration of Aβ in areas of the brain may rise and cause polymerization and fiber formation. In the presence of apoE2 or apoE3 a soluble apoE-Aβ complex may be formed which may be removed subsequently by cell receptors. Candidates for such removal are the LDL receptor (94–96), the LRP (97,98) and possibly the VLDL receptor (99–101). Reduction in Aβ or increase in the apoE2 or apoE3 concentration in the brain may then have a protective role against AD (Figure 19). Such hypotheses may be addressed with existing or new animal models expressing appropriate forms of apoE and APP isoforms (66,192,193).

Other therapeutic approaches could include: a) inhibition of the activity of the β secretase with concomitant stimulation of the a secretase. This will increase the concentration of the neuroprotective APPsα form (39) and will reduce the concentration of Aβ which is neurotoxic. b) prevention of the maturation of Aβ that leads to fibrillogenesis; c) prevention of the binding of Aβ associated proteins or inhibiting the inflammatory response that is associated with the Aβ deposit.

Potential functions of the brain endothelium in preventing the pathogenesis or protection from AD are not known. Smooth muscle cells have been shown to synthesize apoE (88,89). Thus apoE secreted by the smooth muscle cells could prevent the formation of amyloid fibers which are deposited between endothelial cells and smooth muscle cells. In addition, receptor-mediated transcytosis of apoE might allow apoE synthesized by smooth muscle cells or other tissues to enter the brain compartment.

Both the genetic and the biochemical data suggest an important role of apoE in the pathogenesis of Alzheimer's disease. Systematic analysis of the interactions of apoE with its intra- and extracellular protein ligands as well as with cell membranes and cell receptors may provide new insights into the molecular events which lead to the pathogenesis of AD.

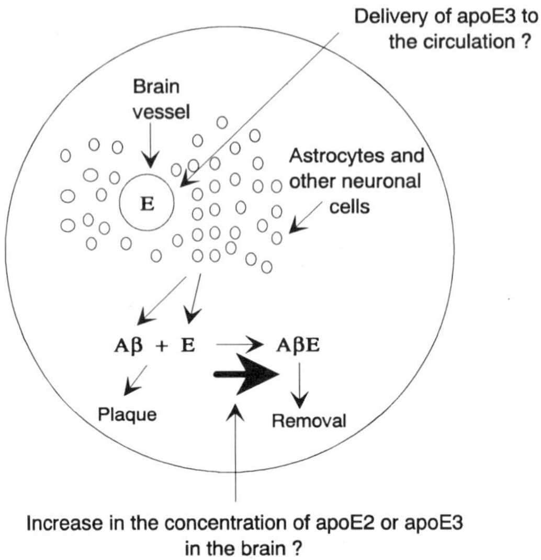

Figure 19. Schematic representation showing potentially beneficial effects of increase in the levels of apoE2 and/or apoE3 in the cerebrospinal fluid or the plasma for the protection against AD.

ACKNOWLEDGMENTS

This work was supported by grants from the National Institutes of Health (AG12717) and grants from the Alzheimer's Association (TRGM 95-016), the RTD Program (BMH4-CT96-0898), and Kos Pharmaceuticals of Miami, Florida. We thank Anne Plunkett and Sophia Lavrentiadou for excellent technical assistance, and Dr. Carmela Abraham for carefully reading the manuscript.

REFERENCES

1. Myers, R.H., Schaefer, E.J., Wilson, P.W.F., D'Agostino, R., Ordovas, J.M., Espino, A., Au, R., White, R.F., Knoefel, J.E., Cobb, J.L., McNulty, K.A., Beiser, A., Wolf, P.A. 1996. Apolipoprotein E,4 association with dementia in a population-based study: The Framingham Study. *Amer. Acad. Neur.* 46:673–677.
2. Bachman, D.L., Wolf, P.A., Linn, R., Knoefel, J.E., Cobb, J., Belanger, A., D'Agostino, R.B., White, L.R. 1992. Prevalence of dementia and probable senile dementia of the Alzheimer type in the Framingham Study. *Neurology* 42:115–119.
3. Price, D.L., Sisodia, S.S. 1994. Cellular and molecular biology of Alzheimer's disease and animal models. *Annu. Rev. Med.* 45:435–446.
4. Goate, A., Chartier-Harlin, M.-C., Mullan, M., et al. 1991. Segregation of a missense mutation in the amyloid precursor protein with familial Alzheimer's disease. *Nature* 349:704–706.
5. Chartier-Harlin, M.-C., Crawford, F., Houlden, H., et al. 1991. Early-onset Alzheimer's disease caused by mutations at codon 717 of the β amyloid precursor protein gene. *Nature* 353:844–846.
6. Murrell, J., Farlow, M., Ghetti, B., Benson, M.D. 1991. A mutation in the amyloid precursor protein associated with hereditary Alzheimer's disease. *Science* 254:97–99.
7. Mullan, M., Crawford, F., Axelman, K., Houlden, H., Lilius, L., Winblad, B., Lannfeldt., L. 1993. A pathogenic mutation for probable Alzheimer's disease in the APP gene at the N-terminus of β-amyloid. *Nature Genet.* 1:345–347.
8. Levy, E., Carman, M.D., Fernandez-Madrid, I.J., Power, M.D., Lieberburg, I., van Duinen, S.G., Bots, G.T.A.M., Luyendijk, W., Frangione, B. 1990. Mutation of the Alzheimer's disease amyloid gene in hereditary cerebral hemorrhage, Dutch-type. *Science* 248:1124–1126.

9. Hyman, B.T., West, H.L., Rebeck, G.W., Buldyrev, S.V., Mantegna, R.N., Ukleja, M., Havlin, S., Stanley, H.E. 1995. Quantitative analysis of senile plaques in Alzheimer disease: Observation of log-normal size distribution and molecular epidemiology of differences associated with apolipoprotein E genotype and trisomy 21 (Down syndrome). *Proc. Natl. Acad. Sci. USA* 92:3586–3590.
10. Sherrington, R., Rogaev, E.I., Liang, Y., Rogaeva, E.A., Levesque, G., Ikeda, M., Chi, H., Lin, C., Li, G., Holman, K., Tsuda, T., Mar, L., Foncin, J.-F., Bruni, A.C., Montesi, M.P., Sorbi, S., Rainero, I., Pinessi, L., Nee, L., Chumakov, I., Pollen, D., Brookes, A., Sanseau, P., Polinsky, R.F., Wasco, W., DaSilva, H.A.R., Haines, J.L., Pericak-Vance, M.A., Tanzi, R.E., Roses, A.D., Fraser, P.E., Rommens, J.M., St. George-Hyslop, P.H. 1995. Cloning of a gene bearing missense mutations in early-onset familial Alzheimer's disease. *Nature* 375:754–760.
11. Levy-Lahad, E., Wasco, W., Poorkaj, P., Romano, D.M., Oshima, J., Pettingell, W.H., Yu, C-E., Jondro, P.D., Schmidt, S.D., Wang, K., Crowley, A.C., Fu, Y-H., Guenette, S.Y., Galas, D., Nemens, E., Wijsman, E.M., Bird, T.D., Schellenberg, G.D., Tanzi, R.E. 1995. Candidate gene for the chromosome 1 familial Alzheimer's disease locus. *Science* 269:973–977.
12. Selkoe, D.J. 1996. Amyloid β-protein and the genetics of Alzheimer's disease. *J. Biol. Chem.* 271:18295–18298.
13. Selkoe, D.J. 1994. Cell biology of the amyloid β-protein precursor and the mechanism of Alzheimer's disease. *Annu. Rev. Cell Biol.* 10:373–403.
14. Strittmatter, W.J., Saunders, A.M., Schmechel, D., Pericak-Vance, M., Enghild, J., Salvesen, G.S., Roses, A.D. 1993. Apolipoprotein E: High-avidity binding to β-amyloid and increased frequency of type 4 allele in late-onset familial Alzheimer disease. *Proc. Natl. Acad. Sci. USA* 90:1977–1981.
15. Selkoe, D.J. 1991. The molecular pathology of Alzheimer's disease. *Neuron* 6:487–498.
16. Kidd, M. 1963. Paired helical filaments in electron microscopy of Alzheimer's disease. *Nature* 197:192–193.
17. Terry, R.D. 1963. The fine structure of neurofibrillary tangles in Alzheimer's disease. *J. Neuropathol. Exp. Neurol.* 22:629–641.
18. Nukina, N., Ihara, Y. 1986. One of the antigenic determinants of paired helical filaments is related to a tau protein. *J. Biochem.* 99:1541–1544.
19. Kosik, K.S., Joachim, C.L., Selkoe, D.J. 1986. Microtubule-associated protein, tau, is a major antigenic component of paired helical filaments in Alzheimer's disease. *Proc. Natl. Acad. Sci. USA* 83:4044–4048.
20. Glenner, G.G. 1980. Amyloid deposits and amyloidosis. *New Engl. J. Med.* 203:1283–1291.
21. Glenner, G.G., Wong, C.W. 1984. Alzheimer's disease: Initial report of the purification and characterization of a novel cerebrovascular amyloid protein. *Biochem. Biophys. Res. Commun.* 120:885–890.
22. Masters, C.L., Multhaup, G., Simms, G., Pottigiesser, J., Martins, R.N., Beyreuther, K. 1985a. Neuronal origin of a cerebral amyloid: Neurofibrillary tangles of Alzheimer's disease contain the same protein as the amyloid of plaque cores and blood vessels. *EMBO J.* 4:2757–2763.
23. Kang, J., Lemaire, H.G., Unterbeck, A., Salbaum, J.M., Masters, C.L., Grezeschik, K.-H., Multhaup, G., Beyreuther, K., Muller-Hill, B. 1987. The precursor of Alzheimer's disease amyloid Aβ protein resembles a cell-surface receptor. *Nature* 325:733–736.
24. Ponte, P., Gonzalez-DeWhitt, P., Schilling, J., Miller, J., Hus, D., Greenberg, B., Davis, K., Wallace, W., Lieberburg, I., Fuller, F., Cordell, B. 1988. A new A4 amyloid mRNA contains a domain homologous to serine proteinase inhibitors. *Nature* 331:525–527.
25. Tanzi, R.E., McClatchey, A.I., Lamperti, E.D., Villa-Komaroff, L.L., Gueslla, J.F., Neve, R.L. 1988. Protease inhibitor domain encoded by an amyloid protein precursor mRNA associated with Alzheimer's disease. *Nature* 331:528–530.
26. Kitaguchi, N., Takahashi, Y., Tolkushima, Y., Shiojiri, S., Ito, H. 1988. Novel precursor of Alzheimer's disease amyloid protein shows protease inhibitory activity. *Nature* 331:530–532.
27. Van Nostrand, W.E., Wagner, S.L., Suzuki, M., Choi, B.H., Farrow, J.S., Geddes, J.W., Cotman, C.W., Cunningham, D.D. 1989. Protease nexin-II, a potent anti-chymotrypsin, shows identity of amyloid β-protein precursor. *Nature* 341:546–549.
28. Oltersdorf, T., Fritz, L.C., Schenk, D.B., Lieberburg, I., Johnson-Wood, K.L., Beattie, E.C., Ward, P.J., Blacher, R.W., Dovey, H.F., Sinha, S. 1989. The secreted form of the Alzheimer's amyloid precursor protein with the Kunitz domain is protease nexin-II. *Nature* 341:144–147.
29. Smith, R.P., Higuchi, D.A., Broze, Jr., G.J. 1990. Platelet coagulation factor Xia-inhibitor, a form of Alzheimer amyloid precursor protein. *Science* 248:1126–1128.
30. Schmaier, A., Dahl, L., Hasan, A., Cines, D., Bauer, K., Van Nostrand, W. 1995. Factor IXa inhibition by protease nexin-2/amyloid β-protein precursor on phospholipid vesicles and cell membranes. *Biochemistry* 34:1171–1178.

31. Kounnas, M.Z., Moir, R.D., Rebeck, G.W., Bush, A.I., Argraves, W.S., Tanzi, R.E., Hyman, B.T., Strickland, D.K. 1995. LDL receptor-related protein, a multifunctional apoE receptor, binds secreted β-amyloid precursor protein and mediates its degradation. *Cell* 82:331–340.
32. Saitoh, T., Sundsmo, M., Roch, J.-M., Kimura, N., Cole, G., Schubert, D., Oltersdorf, T., Schenk, D.B. 1989. Secreted form of amyloid β protein precursor is involved in the growth regulation of fibroblasts. *Cell* 58:615–622.
33. Van Nostrand, W.E., Schmaier, A.H., Farrow, J.S., Cunningham, D.D. 1990. Protease nexin-II (amyloid beta-protein precursor): A platelet alpha-granule protein. *Science* 248:745–748.
34. Klier, F.G., Cole, G., Stallcup, W., Schubert, D. 1990. Amyloid β-protein precursor is associated with extracellular matrix. *Brain Res.* 515:336–342.
35. Selkoe, D.J., Podisny, M.B., Joachim, C.L., Vickers, E.A., Lee, G., Oltersdorf, T. 1988. β-Amyloid precursor protein of Alzheimer disease occurs as 110- to 135-kilodalton membrane-associated proteins in neural and non-neural tissues. *Proc. Natl. Acad. Sci. USA* 85:7341–7345.
36. Esch, F.S., Keim, P.S., Beattie, E.C., Blacher, R.W., Culwell, A.R., Oltersdorf, T., McClure, D., Ward, P.J. 1990. Cleavage of amyloid β peptide during constitutive processing of its precursor. *Science* 248:1122–1124.
37. Sisodia, S.S., Koo, E.H., Beyreuther, K., Unterbeck, A., Price, D.L. 1990. Evidence that beta-amyloid protein in Alzheimer's disease is not derived by normal processing. *Science* 248:492–495.
38. Seubert, P., Oltersdorf, T., Lee, M.G., et al. 1993. Secretion of β-amyloid precursor protein cleaved at the amino terminus of the β-amyloid peptide. *Nature* 361:260–262.
39. Furukawa, K., Sopher, B.L., Rydel, R.E., Begley, J.G., Pham, D.G., Martin, G.M., Fox, M., Mattson, M.P. 1996. Increased activity-regulating and neuroprotective efficacy of "-secretase-derived secreted amyloid precursor protein conferred by a C-terminal heparin-binding domain. *J. Neurochem.* 67:1882–1896.
40. Haass, C., Koo, E.H., Mellon, A., Hung, A.Y., Selkoe, D.J. 1992. Targeting of cell-surface β-amyloid precursor protein to lysosomes: alternative processing into amyloid-bearing fragments. *Nature* 357:500–503.
41. Golde, T.E., Estus, S., Younkin, L.H., Selkoe, D.J., Younkin, S.G. 1992. Processing of the amyloid protein precursor to potentially amyloidogenic derivatives. *Science* 255:728–730.
42. Ripellino, J.A., Vassilacopoulou, D., Robakis, N.K. 1994. Solubilization of full-length amyloid precursor proteins from PC12 cell membranes. *J. Neurosci. Res.* 39:211–218.
43. Bhasin, R., Gregori, L., Morozov, I., Goldgaber, D. 1994. A processing pathway for amyloid β-protein precursor generates a secreted form with an intact carboxyl terminus. *Amyloid: Int. J. Exp. Clin. Invest.* 1:221–231.
44. Conn, K.J., Papastoitsis, G., Meckelein, B., Abraham, C.R. 1994. Identification of full length β-amyloid precursor protein in human neuronal and non-neuronal cell culture supernatant: A possible extracellular source for the generation of Aβ. *Amyloid: Int. J. Exp. Clin. Invest.* 1:232–239.
45. Haass, C., Schlossmacher, M.G., Hung, A.Y., et al. 1992. Amyloid β-peptide is produced by cultured cells during normal metabolism. *Nature* 359:322–325.
46. Shoji, M., Golde, T.E., Ghiso, J., et al. 1992. Production of the Alzheimer amyloid β protein by normal proteolytic processing. *Science* 258:126–129.
47. Seubert, P., Vigo-Pelfrey, C., Esch, F., et al. 1992. Isolation and quantification of soluble Alzheimer's β-peptide from biological fluids. *Nature* 359:325–327.
48. Busciglio, J., Gabuzda, D.H., Matsudaira, P., Yankner. B.A. 1993. Generation of β-amyloid in the secretory pathway in neuronal and non-neuronal cells. *Proc. Natl. Acad. Sci. USA* 90:2092–2096.
49. Abraham, C.R., Selkoe, D.J., Potter, H. 1988. Immunochemical identification of the serine protease inhibitor, a_r-antichymotrypsin in the brain amyloid deposits of Alzheimer's disease. *Cell* 52:487–501.
50. Abraham, C.R., Shirahama, T., Potter, H. 1990. a1-antichymotrypsin is associated solely with amyloid deposits containing the β-protein. Amyloid and cell localization of a1- antichymotrypsin. *Neurobiol. Aging* 11:123–129.
51. Pasternak, J.M., Abraham, C.R., Van Dyke, B.J., Potter, H., Younkin, S.G. 1989. Astrocytes in Alzheimer's disease gray matter express a1-antichymotrypsin mRNA. *Am. J. Pathol.* 135:827–833.
52. Koo, E.H., Abraham, C.R., Potter, H., Cork, L.C., Price, D.L. 1991. Developmental expression of a1-antichymotrypsin in brain may be related to astrogliosis. *Neurobiol. Aging* 12:495–501.
53. Eikelenboom, P., Stam, F.C. 1982. Immunoglobulins and complement factors in senile plaques: An immunoperoxidase study. *Acta Neuropathol.* 57:239–242.
54. Itagaki, S., McGeer, P.L., Akiyama, H., Zhu, S., Selkoe, D.J. 1989. Relationship of microglia and astrocytes to amyloid deposits of Alzheimer's disease. *J. Neuroimmunol.* 24:173–182.
55. Eddleston, M., Mucke, L. 1993. Molecular profile of reactive astrocytes-implications for their role in neurologic disease. *Neuroscience* 54:15–36.

56. Siman, R., Card, J.P., Nelson, R.B., Davis, L.G. 1989. Expression of β-amyloid precursor protein in reactive astrocytes. *Neuron* 3:275–285.
57. Abraham, C.R., Kanemaru, K., Mucke, L. 1993. Expression of cathepsin G-like and a1-antichymotrypsin-like proteins in reactive astrocytes. *Brain Res.* 621:222–232.
58. Razzaboni, B.L., Papastoitsis, G., Koo, E.H., Abraham, C.R. 1992. A calcium-stimulated serine protease from monkey brain degrades the β-amyloid precursor protein. *Brain Res.* 589:207–216.
59. Whitson, J.S., Selkoe, D.J., Cotman, C.W. 1989. Amyloid β protein enhances the survival of hippocampal neurons in vitro. *Science* 243:1488–1490.
60. Pike, C.J., Burdick, D., Walencewicz, A.J., Glabe, C.G., Cotman, C.W. 1993. Neurodegeneration induced by beta-amyloid peptides in vitro: The role of peptide assembly state. *J. Neurosci.* 13:1676–1687.
61. Pike, C.J., Cummings, B.J., Monzavi, R., Cotman, C.W. 1994. β amyloid-induced changes in cultured astrocytes parallel reactive astrocytosis associated with senile plaques in Alzheimer's disease. *Neuroscience* 63:517–531.
62. Lorenzo, A., Yankner B. 1994. β amyloid-induced changes in cultured astrocytes parallel reactive astrocytosis associated with senile plaques in Alzheimer's disease. *Proc. Nat. Acad. Sci. USA* 91:12243–12247.
63. Meda, L., Lasatella, M.A., Szendru, G.I., Otvos, L., Baron, P., Villalba, M., Ferrari, D., Rossi, F. 1995. Activation of microglial cells by β amyloid protein and interferon-(. *Nature* 374:647–650.
64. Thomas, T., Thomas, G., McLendon, C., Sutton, T., Mullan, M. 1996. β amyloid-mediated vasoactivity and vascular endothelial damage. *Nature* 380:168–171.
65. Yankner, B.A., Duffy, L.K., Kirschner, D.A. 1990. Neurotrophic and neurotoxic effects of amyloid beta protein: Reversal by tachykinin neuropeptides. *Science* 250:279–282.
66. Games, D., Adams, D., Alessandrini, R., Barbour, R., Berthelette, P., Blackwell, C., Carr, T., Clemens, J., Donaldson, T., Gillespie, F., Guido, T., Hagopian, S., Johnson-Wood, K., Khan, K., Lee, M., Leibowitz, P., Lieberburg, I., Little, S., Masliah, E., McConlogue, L., Montoya-Zavala, M., Mucke, L., Paganini, L., Penniman, E., Power, M., Schenk, D., Seubert, P., Snyder, B., Soriano, F., Tan, H., Vitale, J., Wadsworth, S., Wolozin, B., Zhao, J. 1995. Alzheimer-type neuropathology in transgenic mice overexpressing V717F β-amyloid precursor protein. *Nature* 373:523–527.
67. Zannis, V.I., Kardassis, D., Zanni, E.E. 1993. Genetic mutations affecting human lipoproteins, their receptors and their enzymes. *Adv. Hum. Genetics* (H. Harris and K. Hirschorn, eds.) Plenum Press, NY. Vol. 21, pp. 145–319.
68. Rall, S.C., Weisgraber, K.H., Mahley, R.W. 1981. Human apolipoprotein E: The complete amino acid sequence. *J. Biol. Chem.* 257:4171–4178.
69. Zannis, V.I., McPherson, J., Goldberger, G., Karathanasis, S.K., Breslow, J.L. 1984. Synthesis, intracellular processing and signal peptide of human apoE. *J. Biol. Chem.* 259:5495–5499.
70. Paik, Y.K., Chang, D.J., Reardon, C.A., Davies, G.E., Mahley, R.W., Taylor, J.M. 1985. Nucleotide sequence and structure of the human apolipoprotein E gene. *Proc. Natl. Acad. Sci USA* 82:3445–3451.
71. Das, H.K., McPherson, J., Bruns, G.A.P., Karathanasis, S.K., Breslow, J.L. 1985. Isolation, characterization and mapping to chromosome 19 of the human apolipoprotein E gene. *J. Biol. Chem.* 260:6240–6246.
72. McLean, J.W., Elshourbagy, N.A., Chang, D.J., Mahley, R.W., Taylor, J.M. 1984. Human apolipoprotein E mRNA. cDNA cloning and nucleotide sequencing of a new variant. *J. Biol. Chem.* 259:6498–7504.
73. Wilson, C., Wardell, M.R., Weisgraber, K.H., Mahley, R.W., Agard, D.A. 1991. Three-dimensional structure of the LDL receptor-binding domain of human apolipoprotein E. *Science* 252:1817–1822.
74. Wilson C., Agard A., Mahley, R.W., Weisgraber, K.H. 1993. Crystal structures of the common apolipoprotein E variants: Insights into functional mechanisms. *NATO ASI Series.* Vol. II, 73:81–96.
75. Wilson, C. et al. 1994. Salt bridge relay triggers defective LDL receptor binding by a mutant apolipoprotein. *Structure* 2:713–718.
76. Dong, L-M., Wilson, C., Wardell, M.R., Simmons, T., Mahley, R.W., Weisgraber, K.H., Agard, D.A. 1994. Human apolipoprotein E. Role of arginine 61 in mediating the lipoprotein preferences of the E3 and E4 isoforms. *J. Biol. Chem.* 269:22358–22365.
77. Nolte, R.T., Atkinson, D. 1992. Conformational analysis of apolipoprotein A-I and E-3 based on primary sequence and circular dichroism. *Biophys. J.* 63:1221.
78. Newman, T.C., Dawson, P.A., Rudel, L.L., Williams, D.L. 1985. Quantitation of apolipoprotein E mRNA in the liver and peripheral tissues of nonhuman primates. *J. Biol. Chem.* 260:2452–2457.
79. Skene, J.H.P., Shooter, E.M. 1983. Denervated sheath cells secrete a new protein after nerve injury. *Proc. Natl. Acad. Sci. USA* 80:4169–4173.
80. Muller, H.W., Gebicke-Harter, P.J., Hangen, D.H., Shooter, E.M. 1985. A specific 37,000-dalton protein that accumulates in regenerating but not in non-regenerating mammalian nerves. *Science* 228:499–501.

81. Ignatius, M.J,. Gebicke-Harter, P.J., Skene, J.H.P., et al. 1986. Expression of apolipoprotein E during nerve degeneration and regeneration. *Proc. Natl. Acad. Sci. USA* 83:1125–1129.
82. Handelmann, G.E., Boyles, J.K., Weisgraber, K.H., Mahley, R.W., Pitas, R.E. 1992. Effects of apolipoprotein, β-very low density lipoproteins, and cholesterol on the extension of neurites by rabbit dorsal root ganglion neurons in vitro. *J. Lipid Res.* 33:1677–1688.
83. Murakami, M., Yshio, Y., Morino, Y., Ohta, T., Matsukado, Y. 1988. Immunohistochemical localization of apolipoprotein E in human glial neoplasm. *J. Clin. Invest.* 82:177–188.
84. Poirier, J., Hess, M., May, P.C., Finch, C.E. 1991. Astrocytic apolipoprotein E mRNA and GFAP mRNA in hippocampus after entorhinal cortex lesioning. *Mol. Brain Res.* 11:97–106.
85. Pitas, R.E., Boyles, J.K., Lee, S.H., Foss, D., Mahley, R.W. 1987. Astrocytes synthesize apolipoprotein E and metabolize apolipoprotein E-containing lipoproteins. *Biochim. Biophys. Acta* 917:148–161.
86. Boyles, J., Zoellner, C.D., Anderson, L.J., Kosik, L.M., Pitas, R.E., Weisgraber, K.H., Hui, D.Y., Mahley, R.W., Gebicke-Harter, P.J., Ignatius, M.J., Shooter, E.M.. 1989. A role of apolipoprotein E, apolipoprotein A-I, and low density lipoprotein receptors in cholesterol transport during regeneration and remyelination of the rat sciatic nerve. *J. Clin. Invest.* 83:1015–1031.
87. Boyles, J.K., Pitas, R.E., Wilson, E., Mahley, R.W., Taylor, J.M. 1985. Apolipoprotein E associated with astrocytic glia of the central nervous system and with non-myelinating glia of the peripheral nervous system. *J. Clin. Invest.* 76:1501–1513.
88. Driscoll, D.M., Getz, G.S. 1984. Extrahepatic synthesis of apolipoprotein E. *J. Lipid Res.* 25:1368–1374.
89. Hussain, M.M., Bucher, N.L.R., Faris, B., Franzblau, C., Zannis, V.I. 1988. Tissue-specific post-translational modification of rat apoE synthesis of sialated apoE forms by neonatal rat aortic smooth muscle cells. *J. Lipid Res.* 29:915–923.
90. Zanni, E.E., Kouvatsi, A., Hadzopoulou-Cladaras, M., Krieger, M., Zannis, V.I. 1989. Expression of apoE gene in Chinese hamster cells with a reversible defect in O-glycosylation. Glycosylation is not required for apoE secretion. *J. Biol. Chem.* 264:9137–9140.
91. Wernette-Hammond, M.E., Lauer, S.J., Corsini, A., Walker, D., Taylor, J.M., Rall, Jr., S.C. 1989. Glycosylation of human apolipoprotein E. The carbohydrate attachment site is threonine 194. *J. Biol. Chem.* 264:9094–9101.
92. Hussain, M.M., Zanni, E.E., Kelly, M., Zannis, V.I. 1989. Synthesis, modification and flotation properties of rat hepatocyte apolipoproteins. *Biochim. Biophys. Acta* 101, 90–101.
93. Hussain, M.M., Roghani, A., Cladaras, C., Zanni, E.E., Zannis, V.I. 1991. Secretion of lipid poor nascent human apoA-I, apoCIII and apoE by cell clones expressing the corresponding genes. *Electrophoresis* 12:273–283.
94. Pitas, R.E., Innerarity, T.L., Mahley, R.W. 1980. Cell surface receptor binding of phospholipid-protein complexes containing different ratios of receptor-active and inactive E apoprotein. *J. Biol. Chem.* 255:5454–5460.
95. Rebeck, W.G., Reiter, J.S., Strickland, D.K., Hyman, B.T. 1993. E in sporadic Alzheimer's disease: Allelic variation and receptor interactions *Neuron* 11:575–580.
96. Handelmann, G.E., Boyles, J.K., Weisgraber, K.H., Mahley, R.W., Pitas, R.E. 1992. Effects of apolipoprotein E, β-very low density lipoproteins, and cholesterol on the extension of neurites by rabbit dorsal root ganglion neurons *in vitro*. *J. Lipid Res.* 33:1677–1688.
97. Wolf, B.B., Lopes, M.B., VandenBerg, S.R., Gonias, S.L. 1992. Characterization and immunohistochemical localization of a-2-macroglobulin receptor (low density lipoprotein receptor-related protein) in human brain. *Am. J. Pathol.* 141:37–42.
98. Tooyama, I., Kawamata, T., Akiyama, H., Moestrup, S.K., Gliemann, J., McGeer, P. 1993. Immunohistochemical study of a_2-macroglobulin receptor in Alzheimer and control postmortem human brain. *Mol. Chem. Neuropathol.* 18:153–159.
99. Takahashi, S., Kawarabayasi, Y., Nakai, T., Sakai, J., Yamamoto, T. 1992. Rabbit very low density lipoprotein receptor: A low density lipoprotein receptor-like protein with distinct ligand specificity. *Proc. Natl. Acad. Sci. USA* 89:9252–9256.
100. Okuizumi, K., Onodera, O., Namba, Y., Ikeda, K., Yamamoto, T., Seki, K., Ueki, A., Nanko, S., Tanaka, H., Takahashi, H., Oyanagi, K., Mizusawa, H., Kanazawa I., Tsuji, S. 1995. Genetic association of the very low density lipoprotein (VLDL) receptor gene with sporadic Alzheimer's disease. *Nature Genet.* 11:207–209.
101. Wyne, K.L., Pathak, K., Seabra, M.C., Hobbs, H.H. 1996. Expression of the VLDL receptor in endothelial cells. *Arterioscler. Thromb. Vasc. Biol.* 16:407–415.
102. Innerarity, T.L., Friedlander, E.J., Rall, Jr., S.C., Weisgraber, K.H., Mahley, R.W. 1983. The receptor-binding domain of human apolipoprotein E. Binding of apolipoprotein E fragments. *J. Biol. Chem.* 258:12341–12347.

103. Weisgraber, K.H., Innerarity, T.L., Harder, K.J., Mahley, R.W. Milne, R.W., Marcel, Y.L. J.T. Sparrow, J.T. 1983. The receptor-binding domain of human apolipoprotein E. *J. Biol. Chem.* 258:12348–12354.
104. Pitas, R.E., Innerarity, T.L., Mahley, R.W. 1980. Cell surface receptor binding of phospholipid-protein complexes containing different ratios of receptor-active and inactive E apoprotein. *J. Biol. Chem.* 255:5454–5460.
105. Lalazar, A., Weisgraber, K.H., Rall Jr., S.C., Giladi, H., Innerarity, T.L., Levanon, A.Z., Boyles, J.K., Amit, B., Gorecki, M., Mahley, R.W. 1988. Site-specific mutagenesis of human apolipoprotein E. Receptor binding activity of variants with single amino acid substitutions. *J. Biol. Chem.* 263:3542–3545.
106. Weisgraber, K.H., Rall, Jr., S.C., Mahley, R.W., Milne, R.W., Marcel, Y.L., Sparrow, J.T. 1986. Human apolipoprotein E. *J. Biol. Chem.* 261:2068–2076.
107. Cardin, A.D., Hirose, N., Blankenship, D.T., Jackson, R.L., Harmony, J.A.K. 1986. Binding of a high reactive heparin to human apolipoprotein E: Identification of two heparin-binding domains. *Biochem. Biophys. Res. Commun.* 134:783–789.
108. Ji, Z-S., Fazio, S., Lee, Y-L., Mahley, R.W. 1994. Secretion-capture role for apolipoprotein E in remnant lipoprotein metabolism involving cell surface heparan sulfate proteoglycans. *J. Biol. Chem.* 269:2764–2772.
109. Aggerbeck, L.P., Wetterau, J.R., Weisgraber, K.H., Wu, C.S.C., Lindgren, F.T. 1988. Human apolipoprotein E3 in aqueous solution. II. Properties of the amino- and carboxyl-terminal domains. *J. Biol. Chem.* 263:6249–6258.
110. Weisgraber, K.H. 1990. Apolipoprotein E distribution among human plasma lipoproteins: Role of the cysteine-arginine interchange at residue 112. *J. Lipid Res.* 31:1503–1511.
111. Dong, L.-M., Weisgraber, K.H. 1996. Human apolipoprotein E4 domain interaction. *J. Biol. Chem.* 271:19053–19057.
112. Westerlund, J.A., Weisgraber, K.H. 1994. Discrete carboxyl-terminal segments of apolipoprotein E mediate lipoprotein association and protein oligomerization. *J. Biol. Chem.* 268:15745–15750.
113. Zannis, V.I., Just, P.W., Breslow, J.L. 1981. Human apolipoprotein E isoprotein subclasses are genetically determined. *Am. J. Hum. Genet.* 33:11–24.
114. Zannis, V.I., Breslow, J.L., Utermann, G., Mahley, R.W., Weisgraber, K.H., Havel, R.J., Goldstein, J.L., Brown, M.S., Schonfeld, G., Hazzard, W.R., Blum, C.B. 1982. Proposed nomenclature of apoE isoprotein genotypes and phenotypes. *J. Lipid Res.* 23:911–914.
115. Zannis, V.I., Breslow, J.L. 1981. Human very low density lipoprotein apolipoprotein E isoprotein polymorphism is explained by genetic variation and post-translational modification. *Biochemistry* 20:1033–1041.
116. Davignon, J., Gregg, R.E., Sing, C.F. 1988. Apolipoprotein E polymorphism and atherosclerosis. *Arteriosclerosis* 8:1–21.
117. Rall, Jr., S.C., Weisgraber, K.H., Innerarity, T.L., R.W. 1982. Structural basis for receptor binding heterogeneity of apolipoprotein E from type III hyperlipoproteinemic subjects. *Proc. Natl. Acad. Sci. USA* 79:4696–4703.
118. Weisgraber, K.H., Innerarity, T.L., Mahley, R.W. 1982. Abnormal lipoprotein receptor-binding activity of the human E apoprotein due to cysteine–arginine interchange at a single site. *J. Biol. Chem.* 257:2518–2521.
119. Mahley, R.W., Innerarity, T.L., Rall, Jr., S.C., Weisgraber, K.H., Taylor, J.M. 1990. Apolipoprotein E: genetic variants provide insights into its structure and function. *Curr. Opin. Lipidol.* 1:87–95.
120. Dong, L.-M., Parkin, S., Trakhanov, S.D., Rupp, B., Simmons, T., Arnold, K.S., Newhouse, Y.M., Innerarity, T.L., Weisgraber, K.H. 1996. Novel mechanism for defective receptor binding of apolipoprotein E2 in type III hyperlipoproteinemia. *Nature Struc. Biol.* 3:718–722.
121. Wardell, M.R., Brennan, S.O., Janus, E.D., R. Fraser, R., Carrell, R.W. 1987. Apolipoprotein E2-Christchurch (136 Arg—Ser). New variant of human apolipoprotein E in a patient with type III hyperlipoproteinemia. *J. Clin. Invest.* 80:483–490.
122. Rall, S.C. Jr., Newhouse, Y.M., Clarke, H.R.G., Weisgraber, K.H., McCarthy, B.J., Mahley, R.W., Bersot, T.P. 1989. Type III hyperlipoproteinemia associated with apolipoprotein E phenotype E3/3: structure and genetics of an apolipoprotein E3 variant. *J. Clin. Invest.* 83:1095–1101.
123. Mann, W.A., Gregg, R.E., Sprecher, D.L., Brewer, Jr., H.B. 1989. Apolipoprotein E-1 Harrisburg: a new variant of apolipoprotein E dominantly associated with type III hyperlipoproteinemia. *Biochim. Biophys. Acta* 1005:239–244.
124. Wardell, M.R., Weisgraber, K.H., Havekes, L.M., Rall, Jr., S.C. 1989. Apolipoprotein E3-Leiden contains a seven-amino acid insertion that is a tandem repeat of residues 121–127. *J. Biol. Chem.* 264:21205–21210.
125. van den Maagdenberg, A.M., de Knijff, P., Stalenhoef, A.F., Gevers Leuven, J.A., Havekes, L.M., Frants, R.R. 1989. Apolipoprotein,3-Leiden allele results from a partial gene duplication in exon 4. *Biochem. Biophys. Res. Commun.* 165:851–857.

126. Smit, M., de Knijff, P., van der Kooij-Meijs, E., Groenendijk, C., van den Maagdenberg, A.M., Gevers Leuven, J.A., Stalenhoef, A.F., Stuyt, P.M., Frants, R.R., Havekes, L.M. 1990. Genetic heterogeneity in familial dysbetalipoproteinemia. The E2 (lys146----gln) variant results in a dominant mode of inheritance. *J. Lipid Res.* 31:45–53.
127. Ghiselli, G., Schaefer, E.J., Gascon, P., Brewer Jr., H.B. 1981. Type III hyperlipoproteinemia associated with apolipoprotein E deficiency. *Science* 214:1239–1241.
128. Schaefer, E.J., Gregg, R.E., Ghiselli, G., Forte, T.M., Ordovas, J.M., Zech, L.A., Brewer, H.B. 1986. Familial apolipoprotein E deficiency. *J. Clin. Invest.* 78:1206–1219.
129. Cladaras, C., Hadzopoulou-Cladaras, M., Felber, B.K., Pavlakis, G., Zannis, V.I. 1987. The molecular basis of a familial apoE deficiency. An acceptor splice site mutation in the third intron of the deficient apoE gene. *J. Biol. Chem.* 262:2310–2315.
130. Plump, A.S., Smith, J.D., Hayek, T., Aalto-Setala, K., Walsh, A., Verstuyft, J.D., Rubin, E.M., Breslow, J.L. 1992. Severe hypercholesterolemia and atherosclerosis in apolipoprotein E-deficient mice created by homologous recombination in ES cells. *Cell* 71:343–353.
131. Piedrahita, J.A., Zhang, S.H., Hagaman, J.R., Oliver, P.M., Maeda, N. 1992. Generation of mice carrying a mutant apolipoprotein E gene inactivated by gene targeting in embryonic stem cells. *Proc. Natl. Acad. Sci. USA* 89:4471–4475.
132. Zhang, S.H., Reddick, R.L., Piedrahita, J.A., Maeda, N. 1992. Spontaneous hypercholesterolemia and arterial lesions in mice lacking apolipoprotein E. *Science* 2588:468–471.
133. Reddick, R.L., Zhang, S.H., Maeda, N. 1994. Atherosclerosis in mice lacking apoE. Evaluation of lesional development and progression. *Arterioscler. Thromb.* 14:141–147.
134. Vanden Maagdenberg, A.M.J., Hofker, M.H., Krimpenfort, P.J.A., DeBruijn, I., Van Vlijmen, B., vander Boom, H., Havekes, L.M., Frants, R.R. 1993. Transgenic mice carrying the apolipoprotein E3-Leiden gene exhibit hyperlipoproteinemia. *J. Biol. Chem.* 268:10540–10545.
135. Fazio, S., Lee, Y., Sheng, X., Rall, Jr., S.C. 1993. Type III hyperlipoproteinemic phenotype in transgenic mice expressing dysfunctional apolipoprotein. *J. Clin. Invest.* 92:1497–1503.
136. Fazio, S., Horie, Y., Simonet, W.S., Weisgraber, K.H., Taylor, J.M., Rall, Jr., S.C. 1994. Altered lipoprotein metabolism in transgenic mice expressing low levels of a human receptor-binding-defective apolipoprotein E variant. *J. Lipid Res.* 35:408–416.
137. Fazio, S., Sanan, D.A., Lee, Y.-L., Ji, Z.-S., Mahley, R.W., Rall, Jr., S.C. 1994. Susceptibility to diet-induced atherosclerosis in transgenic mice expressing a dysfunctional human apolipoprotein E (Arg112, Cys142). *Arterioscler. Thromb.* 14:1873–1879.
138. Huang, Y., Schwendner, S.W., Rall, Jr., S.C., Mahley, R.W. 1996. Hypolipidemia and hyperlipidemic phenotypes in transgenic mice expressing human apolipoprotein E2*. *J. Biol. Chem.* 271:29146–29151.
139. Shimano H., Yamada, N., Katsuki, M., Shimada, M., Gotoda, T., Harada, K., Murase, T., Fukazawa, C.., Takaku, F., Yazaki, Y. 1992. Overexpression of apolipoprotein E in transgenic mice: Marked reduction in plasma lipoprotein except high density lipoprotein and resistance against diet-induced hypercholesterolemia. *Proc. Natl. Acad. Sci. USA* 89:1750–1754.
140. Shimano H., N. Yamada, N., Katsuki, M., Yamamoto, K., Gotoda, T., Harada, K., Shimada, M., Yazaki, Y. 1992. Plasma lipoprotein metabolism in transgenic mice overexpressing apolipoprotein E. *J. Clin. Invest.* 90:2084–2091.
141. Pericak-Vance, M.A., Bebout, J.L., Gaskell Jr., P.C., Yamaoka, L.H., Hung, W.-Y., Alberts, M.J., Walker, A.P., Bartlett, R.J., Haynes, C.A., Welsh, K.A., Earl, N.L., Heyman, A., Clark, C.M., Roses, A.D. 1991. Linkage studies in familial Alzheimer disease: Evidence for chromosome 19 linkage. *Am. J. Hum. Genet.* 48:1034–1050.
142. Namba, Y., Tomonaga, M., Kawasaki, H., Otomo, E., Ikeda, K. 1991. Apolipoprotein E immunoreactivity in cerebral amyloid deposits and neurofibrillary tangles in Alzheimer's disease and kuru plaque amyloid in Creutzfeldt-Jakob disease. *Brain Res.* 541:163–166.
143. Wisniewski, T., Frangione, B. 1992. Apolipoprotein E: A pathological chaperon protein in patients with cerebral and systemic amyloid. *Neuroscience Letters* 135:235–238.
144. Corder, E.H., Saunders, A.M., Risch, N.J., Strittmatter, W.J., Schmechel, D.E., Gaskell Jr., P.C., Rimmler, J.B., Locke, P.A., Conneally, P.M., Schmader, K.E., Small, G.W., Roses, A.D., Haines, J.L., Pericak-Vance, M.A. 1994. Protective effect of apolipoprotein E type 2 allele for late-onset Alzheimer disease. *Nature Genetics* 7:180–184.
145. Saunders, A.M., Strittmatter, W.J., Schmechel, D., St. George-Hyslop, P.H., Pericak-Vance, M.A., Joo, S.H., Rosi, B.L., Gusella, J.F., Crapper-MacLachlan, D.R., Alberts, M.J., Hulette, C., Crain, B., Goldgaber, D., Roses, A.D. 1993. Association of apolipoprotein E allele,4 with late-onset familial and sporadic Alzheimer's disease. *Neurology* 43:1467–1472.

146. Corder, E.H., Saunders, A.M., Strittmatter, W.J., Schmechel, D.E., Gaskell, P.C., Small, G.W., Roses, A.D., Haines, J.L., Pericak-Vance, M.A. 1993. Gene dose of apolipoprotein E type 4 allele and the risk of Alzheimer's diseases in late onset families. *Science* 261:921–923.
147. Naslund, J., Thyberg, J., Tjernberg, L.O., Wernstedt, C., Karlstrom, A.R., Bodganovic, N., Gandy, S.E., Lannfelt, L. Terenius, L., Nordstedt, C. 1995. Characterization of stable complexes involving apolipoprotein E and the amyloid β peptide in Alzheimer's disease brain. *Neuron* 15:219–228.
148. Strittmatter, W.J., Weisgraber, K., Huang, D.Y., Dong, L.-M., Salvesen, G.S, Pericak-Vance, M., Schmechel, D., Saunders, A.M., Goldgaber, D., Roses, A.D. 1993. Binding of human apolipoprotein E to synthetic amyloid beta peptide: Isoform-specific effects and implications for late-onset Alzheimer disease. *Proc. Natl. Acad. Sci. USA* 90:8098–8102.
149. Zhou, Z., Smith, J.D., Greengard, P., Gandy, S. 1996. Alzheimer amyloid-β peptide forms denaturant-resistant complex with type E3 but not type E4 isoform of native apolipoprotein E. *Molecul. Med.* 2:175–180.
150. LaDu, M.J., Falduto, M.T., Manelli, A.M., Reardon, C.A., Getz, G.S., Frail, D.E. 1994. Isoform-specific binding of apolipoprotein E to β-amyloid. *J. Biol. Chem.* 269:23403–23406.
151. LaDu, M.J., Pederson, T.M., Frail, D.E., Reardon, C.A., Getz, G.S., Falduto, M.T. 1995. Purification of apolipoprotein E attenuates isoform-specific binding to β-amyloid. *J. Biol. Chem.* 9039–9042.
152. Strittmatter, W.J., Saunders, A.M., Goedert, M., Weisgraber, K.H., Dong, L.-M., Jakes, R., Huang, D., Pericak-Vance, M., Schmechel, D., Roses, A.D. 1994. Isoform-specific interactions of apolipoprotein E with microtubule-associated protein tau: Implications for Alzheimer's disease. *Proc. Natl. Acad. Sci. USA* 91:11183–11186.
153. Strittmatter, W.J., Roses, A.D. 1995. Apolipoprotein E and Alzheimer disease. *Proc. Natl. Acad. Sci. USA* 92:4725–4727.
154. Huang, D.Y., Goedert, M., Jakes, R., Weisgraber, K.H., Garner, C.C., Saunders, A.M., Pericak-Vance, M.A., Schmechel, D.E., Roses, A.D., Strittmatter, W.J. 1994. Isoform-specific interactions of apolipoprotein E with the microtubule-associated protein MAP2C: Implications for Alzheimer's disease. *Neurosci. Lett.* 182:55–58.
155. Wisniewski, T., Golabek, A., Matsubara, E., Ghiso, J., Frangione, B. 1993. Apolipoprotein E: Binding to soluble Alzheimer's β-amyloid. *Biochem. Biophys. Res. Commun.* 192:359–365.
156. Strittmatter, W.J., Weisgraber, K.H., Godert, M., Saunders, A.M., Huang, D., Corder, E.H., Dong, L.-M., Jakes, R., Alberts, M.J., Gilbert, J.R., Han, S.-H., Hulette, C., Einstein, G., Schmechel, D.E., Pericak-Vance, M.A., Roses, A.D. 1994. Hypothesis: Microtubule instability and paired helical filament formation in the Alzheimer disease brain are related to apolipoprotein E genotype. *Exper. Neurol.* 125:163–171.
157. Weisgraber, K.H., Mahley, R.W. 1996. Human apolipoprotein E: The Alzheimer's disease connection. *FASEB J.* 10:1485–1494.
158. Matthews, D.A., Cotman, C., Lynch, G. 1976. An electron microscopic study of lesion-induced synaptogenesis in the dentate gyrus of the adult rat. I. Magnitude and time course of degeneration. *Brain Research* 115:1–21.
159. Matthews, D.A., Cotman, C., Lynch, G. 1976. An electron microscopic study of lesion-induced synaptogenesis in the dentate gyrus of the adult rat. II. Reappearance of morphologically normal synaptic contacts. *Brain Research* 115:23–41.
160. Poirier, J., Baccichet, A., Dea, D., Gauthier, S. 1993. Cholesterol synthesis and lipoprotein reuptake during synaptic remodelling in hippocampus in adult rats. *Neuroscience* 55:81–90.
161. Flood, D.G., Coleman, P.D. 1986. Failed compensatory dendritic growth as a pathophysiological process in Alzheimer's disease. *Can. J. Neurol. Sci.* 13:475–479.
162. Ransmayr, G., Cervera, P., Hirsch, E., Ruberg, M., Hersh, L.B., Duyckaerts, C., Hauw, J.-J., Delumeau, C., Agid, Y. 1989. Choline acetyltransferase-like immunoreactivity in the hippocampal formation of control subjects and patients with Alzheimer's disease. *Neuroscience* 32:701–714.
163. Nathan, B.P., Bellosta, S., Sanan, D.A., Weisgraber, K.H., Mahley, R.W., Pitas, R.E. 1994. Differential effects of apolipoproteins E3 and E4 on neuronal growth in vitro. *Science* 264:850–852.
164. Nathan, B.P., Chang, K.-C., Bellosta, S., Brisch, E., Ge, N., Mahley, R.W., Pitas, R.E. 1995. The inhibitory effect of apolipoprotein E4 in neurite outgrowth is associated with microtubule depolymerization. *J. Biol. Chem.* 270:19791–19799.
165. Bellosta, S., Nathan, B.P., Orth, M., Dong, L.-N., Mahley, R.W., Pitas, R.E. 1995. Stable expression and secretion of apolipoproteins E3 and E4 in mouse neuroblastoma cells produces differential effects on neurite outgrowth. *J. Biol. Chem.* 270:27063–27071.
166. Beisiegel, U., Weber, W., Heeren, J., Hilpert, J. 1995. Intracellular consequences of chylomicron uptake. *Circulation* 92:691.

167. Han, S.-H., Einstein, G., Weisgraber, K.H., Strittmatter, W.J., Saunders, A.M., Pericak-Vance, M., Roses, A.D., Schmechel, D.E. 1994. Apolipoprotein E is localized to the cytoplasm of human cortical neurons: A light and electron microscopic study. *J. Neuropath. Exper. Neurol.* 53:535–544.
168. Poirier, J., Delisle, M.C., Quirion, R., Aubert, I., Farlow, M., Lahiri, D., Hui, S., Bertrand, P., Nalbantoglu, J., Gilfix, B.M., Gauthier, S. 1995. Apolipoprotein E4 allele as a predictor of cholinergic deficits and treatment outcome in Alzheimer disease. *Proc. Natl. Acad. Sci. USA* 92:12260–12264.
169. Poirier, J. 1994. Apolipoprotein E in animal models of CNS injury and in Alzheimer's disease. *Trends Neurosci.* 17:525–530.
170. Soininen, H., Kosunen, O., Helisalmi, S., Mannermaa, A., Paljdrvi, L., Talasniemi, S., Ryyndnen, M., Sr, P.R. 1995. A severe loss of choline acetyltransferase in the frontal cortex of Alzheimer patients carrying apolipoprotein,4 allele. *Neurosci. Lett.* 187:79–82.
171. Etienne, P., Robitaille, Y., Wood, P., Gauthier, S., Nair, N.P.V., Quirion, R. 1986. Nucleus basalis neuronal loss, neuritic plaques and choline acetyltransferase activity in advanced Alzheimer's disease. *Neuroscience* 19:1279–1291.
172. Whitehouse, P.J., Price, D.L., Struble, R.G., Clark, A.W., Coyle, J.T., DeLong, M.R. 1982. Alzheimer's disease and senile dementia: loss of neurons in the basal forebrain. *Science* 215:1237–1239.
173. de Ruiter, J.P., Uylings, J.B.M. 1987. Morphometric and dendritic analysis of fascia dentata granule cells in human aging and senile dementia. *Brain Res.* 402:217–229.
174. Bertrand, P., Poirier, J., Oda, T., Finch, C.E., Pasinetti, G.M. 1995. Association of apolipoprotein E genotype with brain levels of apolipoprotein E and apolipoprotein J (clusterin) in Alzheimer disease. *Mol. Brain Res.* 33:174–178.
175. Roheim, P.S., Carey, M., Forte, T., Vega, G.L. 1979. Apolipoproteins in human cerebrospinal fluid. *Proc. Natl. Acad. Sci. USA* 76:4646–4649.
176. Nitsch, R.M., Blusztajn, J.K., Pitas, A.G., Slack, B.E., Growdon, J.H., Wurtman, R.J. 1992. Evidence for a membrane defect in Alzheimer disease brain. *Proc. Natl. Acad. Sci. USA* 89:1671–1675.
177. Wurtman, R.J. 1992. Choline metabolism as a basis for the selective vulnerability of cholinergic neurons. *Trends Neurosci.* 15:117–122.
178. Arendt, T., Schindler, C., Bruckner, M.K., Eschrich, K., Bigl, V., Zedlick, D., Marcova, L. 1997. Plastic neuronal remodeling is impaired in patients with Alzheimer's disease carrying apolipoprotein E4 allele. *J. Neurosci.* 17:516–529.
179. Pillott, T., Goethals, M., Vanloo, B., Talussot, C., Brasseur, R., Vandekerckhove, J., Rosseneu, M., Lins, L. 1996. Fusogenic properties of the C-terminal domain of the Alzheimer β-amyloid peptide*. *J. Biol. Chem.* 271:28757–28765.
180. Pillot, T., Goethals M., Vanloo B., Lins L., Brasseur R.,, L., Vandekerckhove, J., Rosseneu, M. 1996. Specific modulation of the fusogenic properties of the Alzheimer β-amyloid peptide by apolipoprotein E isoforms. *Eur. J. Biochem.* in press, 1997.
181. McLaurin, J., Chakrabarty, A. 1996. Membrane disruption by Alzheimer β-amyloid peptides mediated through specific binding to either phospholipids or gangliosides. *J. Biol. Chem.* 271:26482–26489.
182. Miyata, M., Smith, J.D. 1996. Apolipoprotein E allele-specific antioxidant activity and effects on cytotoxicity by oxidative insults and β-amyloid peptides. *Nature Genet.* 14:55–61.
183. Chartier-Harlin, M.-C., Parfitt, M., Legrain, S., Perez-Tur, J., Brousseau, T., Evans, A., Berr, C., Vidal, O., Roques, P., Gourlet, V., Fruchart, J.-C., Delacourte, A., Rossor, M., Amouyel, P. 1994. Apolipoprotein E, ε4 allele as a major risk factor for sporadic early and late-onset forms of Alzheimer's disease: Analysis of the 19q13.2 chromosomal region. *Hum. Molec. Genet.* 3:569–574.
184. Corder, E.H., Saunders, A.M., Risch, N.J., Strittmatter, W.J., Schmechel, D.E., Gaskell, P.C., Jr., Rimmler, J.B., Locke, P.A., Conneally, P.M., Schmader, K.E., Small, G.W., Roses, A.D., Haines, J.L., Pericak-Vance, M.A. 1994. Protective effect of apolipoprotein E type 2 allele for late onset Alzheimer disease. *Nature Genet.* 7:180.
185. Aleshkov, S.A., Abraham, C.R., Zannis, V.I. 1996. Complex formation of apoE produced by mammalian cells with amyloid peptide β (Aβ) follows the order apoE3>apoE3>>apoE4. 26th Annual Meeting, *Society for Neuroscience* 22:1696.
186. Zannis, V.I., Kurnit, D., Breslow, J.L. 1982. Hepatic apoA-I and apoE and intestinal apoA-I are synthesized in precursor isoprotein forms by organ cultures of human fetal tissues. *J. Biol. Chem.* 257:536–544.
187. Zannis, V.I., Ordovas, J.M., Cladaras, C., Cole, F.S., Forbes, G., Schaefer, E.J., 1985. mRNA and apolipoprotein E synthesis abnormalities in peripheral blood monocyte macrophages in familial apolipoprotein E deficiency. *J. Biol. Chem.* 24:12891–12894.
188. Zannis, V.I., vanderSpek, J., Silverman, D. 1986. Intracellular modifications of human apolipoprotein E. *J. Biol. Chem.* 261:13415–13421.

189. Li, W.H. Tanimura, M., Luo, C.C., Datta, S., Chan, L. 1988. The apolipoprotein multigene family: Biosynthesis, structure, structure-function relationships and evolution. *J. Lipid Res.* 29:245–271.
190. Innerarity, T.L., Pitas, R.E., Mahley, R.W. 1979. Binding of arginine-rich (E) apoprotein after recombination with phospholipid vesicles to the low density lipoprotein receptors of fibroblasts. *J. Biol. Chem.* 254:4186–4190.
191. Laggner, P., Gotto, Jr., A.M., Morrisett, J.D. 1977. Physical properties of the dimyristoylphosphatidylcholine vesicle and the complex formed by its interaction with apolipoprotein CIII: X-ray single angle scattering studies. *Biochemistry* 18:164–171.
192. Higgins, L.S., Rodems, J.M., Catalano, R., Quon, D., Cordell, B. 1995. Early Alzheimer disease-like histopathology increases in frequency with age in mice transgenic for b-APP751. *Proc. Natl. Acad. Sci. USA* 92:4402–4406.
193. Hsiao, K., Chapman, P., Nilsen, S., Eckman, C., Harigaya, Y., Younkin, S., Yang, F., Cole, G. 1996. Correlative memory deficits and amyloid plaques in transgenic mice. *Science* 274:99.

VI. GENE TRANSFER IN THE VASCULATURE

14

GENE MEDIATED ANTI-CANCER STRATEGIES INVOLVING THE VASCULATURE

Inhibition of Angiogenesis and Vascular Targeting of Gene Therapy

Rhys T. Jaggar, Hock-Ye Chan, Adrian L. Harris, and Roy Bicknell

Molecular Angiogenesis Group
Imperial Cancer Research Fund
Institute of Molecular Medicine
University of Oxford
John Radcliffe Hospital
Oxford, United Kingdom

Recently, novel approaches to anti-cancer therapy have been described that involve gene transfer. These include anti-angiogenic strategies and vascular targeting of gene therapy.

1. ANTI-ANGIOGENIC GENE THERAPY

There now exists much evidence that growth of solid tumors is angiogenesis dependent. In 1994, the groups of Risau and Ulirich published a paper that described for the first time an approach to anti-angiogenic gene therapy (Millauer et al., 1994). In this study a cDNA coding for the extracellular domain and transmembrane region of the mouse vascular endothelial growth factor (VEGF) receptor flk- 1 (the mouse homolog of human KDR) was packaged into a retrovirus. The extracellular domain of the receptor was shown to act as a dominant negative receptor, that is one which will bind VEGF but not transduce a signal. The dominant negative receptor blocks VEGF signal transduction. A cell line producing this virus was then co-xenografted with a glioblastoma line in which the angiogenic activity was known to be VEGF. The retroviral producing line inhibited growth of the glioblastoma and it was reasoned that the retrovirus infected the tumor endothelium which then expressed the dominant negative flk- 1 receptor and became unresponsive to VEGF secreted by the glioblastoma. Subsequent studies have shown that the approach is widely applicable to the inhibition of different xenografted tumor types (Millauer et at., 1996).

Vascular Endothelium, edited by Catravas *et al.*
Plenum Press, New York, 1998

2. VASCULAR TARGETING OF ANTI-CANCER GENE THERAPY

Vascular targeting as an anti-cancer strategy has attempted to exploit differences between tumor and normal vasculature. The idea was originally put forward by Juliana Denekamp of the Gray Laboratories in London (Denekamp, 1982). It followed from the observation that endothelial proliferation in tumors is at least 10-fold greater than that in comparative normal tissues (Denekamp and Hobson, 1982). Vascular targeting should be distinguished from the inhibition of angiogenesis, since the former aims to destroy the existing tumor vasculature with toxins while the latter involves the inhibition of endothelial migration and proliferation. One of the most significant studies in vascular targeting was that described by Burrows and Thorpe in 1993. Despite attempts by several groups over many years, it had proven difficult to obtain monoclonal antibodies that specifically recognized tumor endothelium but not normal endothelium. In the light of this Burrows and Thorpe developed a strategy to test the proof of principle of vascular targeting. In their model, γ-interferon was transfected into a neuroblastoma line that forms sub-cutaneous tumors. Animals bearing xenografted transfectants were then given intravenously an anti-MHC class II-ricin conjugate (an immunotoxin). The immunotoxin damaged the tumor but not the normal vasculature resulting in ischemic necrosis followed by complete eradication of the tumors.

In anti-cancer gene therapy we have the opportunity to be destructive rather than correctional. The former is far easier than the latter and for this reason anti-cancer studies are likely to comprise a major area of the gene therapy field for the foreseeable future. The greatest problem facing gene therapy at present is the one of how to deliver the gene of interest to the cell in which its function is desired. The vascular endothelium comprises a single monolayer of cells throughout the vascular tree with each cell in intimate contact with the blood. As such it is a uniquely accessible tissue and an ideal one with which to examine strategies of gene delivery.

A major consideration in vascular targeting is the choice of toxic gene. We have worked with two genes, namely tumor necrosis factor-α (TNF-α) and the pro-drug activating enzyme thymidine phosphorylase. TNF-α is a 17 kDa homotrimeric polypeptide which is released from the cell surface after cleavage of a plasma membrane-associated 26 kDa propolypeptide. TNF-α has the desirable property for vascular targeting of being toxic to proliferating but not quiescent microvascular endothelial cells *in vitro* (Schweigerer et al., 1987; Sato et al., 1986). Local administration of high doses of TNF-α by isolated limb perfusion damages the tumor vasculature leading to rapid necrosis of metastatic melanoma in patients (Lienard et al., 1992). This has permitted the salvage of limbs of patients with unresectable melanoma (Lienard et al., 1992). Systemic therapy with TNF-α is not possible because of toxic shock. Thus, the delivery of TNF-α locally to the tumor by gene therapy is an attractive option.

An alternative strategy involves delivery of a pro-drug activating gene to the endothelium. The two most studied pro-drug activating enzymes are Herpes Simplex Virus Thymidine Kinase (HSV tk) activation of gangcyclorin (Borelli et al., 1988) and *E. coli* cytosine deaminase conversion of 5-fluorocytosine to 5-fluorouracil (Mullen, et al., 1992). In our laboratory we have, however, been developing a novel system that uses human thymidine phosphorylase (TP). TP converts the non-toxic pro-drug furtulon to the toxic 5-FU. The advantage of this system over that of, for example, HSV tk and gangcyclorin is that the toxic products are freely cell permeant which enables a wider cell kill around the cell that has been transduced (what is known as the bystander effect) (Patterson et al., 1995a). Pro-drug activating systems are particularly attractive for systemic therapy, since possible toxic side effects may be reversed upon withdrawal of the pro-drug.

3. RETROVIRAL DELIVERY OF GENES TO ENDOTHELIUM

Our work has concentrated on delivery of genes via retroviruses to endothelium. We have found this to be a highly efficient means of introducing genes into endothelial cells *in vitro*. (Endothelial cells have long been known to be a problematical cell lineage to transfect. The most efficient *in vitro* method, aside from retroviral delivery, is currently electroporation.) Retroviruses are single stranded RNA viruses which replicate via a double-stranded DNA intermediate (the provirus) which integrates into the genome of the infected cell. With commonly used vectors integration only occurs in a dividing cell. This provides a selective means of targeting tumor endothelium which is proliferating, unlike the endothelium in normal tissues. The average endothelial cell divides twice in a lifetime and endothelial proliferation in tumors is at the very least 10-fold greater than that in normal tissues.

Retroviruses are produced from so-called packaging cell lines. A packaging cell line normally expresses all of the necessary retroviral proteins (gag, pol and env), but contains no packageable viral RNA. A retroviral vector, when stably transfected into a packaging cell line, produces the retroviral RNA which is packaged but contains no coding sequence for retroviral proteins. The highest viral titres achieved using standard vectors and packaging lines is $1–4 \times 10^7$ CFU/ml (Miller et al., 1986), however, titres are more usually in the range of $10^5–10^6$ CFU/ml.

To date the vast majority of retroviral vectors have been based on Moloney Murine Leukemia Virus. Amphotrophic virions that readily infect human cells are easily produced, although they do have the problem of being rapidly inactivated by human serum (Welsh et al., 1975) although not by human cerebrospinal fluid (Russell, et al., 1995).

The most commonly used approach to the development of tissue-specific gene expression vectors involves the use of transcriptional control elements. For many cell types, the use of tissue culture systems and transgenic mice have delineated sequences necessary for either induced or constitutive gene expression. On a cautionary note, however, the use of transgenic mice has shown that sequences sufficient to generate physiological levels of tissue specific gene expression *in vitro* can sometimes be very poor in inducing efficient gene expression *in vivo*.

Endothelial-specific promoters have only been identified comparatively recently. They include among others those for the genes coding for von Willebrand factor (Factor VIII related antigen) (Aird et al., 1995), E-selectin (Collins, et al., 1991; Montgomery et al., 1991), and most recently the VEGF receptor KDR (Patterson et al., 1995b). Our initial studies have been with the KDR receptor promoter. Three generations of retrovirus were constructed with which to examine the feasibility of endothelial specific gene targeting (Figure 1). The first generation virus had the KDR promoter cloned behind the 5′LTR driving the toxic gene (TNF-α or TP). Second generation virus had a deletion in the 3′LTR. In the process of integration the 3′LTR ends up 5′ of the KDR promoter and so in effect this virus has the LTR promoter inactivated. These viruses are known as self-inactivating vectors (Yu et al., 1986). Third generation vectors have the endothelial specific promoter cloned into the deletion in the 3′ promoter. As this promoter ends up in front of the toxic gene after viral integration, the construct is essentially a generic endothelial specific viral vector.

Figure 2 shows secretion of TNF-α (determined by ELISA) by 3T3 fibroblasts and mouse skin microvascular endothelial (SEND) cells after infection with the first generation virus. It is seen that the LTR promoter leads to strong expression in the 3T3 cells but that from the KDR promoter is very weak. In contrast in SEND cells it is seen that expres-

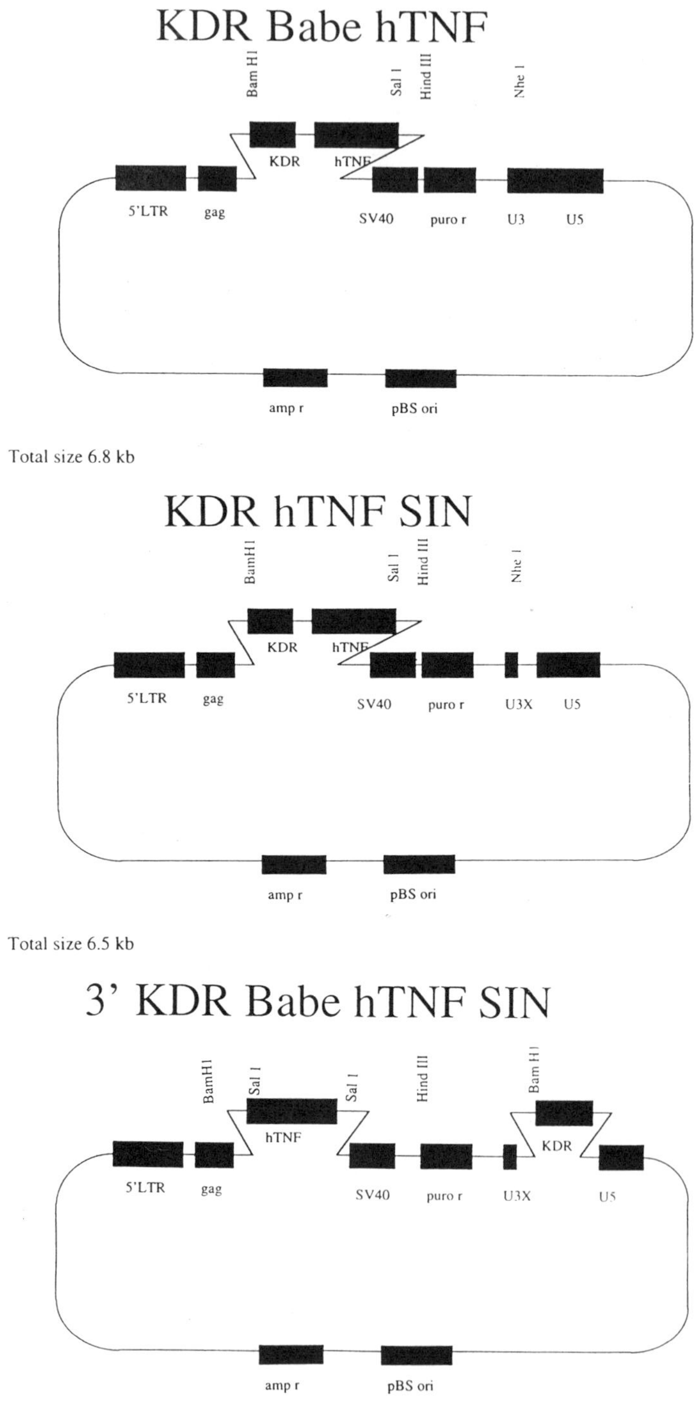

Figure 1. Three generations of retroviral vectors. The second generation (middle) are self-inactivating vectors and the third generation (bottom) have the endothelial specific KDR promoter cloned into the 3′LTR.

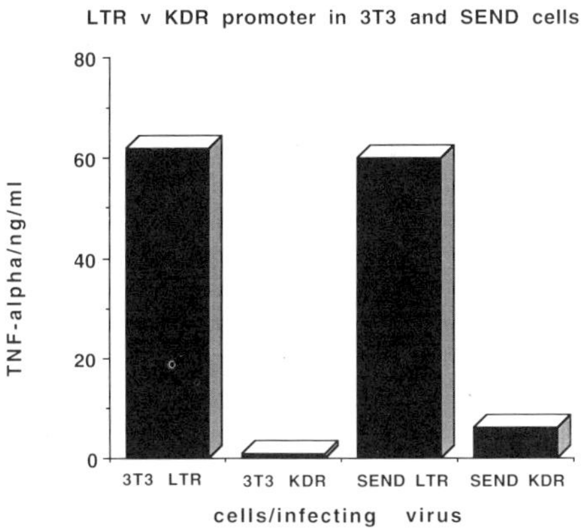

Figure 2. Production of TNF-α (determined by ELISA) by 3T3 fibroblasts and by mouse skin endothelioma cells (SEND) after infection with viruses in which the TNF-α gene is under control of either the viral LTR promoter or the human KDR promoter.

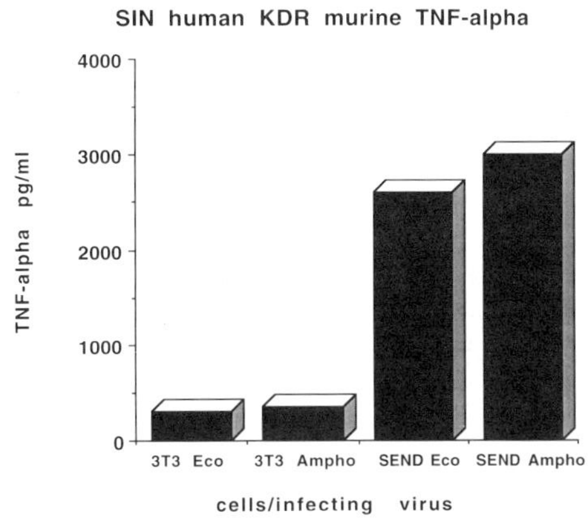

Figure 3. Production of TNF-α by 3T3 fibroblasts and SEND cells following infection with a self-inactivating vector with the human KDR promoter driving expression of the TNF-α.

sion from the KDR promoter is much greater although it still falls short of that from the viral LTR promoter. This was particularly pleasing in view of the fact that we are using the human KDR promoter in mouse endothelium. It was concluded that the KDR promoter is a fairly active constitutive promoter in mouse microvascular endothelium.

Figure 3 shows expression of TNF-α in 3T3 and SEND cells after infection with a second generation self inactivating vector containing the KDR promoter. It is seen that when packaged into either ecotropic or amphotropic virus, the vector gives good endothelial specific expression.

At present it is not possible to obtain sufficiently high retroviral titres to permit systemic delivery of virus and in vivo work is performed with producer cells. This situation will no doubt change as viral technology advances. Co-xenografting of producer cells with e.g. HT-29 colon carcinoma cells leads to a dramatic reduction in tumor growth (Figure 4) and the appearance of side-effects that would be expected from the presence of active

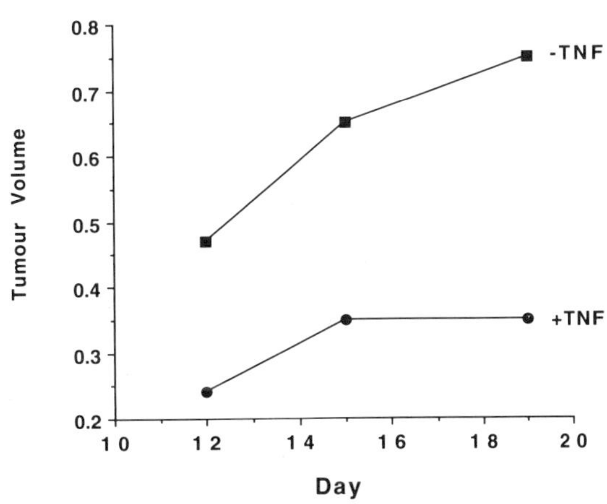

Figure 4. The effect of co-xenografting HT-29 with viral producer cells. The viral producer cells either made a virus that was 'empty' or one that produced TNF-α after infection.

TNF-α (e.g. swollen lymph glands). Co-xenografting is, however, a poor model with which to determine the efficacy of a vascular targeting strategy. Thus, more meaningful models are being sought in which the virus can be delivered within the blood to tumor endothelium. The most promising of these are those based on a paper from Richard Mulligan's group (Hurford et al., 1995). In this model, tumor cells were first injected into the spleen whereupon they colonized the liver, several days later giving rise to liver mets. Viral producer cells were then also injected into the spleen. Virus secreted by the producer cells enters the portal circulation and is delivered to the tumor metastases in the liver. It was shown using this model that the gene β-galactosidase could be effectively delivered to the metastases in the liver. Similar models are being evaluated in our laboratory in order to test vascular targeting strategies. In view of the fact that in these models the virus is delivered via the vasculature, they should be ideal.

Rapid developments in vector technology including viral pseudotyping, envelope targeting, and tissue specific expression should lead to exciting developments in the area over the next two years.

REFERENCES

Aird. W.C., Jahroudi, N., Weiler-Guettlen H., Rayburn, H.B., and Rosenberg. R.D. Human von Willebrand factor gene sequences target expression to a subpopulation of endothelial cells in transgenic mice. *Proc. Natl. Acad. Sci. USA*, 92:4567–4571, 1995.

Borelli. E., Heyman, R., Hsi, M., and Evans, R.M. Targeting of an inducible toxic phenotype in animal cells. *Proc. Natl. Acad. Sci. USA*, 85:7572–7576, 1988.

Burrows, F.J. and Thorpe, P.E. Eradication of large solid tumors in mice with an immunotoxin directed against tumor vasculature. *Proc. Natl. Acad. Sci. USA*, 90:8996–9000, 1993.

Clarke, M.S.F. and West, D.C. The identification of proliferation and tumor-induced proteins in human endothelial cells: a possible target for tumor therapy. *Electrophoresis*, 12:500–508, 1991.

Collins. T., Williams, A., Johnson, G.I., Kim, J., Eddy, R., Shows, T., Gimbrone Jr., M.A., and Bevilacqua, D.L. Structure and chromosomal location of the gene for endothelial-leukocyte adhesion molecule 1. *J. Biol. Chem.*, 266:2466–2473, 1991.

Denekamp, J. Endothelial cell proliferation as a novel approach to targeting tumor therapy. *Br. J. Cancer*, 45:136–139, 1982.

Denekamp, J. and Hobson. B. Endothelial-cell proliferation in experimental tumors. *Br. J. Cancer*, 46:711–720, 1982.

Hurford Jr., R.K., Dranoff G., Mulligan, R.C., and Tepper, R.I. Gene therapy of metastatic cancer by *in vivo* retroviral gene targeting. *Nat. Genet.*, 10:430–435, 1995.

Lienard, D., Ewalenko, P., Delmotte, J.J., Renard, N., and Lejeune, F.J. High dose recombinant tumor necrosis factor α in combination with interferon γ and melphalan in isolation perfusion of the limbs for melanoma and sarcoma. *J. Clin. Oncol.*, 10:52–60, 1992.

Millauer, B., Shawver, L.K., Plate, K.H., Risau, W., and Ullrich, A. Glioblastoma growth inhibited *in vivo* by a dominant-negative Flk-1 mutant. *Nature*, 367:576–579, 1994.

Millauer, B., Longhi, M.P., Plate, K.H., Shawver, L.K., Risau, W., Ullrich, A., and Strawn, L.M. Dominant-negative inhibition of flk-1 suppresses the growth of many tumor types *in vivo*. *Cancer Res.*, 56:1615–1620, 1996.

Miller, A.D., and Buttimore, C. Redesign of retrovirus packaging cell lines to avoid recombination leading to helper virus production. *Mol. Cell. Biol.*, 6:2895–2902, 1986.

Montgomery, K.F., Osborn L., Hession, C., Tizard, R., Goff, D., Vassallo, C., Tarr, P.I., Bomsztyk, K., Lobb, R., Harlan, J.M., and Pohlman, T.H. Activation of endothelial leukocyte adhesion molecule 1 (ELAM- 1) gene transcription. *Proc. Natl. Acad. Sci. USA*, 88:6523–6527, 1991.

Mullen, C.A., Kilstrup, M., and Blaese, R.M. Transfer of the bacterial gene for cytosine deaminase to mammalian cells confers lethal sensitivity to 5-fluorocytosine - a negative selection system. *Proc. Natl. Acad. Sci. USA*, 89:33–37, 1992.

Patterson, A.V., Zhang, H-T., Moghaddam, A., Bicknell, R., Talbot, D.C., Stratford, I.J., and Harris, A.L. Increased sensitivity to the prodrug 5′-deoxy-5-fluorouridine and modulation of 5-fluoro-2′deoxyuridine sensitivity in MCF-7 cells transfected with thymidine phosphorylase. *Br. J. Cancer*, 72:669–675, 1995a.

Patterson, C., Perrella, M.A., Hsieh, C-M., Yoshizumi, M., Lee, M-E., and Haber, E. Cloning and functional analysis of the promoter for KDR/flk-1: a receptor for vascular endothelial growth factor. *J. Biol. Chem.*, 270:23111–23124, 1995b.

Russell, D.W., Berger, M.S., and Miller, A.D. The effects of human serum and cerebrospinal fluid on retroviral vectors and packaging cell lines. *Hum. Gen. Ther.*, 6:635–641, 1995.

Sato, N., Goto, T., Haranaka, K., Satomi, N., Nariuchi, H., Mano-Hirano, Y., and Sawasaki, Y. Actions of tumor necrosis factor on cultured vascular endothelial cells: morphologic modulation, growth inhibition, and cytotoxicity. *J. Natl. Cancer Inst.*, 76:1113–1121, 1986.

Schweigerer, L., Malerstein, B., and Gospodarowicz, D. Tumor necrosis factor inhibits the proliferation of cultured capillary endothelial cells. *Biochem. Biophys Res. Comm.*, 143:997–1004, 1987.

Welsh, R.M., Jensen, F.C., Cooper, N.R., and Oldstone, M.B.A. Inactivation of lysis of oncornaviruses by human serum. *Virology*, 74:432–440, 1975.

Yu, S.-F., Rueden, T., Kantoff, P.W., Garber, C., Seiberg, M., Ruether, U., Anderson, W.F., Wagner, E.F., and Gilboa, E. Self-inactivating retroviral vectors designed for transfer of whole genes into mammalian cells. *Proc. Natl. Acad. Sci. USA*, 83:3194–3198, 1986.

15

NITRIC OXIDE MODULATION OF GENE EXPRESSION

Machelle Manuel,[1] Sha Zhu,[1] Sadis Matalon,[1,2,3] and I. Y. Haddad[3]

[1]Department of Anesthesiology
[2]Department of Physiology and Biophysics
[3]Department of Pediatrics
University of Alabama at Birmingham
Birmingham, Alabama

1. INTRODUCTION

The free radical nitric oxide (·NO) is the major form of the endothelial-derived relaxing factor that is enzymatically synthesized from arginine oxidation by an NADPH-dependent nitric oxide synthase. ·NO causes smooth muscle relaxation by activating soluble guanylate cyclase, and inhibits platelet aggregation and adhesion to endothelium by increasing cGMP[41]. Because of the vasorelaxant properties of ·NO and its rapid inactivation in the blood by its reaction with hemoglobin, ·NO inhalation has been advocated as a means of selectively reducing pulmonary hypertension and improving ventilation/perfusion mismatching in a variety of clinical situations[9,40].

In spite of its therapeutic value, there is reason to believe that in pathological conditions where there is increased production of oxygen free radicals by lung cells, inhalation or overproduction of ·NO may damage alveolar and airway cells. Since ·NO is a free radical, it can undergo a radical-radical reaction with superoxide (O_2^-) at near diffusion-limited rates, to yield peroxynitrite ($ONOO^-$)[4]. Peroxynitrite is a potent oxidizing agent known to initiate lipid peroxidation in biological membranes, at rates that are a thousand fold higher than for hydrogen peroxide (H_2O_2), hydroxylation, and nitration of aromatic amino acid residues and sulfhydryl oxidation of proteins[50,51]. Furthermore, $ONOO^-$ may decompose homolytically to yield nitrogen dioxide ($\cdot NO_2$), and species with hydroxyl radical-like reactivity (·OH) without the need for metal catalysis[4]. Following stimulation by cytokines and lipopolysaccharide, alveolar macrophages and epithelial cells produce large amounts of ·NO and O_2^- for prolonged periods of time[26,45,49] and can act as foci for the intense and localized production of $ONOO^-$ and its toxic intermediates in close proximity to lung cells. Herein, we will: (1) summarize the cellular and molecular mechanisms responsible for the regulation of ·NO synthesis; and (2) review evidence for the beneficial and detri-

Vascular Endothelium, edited by Catravas *et al.*
Plenum Press, New York, 1998

mental functions of ·NO during oxidant stress in the lung. We will point out existing controversies and whenever possible, attempt to reconcile existing differences.

2. ·NO: A FREE RADICAL SIGNAL TRANSDUCING AGENT

·NO is synthesized from the five electron oxidation of either of the two equivalent guanidine nitrogens of L-arginine. The reaction is catalyzed by one of three isozymes of nitric oxide synthase (NOS), using reduced NADPH as the source of electrons and cofactors, including tetrahydrobiopterin (H_4B) and flavin nucleotides (FMN; FAD). Molecular oxygen is a co-substrate and N^G-hydroxy-L-arginine is formed as a short-lived intermediate with L-citrulline as a byproduct. All isoforms of NOS bind calmodulin and contain heme. Electrons are supplied by NADPH, transferred along the flavins and calmodulin and presented to the catalytic heme. Details of the biochemical pathway and electron flow schema appear in several recent reviews.[18]

Our understanding of how the simple, small, diffusible, lipophilic, short-lived molecule, ·NO performs an extraordinarily diverse array of physiologic and pathophysiologic functions remains quite rudimentary. Nonetheless, it appears that regulation of ·NO biosynthesis subserves function, and thus significant advances have been made by studying unique characteristics of NOS isoforms. Collectively, NOS are homodimeric cytochrome P450-like hemeproteins that have oxygenase and reductase domains in the NH_2 and COOH termini, respectively, and are 50% homologous by amino acid sequence. The active dimeric form is hypothesized to be dependent upon H_4B[60]. The domains are separated by a Ca^{2+}/calmodulin binding region. The reductase domain is homologous to NADPH-cytochrome P450 including binding sites for FMN, FAD and NADPH. The oxygenase domain contains similar binding sites amongst the isozymes for heme, L-arginine and H_4B. The constitutive forms have similar phosphorylation sites, while the endothelial cell isozyme has a unique NH_2 terminal myristoylation site.

Differences between the isozymes, however, truly underscore the function of ·NO in various biological systems. The nomenclature evolved from descriptions of cellular sources of the enzyme (e.g. neuronal, macrophage or endothelial) or its expression (constitutive vs. inducible) and by many conventions is now relegated to the chronology in which the enzyme was purified and cloned. NOS I and III are constitutively expressed and their activity is regulated by intracellular calcium; prototypic sources are neuronal and endothelial cell, respectively. NOS II is induced by cytokines, its activity is largely independent of calcium and it is regulated at the transcriptional level. The prototypic source is macrophage. Although the human genes are present on discrete chromosomes, a considerable degree of homology exists between the three isoforms suggesting common ancestral origin with subsequent gene duplication and transposition. Comparison of their calcium dependence and gene regulation are manifested in functional differences. NOS-I and III derived ·NO are produced in small quantities for brief periods of time and underscore intra- and intercellular signaling events such as neurotransmission or vascular homeostasis. In contrast, NOS-II produces large amounts of ·NO for prolonged periods of time (assuming availability of substrate and cofactors) and contributes to more diffuse physiological roles associated with inflammation or infection.

A recent study has shown that normal human and rat airway epithelium in large, cartilaginous airways, unstimulated by cytokines, immunostained in a specific manner with an antibody to inducible NOS, suggesting a constitutive role for a NOS isozyme that shares epitopes with NOS-II[34]. Thus a new methodology for identifying the NOS isoforms has been developed. The constitutive and inducible forms of NOS from a number of species, including smooth muscle and endothelial cells, neurons and macrophages, have been cloned and sequenced.

3. PHYSIOLOGICAL EFFECTS OF ·NO

3.1. Biological Targets

The biologic actions of ·NO are dictated by the reactions it undergoes with different target molecules in cells, membranes, and the extracellular milieu. For each target, depending on the amount and duration of induction, ·NO can exert beneficial or detrimental effects. Known targets for ·NO include:

3.1.1. Guanylate Cyclase. ·NO binds the heme group of soluble guanylate cyclase leading to an increase in cyclic guanosine-3′-5′-monophosphate (cGMP) levels. Many effects of cGMP are mediated via a group of enzymes, cGMP-associated protein kinases PKGs. PKGs act to reduce intracellular calcium causing smooth muscle relaxation. ·NO-mediated increased cGMP levels also prevent platelet aggregation and decreases adhesion of neutrophils[36]. However, excessive ·NO-mediated cGMP production has been implicated in sepsis-induced refractory hypotension and shock[63].

3.1.2. Hemoglobin. The major route for the destruction of ·NO *in vivo* is the fast and irreversible reaction with *oxy*-hemoglobin ($Hb\text{-}Fe^{2}+\text{–}O_2$) or oxy-myoglobin to produce nitrate and methemoglobin according to the following reaction:

$$Hb\text{-}Fe^{2}+\text{–}O_2 + \cdot NO \Rightarrow Hb\text{-}Fe^{2+}ONOO + NO_3^- \quad (1)$$

Because of its vasorelaxant properties and rapid inactivation in the blood by its reaction with hemoglobin, ·NO inhalation has been advocated as a means of selectively reducing pulmonary hypertension and improving systemic oxygenation in a variety of clinical situations including bronchopulmonary dysplasia and the Acute Respiratory Distress Syndrome (ARDS)[54]. However, patients with diminished methemoglobin reductase activity, such as neonates, are unable to efficiently convert methemoglobin to ferrous hemoglobin, and could therefore be at greater risk for developing methemoglobinemia and decreased oxygen transport[43].

3.1.3. Iron/Sulfur (4Fe/4S) Centers of Enzymes. Production of ·NO by activated macrophages defends the host against infectious agents including bacteria[28], parasites[17], and viruses[29] and destroys tumor cells[58]. The proposed mechanisms responsible for these effects involve the reaction of ·NO with the nonheme iron of iron-sulfur complexes, resulting in the inactivation of iron-sulfur containing enzymes including mitochondrial aconitase[13], cytochrome c oxidase[10], and the DNA synthesis rate limiting enzyme, ribonucleotide reductase[37]. Inhibition of these critical enzymes leads to suppression of mitochondrial respiration, energy metabolism, and cell replication.

3.1.4. Other Free Radicals. NO has an unpaired electron, and thus can readily react with other free radicals, as specified below:

a. Reaction with Oxygen. If a tank of ·NO is opened and allowed to leak into the air, a cloud of lethal and highly reactive orange-brown nitrogen dioxide ($\cdot NO_2$) is formed according to the following reaction:

$$2 \cdot NO + O_2 \Rightarrow 2 \cdot NO_2 \quad (2)$$

The production of $\cdot NO_2$ is a third order reaction and the rate of conversion of $\cdot NO$ in the gas phase can be calculated by: $d[\cdot NO]/dt = 4\,k_3[\cdot NO]^2\,[O_2]$ where k_3 is $2 \times 10^6\ M^{-2}\,s^{-1}$.[15] Thus, the rate of $\cdot NO_2$ formation, at a given concentration of oxygen, will be determined by the square of the $\cdot NO$ concentration. At low $\cdot NO$ concentrations ($<5\mu M$), observed *in vivo* at physiologic and most pathologic conditions, the low probability of any two $\cdot NO$ molecules encountering each other makes the formation of $\cdot NO_2$ extremely slow and can, therefore, be discarded. Instead, $\cdot NO$ may react with a single molecule of oxygen in a second order reaction to form a nitrosyldioxyiradical (ONOO·).

b. Reaction with O_2^-. $\cdot NO$ reacts with O_2^- at a near diffusion-limited rate constant of about $7 \times 10^9\ M^{-1}\,s^{-1}$ to form $ONOO^-$. Initially, this reaction was mistakenly thought to be protective, since it decreased the amount of O_2^- detected. $ONOO^-$ has a pK_a of 6.8 at 37°C and thus may remain stable for months in alkaline solutions. The protonated form of $ONOO^-$, peroxynitrous acid (ONOOH), decomposes to form $\cdot NO_2$ and an intermediate with a reactivity equivalent to the hydroxyl radical (·OH), derived from the trans isomerization of ONOOH, as shown in Equation 3:[4]

$$O_2^- + \cdot NO \Rightarrow ONOO^- + H^+ \rightleftharpoons ONOOOH \Rightarrow \text{“}\,OH\cdots\cdot NO_2\text{”} \qquad (3)$$

The ·OH-like reactivity of ONOOH is mediated by a vibrationally excited intermediate that does not physically separate into a free ·OH and $\cdot NO_2$. This intermediate has the reactivity of ·OH and in some cases the toxicity of $ONOO^-$ has been mitigated by ·OH scavengers[21,50].

Under physiological conditions, a minimum of 25% of $ONOO^-$ decomposes to form "$\cdot OH\cdots NO_2$", with the remainder recombining to form nitrate[4]. While being highly reactive, its modest rate of decomposition under physiological conditions allows $ONOO^-$ to diffuse for up to several cell diameters to critical cellular targets before becoming protonated and decomposing. $ONOO^-$ initiates iron-independent lipid peroxidation and oxidizes thiols at rates at least a thousand fold greater than that of hydrogen peroxide (H_2O_2) at pH 7,[51] damages the mitochondrial electron transport chain[52], is a key mediator of macrophage-mediated cytotoxicity towards bacteria and *Trypanosoma cruzi*[55], and causes lipid peroxidation of human low density lipoproteins[16]. In addition, metal ions, such as Fe^{3+} EDTA and copper in the active site of superoxide dismutase (SOD), catalyze the heterolytic cleavage of $ONOO^-$ to form a nitronium ion-like species (NO_2^+) which nitrates phenolics including tyrosine in proteins[5]. A recent report indicates that carbonated species accelerate the oxidation and nitration reactions of $ONOO^-$, via the formation of the nitrosoperoxycarbonate anion ($O{=}N{-}OOCO_2^-$)[61].

Under normal conditions, intracellular O_2^- concentrations are kept at remarkably low levels (10 pM) because eukaryotic cells contain large amounts of SOD (4–10 μM). Under these conditions, $ONOO^-$ formation is prevented. However, pathological conditions can greatly increase the synthesis of both ·NO and O_2^-. When the concentration of ·NO increases to the micromolar range, it can effectively compete with SOD for O_2^- to form $ONOO^-$ (rate constant of the reaction of O_2^- with ·NO is three times faster than that with SOD). Indeed, activated rat alveolar macrophages[26], human neutrophils[8], and agonist-stimulated bovine endothelial cells[35] have been shown to produce $ONOO^-$.

3.2. Effect of ·NO on Gene Expression

In addition to the above considerations regarding regulation of gene expression in L-arginine-dependent ·NO biosynthesis, ·NO itself is now considered to be a modulator of gene expression and also affects the activity of various transactivating factors and DNA binding proteins. ·NO decreases cytokine-induced endothelial cell activation by inhibiting NF-κB expression and, subsequently, either vascular cell adhesion molecule (VCAM)[12] or NOS-II expression[11]. ·NO has been reported to affect not only NF-κB function (in part by altering IκBα expression)[46], but other redox sensitive transcription factors, including AP-1 components[47]. In addition, ·NO has been reported to modulate inflammation by altering the production of soluble cytokines such as TNF -α,[14] IL-8,[3] IL-1 , IL-6,[53] IFN-γ,[62] and chemoattractant factors (MCP-l) and by inhibiting adhesion molecule expression in various cells.

A potential coupling between ·NO-iron interactions, signal transduction and gene expression has been shown by Weiss et al.[64] and Drapier et al.[13]. It is well known that iron is an important target and signaling molecule for ·NO. The iron-sulfur clusters of many proteins are affected by ·NO and in particular, *cis* aconitase serves as a bifunctional protein depending upon its iron-sulfur status. In its 4Fe-4S conformation, *cis* aconitase functions as an enzyme facilitating the interconversion of isocitrate and citrate. While in its 3Fe-4S complex, it functions as the iron-regulatory protein (IRP). In this latter apo-IRP form (after ·NO attack), it can bind to iron responsive elements (IRE) and lead to: (a) inhibition of translation of ferritin and aminolevulinate synthase mRNA by forming an IRE/IRP complex in their 5′ untranslated regions (UTRs); and(b) stabilizing transcription of transferrin receptor mRNA against targeted endonucleolytic degradation by interacting with five IREs in the 3′ UTR[44]. In murine macrophages, endogenous ·NO (after cytokine stimulation), exogenous ·NO (via ·NO donors), or ·NO produced by fibroblasts transfected with an expression vector containing NOS-II activates IREs resulting in repression of ferritin mRNA and enhanced transferrin receptor mRNA. Accordingly, iron acts as both a sensory and regulatory molecule in ·NO signaling and IRP is the switch in affecting gene expression. Although less is known regarding similar mechanisms in the lung, a recent report indicates that exposure of pulmonary artery endothelial cells in culture to S-nitrosoacetylpenicillamine (SNAP) resulted in induction of heme oxygenase, a transient increase in iron, and a repression in the binding activity of IRP as manifested by an enhanced ferritin and EPR-detectable dinitrosyl-iron-sulfur complex.[32]

3.3. ·NO Modulation of Adenovirus-Mediated Gene Transfer

A recent study by Knowles et al.[33] showed that adenovirus-mediated transfer of the cystic fibrosis gene did not result in correction of the functional defect in nasal epithelium and that local inflammatory responses limited the dose of adenovirus that could be administered to overcome inefficient transduction by the adenovirus vector. Endogenous ·NO has been found in exhaled air of both animals and humans with the largest contribution from the upper airway and nasal epithelium in adult humans[57], especially during upper respiratory tract infections[31] and lung inflammation[2]. In addition to modification of the inflammatory process, ·NO may also directly alter the efficiency of adenovirus-mediated gene transfer by several mechanisms. Firstly, ·NO may alter binding and internalization of the adenovirus complex by S-nitrosylation of adenoviral proteins or adenoviral-specific cell receptors. S-nitrosylation of critical thiol groups has been reported to modify protein and receptor functions[39]. Secondly, ·NO may modify gene transfer by regulating the expression of the transgene. We recently reported that ·NO produced by murine fibroblasts, stably transfected with the human NOS-II, decreased the efficiency of reporter gene expression following infection with recombinant replication-deficient adenovirus (E1-deleted) expressing either the luciferase or Lac Z reporter genes (AdCMV-Luc and AdCMV-Lac Z) to approximately 60% of their corresponding control values (Figures 1

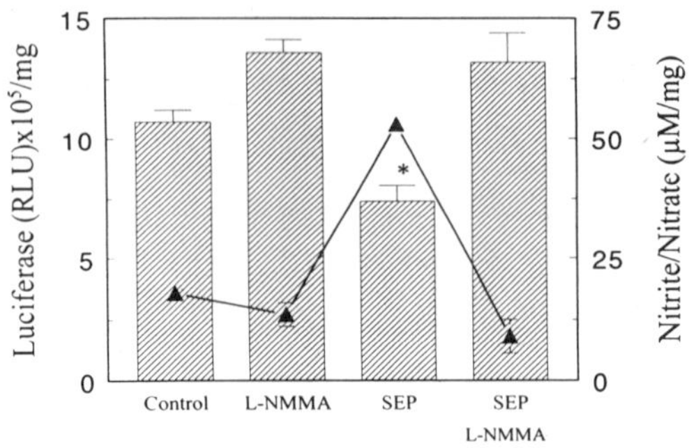

Figure 1. Luciferase expression 24 hours following infection of cultured NIH3T3 fibroblasts, stably transfected with the gene for human NOSII with a recombinant replication-deficient adenovirus expressing the luciferase reporter gene (AdCMV-Luc; 10 pfu/cell). The addition of sepiapterin (SEP; 50 μm) increased nitrite/nitrate in the medium (solid line) and decreased reporter protein expression. L-NMMA is NG-monomethyl-L-arginine at 1 mM. Luciferase activity is expressed in relative light units (RLU) per mg cell proteins (hatched bars). Nitrite/nitrate levels were expressed in μM/mg protein. Values are means + SEM for n= 3 different experiments performed in triplicate. * = $p < 0.05$ compared to control cells. Figure reproduced from reference 20 with permission.

and 2).[20] Pretreatment of cells with N^G-monomethyl-L-arginine (L-NMMA; 1 mM), an inhibitor of NOS, prevented the sepiapterin-mediated induction of ·NO and restored gene transfer to baseline values. Furthermore, parenteral steroid administration to mice enhanced the efficiency of adenoviral gene transfer upon re-administration of the virus, partly by inhibition of NOS-II induction.[59]

3.4. Cytotoxic Potential of ·NO/ONOO

The neutrality of ·NO and its low Stoke's radius, permits facile transmembrane diffusion and its relatively long biological half-life allows it to react with targets distant from its point of origin. Thus, smooth muscle cell and endothelial cell derived ·NO may

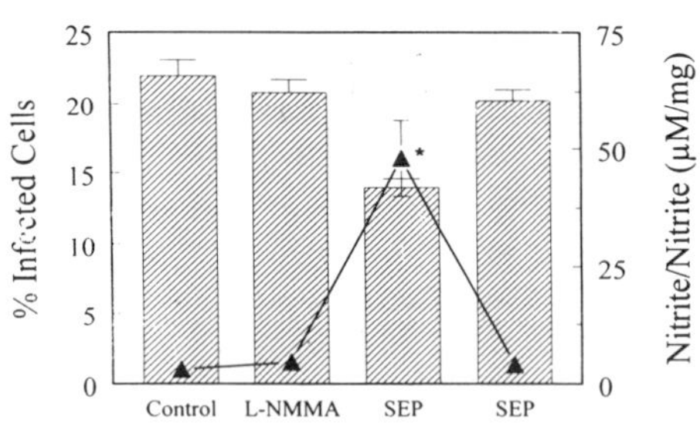

Figure 2. β-galactosidase expression 24 hours following infection of cultured NIH3T3 fibroblasts, stably transfectd with the gene for human NOSII, with a recombinant replication-deficient adenovirus expressing the Lac Z reporter gene (AdCMV-LacZ; 1 pfu/cell). The addition of sepiapterin SEP; 50 : M) increased nitrite/nitrate in the medium (solid line) and decreased reporter protein expression identified by the 4 characteristic histochemical response to the substrate X-gal. Number of b-galactosidase positive cells is shown as percent infected cells (hatched bars). L-NMMA is NG-monomethyl-L-arginine at 1 mM. Nitrite/nitrate levels were expressed in mM/mg protein. Values are means + SEM for n = 3 different experiments in triplicate. * = $p < 0.05$ compared to control cells. Figure reproduced from reference 20 with permission.

combine with epithelial cell-derived O_2^- to form $ONOO^-$ in the lung epithelial lining fluid. The following key observations establish the potential involvement of ·NO, $ONOO^-$ and various reactive nitrogen species in the generation and propagation of pulmonary epithelial injury in a variety of pathological situations including: 1) induction of immune complex alveolitis in rat lungs[42]; 2) paraquat induced injury[7]; 3) ischemia-reperfusion injury; 4) *Bordetella pertussis* infection of airway cells *in vitro*[25]; 5) pathogenesis of influenza virus-induced pneumonia[1]; and 6) injury to pulmonary surfactant and surfactant apoproteins.[19,22–24,38]

4. INHIBITION OF OXIDANT-MEDIATED INJURY

·NO reaction with O_2^- may, in some cases, also serve to protect O_2^--sensitive target molecules. This reaction of ·NO, which kinetically can out compete SOD, forces O_2^- through $ONOO^-$ oxidation and decomposition pathways. At the same time this limits the accumulation of H_2O_2 and decreases the formation of secondary reactive species derived from H_2O_2 reaction with transition metals. For example, it has recently been reported that ·NO donors inhibit reactive oxygen species-mediated cellular damage and cytotoxicity.[65] ·NO can also react with other free radicals, such as lipid alkoxyl (LO·) and peroxyl (LOO·) radicals, decreasing their steady state concentration and terminating free radical lipoprotein oxidation propagation chain reactions. Rubbo et al.[56] showed that when the molar concentration of ·NO exceeds that of O_2^-, ·NO protects against O_2^-- and $ONOO^-$-dependent lipid peroxidation. These reactions are especially important to consider inasmuch as ·NO significantly concentrates m lipophilic cell compartments, having a lipid-water partition coefficient of ˜6.5. ·NO may also ameliorate tissue injury by activating guanylate cyclase with subsequent induction of cGMP-dependent effects, such as reducing platelet and neutrophil adhesion to endothelium[36] and reacting with iron to form iron-nitrosyl compounds. By binding free coordination sites of iron, ·NO can limit Fenton chemistry and iron-dependent electron transfer reactions[27].

In buffer-perfused isolated rabbit lungs, inhaled ·NO (24 ppm) ameliorated the increase in pulmonary vascular permeability produced by the intravascular generation of H_2O_2[48]. Kavanagh et al.[30] investigated the effect of inhaled ·NO on the increase in capillary filtration coefficient ($K_{f,c}$) arising from the perfusion of isolated rabbit lungs with purine and xanthine oxidase. Inhaled ·NO (90–120 ppm) attenuated, and inhibition of endogenous ·NO exacerbated, the oxidant-mediated increase in $K_{f,c}$. In patients with acute lung injury, inhaled ·NO has been shown to decrease permeability pulmonary edema[6] and to suppress pro-inflammatory cytokine production in the lung. We investigated the effects of aminoguanidine (100 mg/kg per dose injected intraperitoneally every 12 h), a nucleophilic compound and presumed selective inhibitor of NOS-II, on indices of hyperoxic lung injury assessed at 48 h. Aminoguanidine significantly increased the total protein concentrations in the BAL. In further experiments, all rats that received aminoguanidine during hyperoxia (n=3) died between 48 and 60 h. In contrast, only one rat that received saline during hyperoxia (n=7) died during that same time period. This data suggests that, in this model of hyperoxic lung injury, upregulation of ·NO production may serve a protective role. However, this conclusion is complicated by the fact that inhibitors of NOS-II activity also affect NOS-III, to varying degrees. Furthermore, aminoguanidine, is known to interfere with glucose metabolism and thus may have diverse biologic functions in addition to blocking NOS-II[66].

5. SUMMARY AND CONCLUSIONS

Studies aimed at identifying regulation of the NOS genes will help us understand the wide range of functional manifestations of NOS-derived ·NO from modulating perfusion to mediating cytotoxicity. The reaction of ·NO with O_2^-, on the other hand, is critical to our understanding of the role of ·NO in oxidant stress and inflammation.

·NO plays a dual role in the modification of oxidant-mediated injury. ·NO acts as a chain reaction-breaking antioxidant to inhibit O_2^--mediated lipid peroxidation by scavenging alkoxyl and peroxyl radicals. However, under similar experimental conditions, ·NO may, at the same time, promote oxidant injury to proteins by rapidly reacting with O_2^- to form a potent oxidizing and nitrating agent, $ONOO^-$. These results clarify, in part, the often confusing literature describing ·NO reactivity during oxidant stress and provide evidence that the target molecule is an important determinant of the final outcome. The efficacy of inhaled ·NO should not be judged only by the physiological parameters noted at the bedside, but also by the variety of other non-cGMP-dependent biochemical reactions including modulation of gene expression.

ACKNOWLEDGMENTS

This work was supported by the grants from the National Institutes of Health (NICHD P30, HD28831, HL31197, HL51173, T32 HL07553), the Cystic Fibrosis Foundation, and a grant from the Office of Naval Research (000014-93-0785).

REFERENCES

Akaike, T., Y. Nogushi, S. Ijiri, K. Setoguchi, M. Suga, Y.M. Sheng, B. Dietzschold, and H. Maeda. Pathogenesis of influenza virus-induced pneumonia: Involvement of both nitric oxide and oxygen radicals. *Proc. Natl. Acad. Sci. USA*, 93:2448–2453, 1996.

Alving, K., E. Weitzberg, and J.M. Lundberg. Increased amount of nitric oxide in exhaled air of asthmatics. *Eur. Respir. J.*, 6:1368–1370, 1993.

Andrew, P.J., H. Harant, and I.J. Lindley. Nitric oxide regulates IL-8 expression in melanoma cells at the transcriptional level. *Biochem. Biophys. Res. Commun.*, 214:949–956, 1995.

Beckman, J.S., T.W. Beckman, J. Chen, P.A. Marshall, and B.A. Freeman. Apparent hydroxyl radical production by peroxynitrite: Implications for endothelial injury from nitric oxide and superoxide. *Proc. Natl. Acad. Sci. USA*, 87:1620–1624, 1990.

Beckman, J.S., H. Ischiropoulos, L. Zhu, M. van der Woerd, C. Smith, J. Chen, J. Harrison, J.C. Martin, and M. Tsai. Kinetics of superoxide dismutase- and iron-catalyzed nitration of phenolics by peroxynitrite. *Arch. Biochem. Biophys.*, 298:438–445, 1992.

Benzing, A., P. Brautigam, K. Geiger, T. Loop, U. Beyer, and E. Moser. Inhaled nitric oxide reduces pulmonary transvascular albumin flux in patients with acute lung injury. *Anesthesiology*, 83:1153–1161, 1995.

Berisha, H.I., H. Pakbaz, A. Absood, and S.I. Said. Nitric oxide as a mediator of oxidant lung injury to paraquat. *Proc. Natl. Acad. Sci. USA*, 91:7445–7449, 1994.

Carreras, M.C., G.A. Pargament, S.D. Catz, J.J. Poderoso, and A. Boveris. Kinetics of nitric oxide and hydrogen peroxide production and formation of peroxynitrite during the respiratory burst of human neutrophils. *FEBS Lett.*, 341:65–68, 1994.

Channick, R.N., R.C. Hoch, J.W. Newhart, F.W. Johnson, and C.M. Smith. Improvement in pulmonary hypertension and hypoxemia during nitric oxide inhalation in a patient with end-stage pulmonary fibrosis. *Am. J. resp. Crit. Care Med.*, 149:811–814, 1994.

Cleeter, M.W., J.M. Cooper, V.M Darley-Usmar, S. Moncada, and A.H. Schapira. Reversible inhibition of cytochrome c oxidase, the terminal enzyme of the mitochondrial respiratory chain, by nitric oxide: Implications for neurodegenerative diseases. *FEBS Lett.*, 345:50–54, 1994.

Colasanti, M., T. Persichini, M. Menegazzi, S. Mariotto, E. Giordano, C.M. Caldarera, V. Sogos, M. Lauro, and H. Suzuki. Induction of nitric oxide synthase mRNA expresson. Suppression by exogeneous nitric oxide. *J. Biol. Chem.* 270:26731–26733, 1995.

De Caterina, R., P. Libby, H.B. Peng, V.J. Thannickal, T.B. Rajavashisth, M.A. Gimbrone, Jr., W.S. Shin, and J.K. Liao. Nitric oxide decreases cytokine-induced endothelial activation. Nitric oxide selectively reduces endothelial expression of adhesion molecules and proinflammatory cytokines. *J. Clin. Invest.*, 96:60–68, 1995.

Drapier, J.C., H. Hirling, J. Wietzerbin, P. Kaldy, and L.C. Kuhn. Biosynthesis of nitric oxide activates iron regulatory factor in macrophages. *EMBO J.*, 12:3643–3649, 1993.

Eigler, A., J. Moeller, and S. Endres. Exogeneous and endogeneous nitric oxide attenuates tumor necrosis factor in the murine macrophage cell line RAW 264.7. *J. Immunol.*, 154:4048–4054, 1995.

Foubert, L., B. Fleming, R. Latimer, M. Jonas, A. Oduro, C. Borland, and T. Higenbottam. Safety guidelines for use of nitric oxide [letter]. *Lancet*, 339:1615–1616, 1992.

Graham, A., N. Hogg, B. Kalyanaraman, V. O'Leary, V. Darley-Usmar, and S. Moncada. Peroxynitrite modification of low density lipoprotein leads to recognition by the macrophage scavenger receptor. *FEBS Lett.*, 330:181–185, 1993.

Green, S.J., L.F. Scheller, M.A. Marletta, M.C. Seguin, F.W. Klotz, M. Slayter, B.J. Nelson, and C.A. Nacy. Nitric oxide: cytokine-regulation of nitric oxide in host resistance to intracellular pathogens. *Immunol. Lett.*, 43:87–94, 1994.

Griffith, O.W. and D.J. Stuehr. Nitric oxide synthases: properties and catalytic mechanism. *Annn. Rev. Physiol.*, 57:707–736, 1995.

Haddad, I.Y., J.P. Crow, P. Hu, Y. Ye, J. Beckman, and S. Matalon. Concurrent generation of nitric oxide and superoxide damages surfactant protein A. *Am. J. Physiol.*, 267:L242-L249, 1994.

Haddad, I.Y., R.I. Garver, Jr., E.J. Sorscher, E. Tzeng, and S. Matalon. Decreased efficiency of adenovirus-mediated gene transfer by nitric oxide. *Am. J. Respir. Crit. Care Med.*, 153:109, 1996.

Haddad, I.Y., H. Ischiropoulos, B.A. Holm, J.S. Beckman, J.R. Baker, and S. Matalon. Mechanisms of peroxynitrite-induces injury to pulmonary surfactants. *Am. J. Physiol.*, 265:L555–64, 1993.

Haddad, I.Y., G. Pataki, P. Hu, C. Galliani, J.S. Beckman, and S. Matalon. Quantitation of nitrotryosine levels in lung sections of patients and animals with acute lung injury. *J. Clin. Invest.*, 94:2407–2413, 1994.

Haddad, I.Y., S. Zhu, J. Crow, E. Barefield, T. Gadilhe, and S. Matalon. Inhibition of alveolar type II cell ATP and surfactant synthesis by nitric oxide. *Am. J. Physiol. Lung Cell. Mol. Physiol.*, In Press.

Haddad, I.Y., S. Zhu, H. Ischiropoulos, and S. Matalon. Nitration of surfactant protein A results in decreased ability to aggregate lipids. *Am. J. Physiol. Lung Cell. Mol. Physiol.*, 270:L281-L288, 1996.

Heiss, L.N., J.R. Lancaster, Jr., J.A. Cobett, and W.E. Goldman. Epithelial autotoxicity of nitric oxide: role in the respiratory cytopathology of pertussis. *Proc. Natl. Acad. Sci. USA*, 91:267–270, 1994.

Ischiropoulos, H., L. Zhu, and J.S. Beckman. Peroxynitrite formation from macrophage-derived nitric oxide. *Arch. Biochem. Biophys.*, 298:446–451, 1992.

Kanner, J., S. Harel, and R. Granit. Nitric oxide as an antioxidant. *Arch. Biochem. Biophys.*, 289:130–136, 1991.

Kaplan, S.S., J.R. Lancaster, Jr., R.E. Basford, and R.L. Simmons. Effect of nitric oxide on staphylococcal killing and interactive effect with superoxide. *Infect. Immun.*, 64:69–76, 1996.

Karupiah, G., Q.W. Xie, R.M. Buller, C. Nathan, C. Duarte, and J.D. MacMicking. Inhibition of viral replication by interferon-gamma-induced nitric oxide synthase. *Science*, 261:1445–1448, 1993.

Kavanaugh, B.P., A. Mouchawar, J. Goldsmith, and R.G. Pearl. Effects of inhaled NO and inhibition of endogeneous NO synthesis in oxidant-induced acute lung injury. *J. Appl. Physiol.*, 76:1324–1329, 1994.

Kharitonov, S.A., D. Yates, and P.J. Barnes. Increased nitric oxide in exhaled air of normal human subjects with upper respiratory tract infections. *Eur. Respir. J.*, 8:295–297, 1995.

Kim, Y.M., H.A. Bergonia, C. Muller, B.R. Pitt, W.D. Watkins, and J.R. Lancaster, Jr. Loss and degradation of enzyme-bound heme induced by cellular nitric oxide synthesis. *J. Biol. Chem.*, 270:5710–5713, 1995.

Knowles, M.R., K.W. Hohneker, Z. Zhou, J.C. Olsen, T.L. Noah, P.C. Hu, M.W. Leigh, J.F. Engelhardt, L.J. Edwards, and K.R. Jones. A controlled study of adenoviral-vector-mediated gene transfer in the nasal epithelium of patients with cystic fibrosis [see comments]. *N. Engl. J. Med.*, 333:823–831, 1995.

Kobzik, L., D.S. Bredt, C.J. Lowenstein, J. Drazen, B. Gaston, D. Sugarbaker, and J.S. Stamler. Nitric oxide synthase in human and rat lung: immunocytochemical and histochemical localization. *Am. J. Resp. Cell Mol. Biol.*, 9:371–377, 1993.

Kooy, N.W. and J.A. Royall. Agonist-induced peroxynitrite production from endothelial cells. *Arch. Biochem. Biophys.*, 310:352–359, 1994.

Kubes, P., M. Suzuki, and D.N. Granger. Nitric oxide: and endogeneous modulator of leukocyte adhesion. *Proc. Natl. Acad. Sci. USA*, 88:4651–4655, 1991.

Kwon, N.S., D.J. Steuher, and C.F. Nathan. Inhibition of tumor cell ribonucleotide reductase by macrophage-derived nitric oxide. *J. Exp. Med.*, 174:761–767, 1991.

Matalon, S., V. DeMarco, I.Y. Haddad, C. Myles, J.W. Skimming, S. Schurch, S. Cheng, and S. Cassin. Inhaled nitric oxide injures the pulmonary surfactant system of lambs *in vivo*. *Am. J. physiol. Lung Cell. Mol. Physiol.*, 270:L273-L280, 1996.

Mohr, S., J.S. Stamler, and B. Brune. Mechanism of covalent modification of glyceraldehyde-3-phosphate dehydrogenase at its active site thiol by nitric oxide, peroxynitrite and related nitrosating agents. *FEBS Lett.*, 348:223–227, 1994.

Moinard, J., G. Manier, O. Pillet, and Y. Castaing. Effect of inhaled nitric oxide on hemodynamics and VA/Q inequalities in patients with chronic obstructive pulmonary disease. *Am. J. Resp. Crit. Care Med.*, 149:1482–1487, 1991.

Moncada, S., R.M. Palmer, and E.A. Higgs. Nitric oxide: physiology, pathophysiology, and pharmacology. *Pharmacol. Rev.*, 43:109–142, 1991.

Mulligan, M.S., J.M. Hevel, M.A. Marletta, and P.A. Ward. Tissue injury caused by deposition of immune complexes is L-arginine dependent. *Proc. Natl. Acad. Sci. USA*. 88:6338–6342, 1991.

Nilsson, A., G. Engberg, S. Henneberg, K. Danielson, and C.H. De Verdier. Inverse relationship between age-dependent erythrocyte activity of methaemoglobin reductase and prilocaine-induced metaemoglobinaemia during infancy. *Br. J. Anest.*, 64:72–76, 1990.

Pantopoulos, K. and M.W. Hentze. Nitric oxide signaling to iron-regulatory protein: direct control of ferritin mRNA translation and transferrin receptor mRNA stability in transfected fibroblasts. *Proc. Natl. Acad. Sci. USA*, 92:1267–1271, 1995.

Pendino, K.J., J.D. Laskin, R.L. Shuler, C.J. Punjabi, and D. L. Laskin. Ehanced production of nitric oxide by alveolar macrophages after inhalation of a pulmonary irritant is associated with increased expression of nitric oxide synthase. *J. Immunol.*, 151:7196–7205, 1993.

Peng, H.B., P. Libby, and J.K. Liao. Induction and stabalization of I kappa B alpha by nitric oxide mediates inhibition of NF-kappa B. *J. Biol. Chem.*, 270:14214–14219, 1995.

Pilz, R.B., M. Suhasini, S. Idriss, J.L. Meinkoth, and G. R. Boss. Nitric oxide and cGMP analogs activate transcription from AP-1 responsive promoters in mammalian cells. *FASEB J.*, 9:552–558, 1995.

Poss, W.B., O.D. Timmons, I.S. Farukh, J.R. Hoidal, and J.R. Michael. Inhaled nitric oxide prevents the increase in pulmonary vascular permeability caused by hydrogen peroxide. *J. Appl. Physiol.*, 79:886–891, 1995.

Punjabi, C.J., J.D. Laskin, K.J. Pendino, N.L. Goller, S.K. Durham, and D.L. Laskin. Production of nitric oxide by rat type II pneumocytes: increased expression of inducible nitric oxide synthase following inhalation of a pulmonary irritant. *Am. J. Respir. Cell Mol. Biol.*, 11:165–172, 1994.

Radi, R., J.S. Beckman, K.M. Bush, and B.A. Freeman. Peroxynitrite-induced membrane lipid peroxidation: the ctyotoxic potential of superoxide and nitric oxide. *Arch. Biochem. Biophys.*, 288:481–487, 1991.

Radi, R., J.S. Beckman, K.M. Bush, and B.A. Freeman. Peroxynitrite oxidation sulfhydryls. The cytotoxic potential of superoxide and nitric oxide. *J. Biol. Chem.*, 266:4244–4250, 1991.

Radi, R. M. Rodriquez, L. Castro, and R. Telleri. Inhibition of mitochondrial electron transport by peroxynitrite. *Arch. Biochem. Biophys.*, 308:89–95, 1994.

Renard, S., N. Voilley, F. Bassilana, M. Lazdunski, and P. Barbry. Localization and regulation by steroids of the alpha, beta, and gamma subunits of the amiloride-sensitive Na+ channel in colon, lung, and kidney. *Pflugers Arch.*, 430:299–307, 1995.

Rossaint, R. K.J. Falke, F. Lopez, K. Slama, U. Pison, and W.M. Zapol. Inhaled nitric oxide for the adult respiratory distress syndrome [see comments]. *N. Engl. J. Med.*, 328:399–405, 1993.

Rubbo, H., A. Denicola, and R. Radi. Peroxynitrite inactivates thiol-containing enzymes of Trypanosoma cruzi energetic metabolism and inhibits cell respiration. *Arch. Biochem. Biophys.*, 308:96–102, 1994.

Rubbo, H. R. Radi, M. Trujillo, R. Telleri, B. Kalyanaraman, S. Barnes, M. Kirk, and B. A. Freeman. Nitric oxide regulation of superoxide and peroxynitrite-dependent lipid peroxidation. Formation of novel nitrogen-containing oxidized lipid derivatives. *J. Biol. Chem.*, 269:26066–26075, 1994.

Schedin, U., C. Frostell, M.G. Persson, J. Jakobsson, G. Andersson, and L.E. Gustafsson. Contribution from upper and lower airways to exhaled endogeneous nitric oxide in humans. *Acta Anesthesiol. Scand.*, 39;327–332,1995.

Stuehr, D.J. and C.F. Nathan. Nitric oxide. A macrophage product responsible for cytostasis and respiratory inhibition in tumor target cells. *J. Exp. Med.*, 169:1543–1555, 1989.

Szabo, C., C. Thiemermann, C.C. Wu, M. Perretti, and J.R. Vane. Attenuation of the induction of nitric oxide synthase by endogeneous glucocorticoids accounts for endotoxin tolerance *in vivo*. *Proc. Natl. Acad. Sci. USA*, 91:271–275, 1994.

Tzeng, E., T.R. Billiar, P.D. Robbins, M. Loftus, and D. Stuehr. Expression of human inducible nitric oxide synthase in a tetrahydrobiopterin (H_4B)-deficient cell line: H_4B promotes assembly of enzyme subunits into an active dimer. *Proc. Natl. Acad. Sci. USA*, 92:11771–11775, 1995.

Uppu, R.M., G.L. Squadrito, and W.A. Pryor. Acceleration of peroxinitrite oxidations by carbon dioxide. *Arch. Biochem. Biophys.*, 327:335–343, 1996.

van der Meide, P.H., M.C. de Labie, C.A. Botman, J. Aten, and J.J. Weening. Nitric oxide suppresses IFN-gamma production in the spleen of mercuric chloride-exposed brown Norway rats. *Cell Immunol.*, 161:195–206, 1995.

Wei, X.Q., I.G. Charles, A. Smith, J. Ure, G.J. Feng, F.P. Huang, D. Xu, W. Muller, S. Moncada, and F.Y. Liew. Altered immune responses in mice lacking inducible nitric oxide synthase. *Nature*, 375:408–411, 1995.

Weiss, G., B. Goossen, W. Doppler, D. Fuchs, K. Pantopoulos, G. Werner-Felmayer, H. Wachter, and M.W. Hentze. Translational regulation via iron-responsive elements by the nitric oxide/NO-synthase pathway. *EMBO J.*, 12:3651–3657, 1993.

Wink, D.A., I. Hanbauer, M.C. Krishna, W. DeGraff, J. Gamson, and J.B. Mitchell. Nitric oxide protects against cellular damage and cytotoxicity form reactive oxygen species. *Proc. Natl. Acad. Sci. USA*, 90:9813–9817, 1993.

Wolff, D.J. and A. Lubeskie. Aminoguanidine is an isoform-selective, mechanism-based inactivator of nitric oxide synthase. *Arc. Biochem. Biophys.*, 316:290–301.

VII. ABSTRACTS OF ORAL AND POSTER PRESENTATIONS

FUNCTIONAL VALIDATION OF TRANSGENES FOR THE DEVELOPMENT OF A TRANSGENIC HYPERTENSIVE RAT ATHEROSCLEROSIS MODEL

Hedy Adari,[1] Xie H. Xiang,[2] Nelson Ruiz-Opazo,[2] Victoria L. M. Herrera,[2] and Savvas C. Makrides[1]

[1]T Cell Sciences, Inc., Needham, Massachusetts
[2]Boston University School of Medicine
Boston, Massachusetts

We have utilized the Dahl-S hypertensive rat in order to develop transgenic rat lines carrying the human cholesteryl ester transfer protein (CETP) gene. The overexpression of the CETP gene, and the hypertensive genetic background of the Dahl-S rat are expected to accelerate the development of atherosclerosis in response to high salt high cholesterol diet, thus providing an animal model of atherosclerosis that would resemble human atherosclerotic lesions. CETP is a component of the reverse cholesterol transport pathway, and it may play a proatherogenic role (1). This is consistent with the observed lipoprotein profile changes in studies of CETP deficiency or overexpression, including: (a) human subjects deficient in CETP (2); (b) overexpression of the human (3) or simian (4) CETP gene in transgenic mice; (c) infusion of CETP into rats (5); (d) the *in vivo* inhibition of CETP in rabbits (6) and hamsters (7) using CETP neutralizing monoclonal antibodies. In addition, species which have relatively high levels of plasma CETP activity (e.g. human, monkey, rabbit) are susceptible to atherosclerosis, whereas species with low or absent CETP activity (e.g. dog, rat, mouse) are relafively resistant to atherosclerosis (8). Pigs are an exception since they lack CETP activity but present atheroma with diet-induced hypercholesterolemia. Moreover, CETP plasma concentrations are directly associated with intima media thickening of carotid arteries (9). However, to date the beneficial or detrimental effects of CETP on atherosclerosis remain to be conclusively established (1), and conflicting observations (10, 11) emphasize the need for further study of the relationship between CETP and development of atherosclerosis.

We placed the CETP cDNA under the control of the human apoCIII promoter, and tested the functionality of the construct by cotransfecting Chinese hamster ovary (CHO) cells with both the CETP construct and a plasmid encoding the hepatic nuclear factor 4

(HNF4) which is necessary for transcriptional activation of the apoCIII promoter. Functionality of the CIII–hCETP transgene was verified by the production of human CETP of the expected size, approximately 74 kd, as determined by Western blot analysis. Control nontransfected CHO cells did not express CETP.

We have also constructed a plasmid encoding the human apoAl gene under the control of the human apoCIII promoter for a second transgenic line. The rationale for the apoAl transgene is to produce a bigenic [CIII–hCETP × CIII–hApoAI] rat model. This model design is based on studies indicating that human CETP has a more pronounced effect on HDL cholesterol levels in transgenic mice expressing human apoAI than single transgenic CETP mice, perhaps due to an enhanced interaction of CETP with human compared to mouse apoAI (3). The apoCIII promoter is expected to provide similar tissue specificity of expression for both the CETP and apoAI genes, both expressed predominantly in the liver and intestine. As with the CIII–hCETP transgene, the functionality of the CIII–hApoAI transgene was verified in double transfectant (CIII–hApoAI and HNF4) CHO cells secreting human apoAI of the expected size, approximately 29 kd, as determined by Western blot analysis.

Microinjection of rat embryos with the CETP construct resulted in the generation of four founder rats which were bred to produce transgenic F1 litter, proving that the CETP transgene is germ-line integrated in all 4 founders. The Fl pups are being bred to homozygosity.

The transgenic CIII–hCETP and the bigenic [CIII–hCETP × CIII–hApoAI] Dahl-S hypertensive rat models should facilitate the testing of antiatherosclerotic agents as well as the dissection of the molecular mechanisms underlying the acceleration of atherosclerosis by hypertension. Comparative analysis of anti-CETP therapeutic agents and other antiatherosclerotic agents versus hypertension control would provide novel insight into the hierarchical interaction of the atherogenesis and genetic hypertension pathogenesis.

ACKNOWLEDGMENTS

We thank Dr. Gerhard Hannig for his help with the apoAI plasmid construction. We are grateful to Dr. Vassilis Zannis for his generous gifts of plasmids and antibodies.

REFERENCES

1. Tall, A. (1995) Ann. Rev. Biochem. 64:235–257.
2. Koizumi, J. et al. (1985) Atherosclerosis 58:175–186.
3. Hayek, T. et al. (1992) J. Clin. Invest. 90:505–510.
4. Marotti, K.R. et al. (1993) Nature 364:73–75.
5. Ha, Y.C., Chang, L.B.F., & Barter, P.J. (1985) Biochim. Biophys. Acta 833:203–210.
6. Abbey, M., & Calvert, G.D. (1989) Biochim. Biophys. Acta 1003:20–29.
7. Evans, G.F. et al. (1994) J. Lipid Res. 35:1634–1645.
8. Ha, Y.C., & Barter, P.3. (1982) Comp. Biochem. Physiol. 71B:265–269.
9. F6ger, B. et al. (1995) J. Mol. Med. 73:369–372.
10. Zhong, S. etal. (1996) J. Clin. Invest. 97:2917–2923.
11. Fielding and Havel (1996) J. Clin. Invest. 97:2687–2688.

EFFECT OF EXTRACORPOREAL CO_2 REMOVAL ON PULMONARY CAPILLARY ENDOTHELIUM-BOUND ANGIOTENSIN CONVERTING ENZYME ACTIVITY IN THE HUMAN

A. Armaganidis,[1] S. E. Orfanos,[1] D. Sfyras,[1] P. Kaltsas,[1] J. D. Catravas,[3] J. Economidou,[2] and Ch. Roussos[1]

[1]Department of Critical Care
[2]Department of Immunology
Evangelismos Hospital
University of Athens Medical School
Athens, Greece
[3]Vascular Biology Center
Medical College of Georgia
Augusta, Georgia

The pulmonary vascular endothelium possesses major metabolic functions such as synthesis and degradation of vasoactive peptides. Enzymes that catalyze these reactions are located on the luminal surface of the capillaries and their activities may be measured *in vivo* by means of *indicator-dilution* techniques. Among these enzymes, angiotensin-converting enzyme (ACE) has been extensively studied in animals and more recently in humans. Pulmonary capillary endothelium-bound (PCEB) ACE activity is reduced in patients with acute lung injury (ALI) and acute respiratory distress syndrome (ARDS), reflecting the underlying endothelial dysfunction (Orfanos et. al., 1996a). In this study we report preliminary results of PCEB ACE activity assessment in two subjects with severe ARDS (NIH ECMO study criteria), who were treated with *Low Frequency Positive Pressure Ventilation* with *Extra-Corporeal CO_2 Removal* (LFPPV with $ECCO_2R$). Our purpose was 1) to investigate if $ECCO_2R$ application, per se, affects PCEB ACE activity, and 2) to assess if the degree of endothelial dysfunction correlates with the clinical severity and the evolution of the syndrome.

Single-pass transpulmonary hydrolysis of the specific ACE substrate ^{3}H-benzoyl–Phe–Ala–Pro (BPAP; 20 μCi, 22.2 Ci/mmol) was measured by means of *indicator-*

dilution techniques, under first order reaction conditions, at three different time-points: (1) pre-$ECCO_2R$, (2) 12 hours (subject 1) and 15 hours (subject 2) post-$ECCO_2R$, and (3) 48 hours post-$ECCO_2R$. Estimated parameters of enzyme activity included percent BPAP metabolism (%M), and the modified first order parameter A_{max}/K_m, an index of functional dynamically perfused capillary surface area (DPCSA).

$ECCO_2R$ application in subject 1 was not associated with any improvement in blood gases or chest x-rays, while subject 2 denoted an improvement in blood gases. All ACE activity parameters were lower than those previously reported in normal volunteers (Orfanos et al., 1996b). Subject 1 showed no changes among the pre- and post-$ECCO_2R$ values: BPAP %M=29% to 28%, and A_{max}/K_m=1268 to 1109 ml/min. In contrast, a slight increase was observed in BPAP %M values of subject 2 (30% to 33%), while A_{max}/K_m decreased (1246 to 896 ml/min) most probably due to a decrease in cardiac output.

These preliminary data confirm that PCEB ACE activity is reduced in ARDS, whereas it is not altered by $ECCO_2R$ application per se, while PCEB ACE activity appears to correlate with the clinical severity of ARDS and could serve as a quantifiable index of the severity of the syndrome an its response to treatment.

Supported by PENED 70/2/1610 and PENED 1997, the Thorax Foundation, and HL31422.

REFERENCES

Orfanos, SE et al., Int. Care Med., 22(supp. 3):S338; 1996a.

Orfanos, SE et al., In: *Vascular Endothelium: Responses to Injury*, JD Catravas, A Callow, and U Ryan eds., Plenum Press, New York, 1996b, pp. 325–326.

INDUCTION OF CYCLO-OXYGENASE IN HUMAN VESSELS *IN VITRO*

David Bishop-Bailey, El B. Haddad,[1] Simon Larkim, John R. Pepper,[2] Magdi Yacoub,[3] Timothy J. Williams, and Jane A. Mitchell[3]

[1]Department of Thoracic Medicine
Department of Applied Pharmacology
The National Heart and Lung Institute
Imperial College
London, United Kingdom
[2]Department of Cardiothoracic Surgery
[3]Departmemt of Anaesthesia and Critical Care Medicine
Royal Brompton Hospital
London, United Kingdom

Cyclo-oxygenase (COX) converts free arachidonic acid to prostaglandin (PG)H_2, the precursor to a group of lipid mediators including prostacyclin (PGI_2), PGE_2, and thromboxane (TX) A_2. The predominant COX metabolite released from vessels is PGI_2, which is a vasodilator, and an inhibitor of cellular proliferation, platelet adhesion and activation, and cholesterol metabolism (see Willis et al., 1986; Hajjar and Pomerantz, 1992). COX is now known to exist in at least 2 different isoforms; a constitutively expressed COX-1, and a cytokine inducible COX-2 (see Mitchell et al., 1995). Within vessels COX-l is normally the dominant isoform, and is concentrated mainly in the endothelial cell layer. The underlying vascular smooth muscle has lower COX activity, but is a rich source of PGI_2 synthetase. In animal models, COX-2 is induced in the vascular smooth muscle layer after mechanical damage to the vessel (Rimarchin et al., 1994). However, the expression of COX-2 in human vessels has not been demonstrated.

Using an *in vitro* organ culture system, we have assessed the effects of surgical preparation on the subsequent induction of COX-2 and release of prostanoids from segments of human saphenous vein (SV) and internal mammary artery (IMA) obtained from the same patient. We have also investigated the ability of interleukin-1β (IL-1) to modulate the induction of COX-2 in this system.

COX-2 mRNA was not detected in freshly prepared SV or IMA. In contrast after 48h in organ culture significant levels of COX-2 mRNA were detected in both SV and IMA. Furthermore the level of COX-2 mRNA were significantly elevated when IL-1β was included in the culture medium. COX-2 metabolites were released in the following rank order PGE_2>= PGI_2 >> thromboxane $(TX)A_2$, with approximately 4 fold greater amounts released from SV treated with IL-1β. The release of prostanoids by IMA was increased, although not significantly when treated by IL-1β. Furthermore, the selective COX-2 inhibitor NS-398, significantly attenuated 6-keto $PGF_{1\alpha}$ (the hydrolysis product of PGI_2) release from SV under all conditions, and from incubations of IMA treated with IL-1β However, NS-398 only inhibited the release of 6-keto $PGF_{1\alpha}$ from the IMA of 2 of the 4 patients under control culture conditions.

These observations show that COX-2 is induced in human SV and IMA, and suggests that this may occur in man following CABG. The induction of COX-2 and subsequent release of PGI_2 may represent an endogenous defense mechanism against endothelial damage incurred during surgical preparation of the these vessels for bypass.

REFERENCES

Willis AL, Smith DL, and Vigo C. *Prog Lipid Res* 1986; 25:645–666.
Rimarachin JA, Jacobson JA, Szabo P, et al. *Arterioscler Thromb* 1994; 14:1021–1031.
Mitchell JA Larkin SW, and Williams TJ. *Biochem Pharmacol* 1995; 5O:1535–1542.
Hajjar DP and Pomerantz KB. *FASEB J* 1992; 6:2933–2941.

EXPRESSION AND FUNCTIONAL CONSEQUENCES OF CYTOKINE- OR LIPOPOLYSACCHARIDE-INDUCED ADHESION MOLECULE EXPRESSION ON HUMAN LUNG MICROVASCULAR ENDOTHELIAL CELLS

Kate Blease, Paul G. Hellewell and Anne C. Burke-Gaffney

Applied Pharmacology
National Heart and Lung Institute
Imperial College
London, United Kingdom

Sepsis and its complications such as lung injury are the most common cause of death in hospital intensive care units (Bone, 1994). It is now generally agreed that Gram-negative sepsis is precipitated when lipopolysaccharide (LPS), released from bacterial cell walls, triggers a cascade of inflammatory mediators including cytokines such as tumor necrosis factor-α (TNF-α) and interleukin (IL)-1. It is thought that cytokine- or LPS-induced adhesion molecule expression on lung microvascular endothelium may play a key role in neutrophil-mediated lung injury associated with sepsis. Although expression of cell adhesion molecules (CAM) on human umbilical vein endothelial cells has been well characterized (Pober et al., 1986; Burke-Gaffney & Hellewell, 1996) little is known about cytokine/LPS-induced CAM expression on human lung microvascular endothelial cells.

In this study we investigated the kinetics and concentration-dependence of TNF-α, IL-1β, LPS- or interferon-γ (IFNγ) induced intercellular adhesion molecule (ICAM-1), E-Selectin and vascular cell adhesion molecule (VCAM-1) expression on normal human lung microvascular cells (HLMVEC: Clonetics, California, USA; purchased from TCS Biologicals Ltd. Buckingham, UK), in culture. CAM were detected on confluent monolayers of HLMVEC grown in 96-well plates, using mouse anti-human ICAM-1 (RR1/1), VCAM-1 (4B2) and E-Selectin (1D2) primary antibodies and a peroxidase-linked goat anti-mouse secondary antibody. Antibody binding was quantified using a peroxidase sensitive substrate 2,2¢-azino-bis (3-ethylbenz-thiazoline-6-sulphonic acid) and chromophore development determined by measuring optical density at 405nm (OD_{405}). Results are expressed as mean OD_{405} ± S.E.M. of 5 experiments. We assessed functional consequences of adhesion molecule expression by measuring adhesion of human peripheral blood neutrophils to HLMVEC monolayers. Neutrophils separated by plasma/Percoll gradient centrifugation (>98% pure), were labeled with a fluorescent dye (Calcein-AM, 10 μM) and incubated (37°C, 30 min) with HLMVEC. Results (mean ± S.D. of a representative experiment) are expressed as a percent of adherent cells over total cells added per well (1.25×10^{-5} cells) as determined by fluorescence.

ICAM-1, but not E-selectin or VCAM-l, was constitutively expressed on HLMVEC monolayers (Table 1). TNFα, IL-1β, or LPS caused a concentration-dependent increase in ICAM-1 (24h), VCAM-1(16h) and E-selectin, (6h) but IFNγ only increased ICAM-1 expression (16h). TNFα, IL-1β, or LPS showed similar kinetics for induction of CAM: ICAM-1 expression increased up to 24h; VCAM-1 expression was maximal at 16h and E-selectin expression peaked at 6h and was significantly (P<0.0l) reduced by 16h (Table 1).

Adhesion of unstimulated neutrophils to unactivated HLMVEC was 9.4 ± 3.2% of total cells added to each well and significantly (P<0.05) increased following activation of HLMVEC for 6h with TNFα- (l0 ng/ml; 16.2 ± 1.4%,), IL-1β- (l0 ng/ml; 33.3 ± 1.6%), LPS (10 μg/ml; 18.0 ± 2.1%) but not IFNγ(6.0 ± 1.5%). Anti-CD18 antibody reduced unstimulated neutrophil adhesion to TNFα-activated HLMVEC but not unactivated HLMVEC. Neutrophil adhesion to unactivated or cytokine/LPS-activated HLMVEC was also increased following neutrophil stimulation for 30 min with PMA (10^{-7}M) but not C5a (10^{-7}M). The results in this study show that TNFα, IL-1β, or LPS increased expression of ICAM-1, E-selectin and VCAM-1 on HLMVEC and also increased adhesion of neutrophils to HLMVEC monolayers. In contrast, IFNγ increased ICAM-1 but not E-selectin or VCAM-1 expression and did not increase neutrophil adhesion. An understanding of the profile of cytokine/LPS-induced lung endothelial cell adhesion molecule expression may allow targeting of specific cytokines or adhesion molecules in sepsis.

KB holds a British Heart Foundation studentship. ABG & PGH are supported by the National Asthma Campaign UK.

REFERENCES

Bone, R. C. (1994). Crit. Care. Med., 22, S8–S11.
Pober, J.S., Gimbrone Jr., M.A., Lapierre, L.A. et al., (1986). J. Immunol., 137, 1893–1896.
Burke-Gaffney, A. & Hellewell, P.O. (1996). Am. J. Physiol., 39, C552–C561.

Table 1. Cytokine-stimulated adhesion molecule expression on human lung microvascular endothelial cells

	ICAM-1 (OD_{405})[1]			E-SELECTIN (OD_{405})[2]			VCAM (OD_{405})[3]		
	6 h	16 h	24 h	6 h	16 h	24 h	6 h	16 h	24 h
TNFα 10 ng/ml	1.17 ± 0.04	1.50 ± 0.04	1.70 ± 0.05*	1.18 ± 0.004	0.60 ± 0.03*	0.40 ± 0.02*	0.50 ± 0.05	0.90 ± 0.06*	0.90 ± 0.01*
IL-1 10 ng/ml	0.97 ± 0.06	1.13 ± 0.03	1.26 ± 0.03*	1.13 ± 0.02	0.59 ± 0.03*	0.34 ± 0.02*	0.32 ± 0.02	0.44 ± 0.04	0.30 ± 0.03
IFNγ 40ng/ml	0.37 ± 0.04	0.51 ± 0.04	0.59 ± 0.04*	0.02 ± 0.01	0.007 ± 0.01	0.005 ± 0.01	0.01 ± 0.01	0.02 ± .008	0.03 ± 0.01
LPS 10 μg/ml	0.58 ± 0.02	0.70 ± 0.03	0.87 ± 0.02*	0.78 ± 0.05	0.20 ± 0.02	0.11 ± 0.02	0.24 ± 0.03	0.22 ± 0.03	0.15 ± 0.04

[1]Basal ICAM-1, 0.02 ± 0.001 OD_{405}.
[2]Basal E-selectin or VCAM-1 were not detected.
[3]P<0.01, denotes a significant difference from CAM expression at 6h.

EFFECTS OF ENDOTOXIN ON ENDOTHELIUM-DEPENDENT AND -INDEPENDENT RESPONSES IN THE ISOLATED RAT HEART

Richard G. Bogle[1] and Patrick Vallance[2]

[1]Clinical Pharmacology Unit
Department of Pharmacology and Clinical Pharmacology
St. George's Hospital Medical School
London, United Kingdom
[2]Centre for Clinical Pharmacology
Cruciform Project for Strategic Medical Research
University College London
The Rayne Institute
London, United Kingdom

Endotoxin is thought to be the initiator of a cascade of responses resulting in septic shock [1]. Its effects are mediated via cytokine-induction that in turn increases expression of enzymes producing pro-inflammatory mediators. It is clear that induction of nitric oxide synthase in the vasculature is important in producing the hypotension and smooth muscle hypo-responsiveness to vasoconstrictors associated with septic shock [2]. Less is known about the effects of endotoxin on endotheliumdependent responses. The aim of this study was to determine the effects of endotoxin on endothelial responses in the microvasculature of the heart.

Hearts were obtained from male Wistar rats (220–300g) and perfused (Langendorff) at a constant flow of 10 ml/min with Krebs solution maintained at 37°C and gassed with 95% O_2/5% CO_2. Coronary perfusion pressure (CPP) was increased in hearts by perfusion with a modified Krebs solution containing 3.2mM potassium [3]. Animals were treated with endotoxin (*E. Coli.*, 055:B5, 2.5 mg/kg, i.p.) for 2, 6 or 24h. Some animals were treated with dexamethasone (l mg/kg, i.p.) 1h prior to endotoxin injection. Plasma samples were obtained following isolation of the heart and nitrate levels measured subsequently by capillary electrophoresis [4].

In control animals CPP was 53.2 ± 2.3 mmHg, rising to 108.3 ± 2.9 mmHg (n=13) after perfusion with 3.2 mM potassium. Hearts obtained from animals treated with endotoxin for 2h or 6h had higher basal CPP 99.4 ± 6.6 mmHg and 83.2 ± 8.5 mmHg respectively (n=5) which were not further increased by perfusion with 3.2 mM potassium. Bradykinin (BK, 10^{-14} –10^{-9} mol) or sodium nitroprusside (SNP, 10^{-13}–10^{-7} mol) caused dose-dependent reductions in CPP with calculated EC_{50} values of –11.5 ± 0.1 and –8.2 ± 0.5 mol respectively (n=13). Endotoxin increased plasma nitrate (μM) from 4.2 ± 1.2 μM in control rats to 30 ± 6 μM at 2h, 426 ± 121 μM at 6h and 48 ± 3 μM at 24h (n=4). Endotoxin also reduced the maximum dilation to BK (10^{-9} mol, n=5) from 56 ± 1 mmHg to 26 ± 1 mmHg ($P < 0.05$) and to SNP (10^{-8} mol, n=5) from 57 ± 5 mmHg to 17 ± 4 mmHg ($P < 0.05$) at 6h. This inhibition was not observed in hearts from animals treated with endotoxin for 24h. LPStreatment (6h) also shortened the response to BK and SNP from 383 ± 17 s and 466 ± 53s in control animals to 144 ± 20 s and 225 ± 26 s respectively ($P < 0.05$). Pretreatment with dexamethasone significantly reduced nitrate concentrations measured at 6h to 120 ± 10 (n=3), however it did not prevent the effects of endotoxin on responses to BK or SNP. Relaxation to nicardipine (10^{-10}–10^{-5} mol) a nitric oxide/endothelium-independent vasodilator was not different in endotoxin-treated animals. These results suggest that endotoxin results in microvascular dysfunction involving both endotheliumdependent and -independent pathways. The dysfunction may lie at the level of guanylate cyclase since responses to endogenous and exogenous NO are impaired.

REFERENCES

1. Bayston, K.F. & Cohen, J. (1990). Bacterial endotoxin and current concepts in the diagnosis and treatment of endotoxaemia. J. Med. Microbiol., 31, 73–83.
2. Bogle, R.G. & Vallance, P. (1996). Regulation of vascular smooth muscle tone in sepsis. In: *Pharmacology of Vascular Smooth Muscle.* Ed. Garland, C. & Allen, J. Oxford University Press.
3. Criddle, D.N., Dewar, G.H., Radniknam, M., Wathey, W.B. & Woodward, B. (1991). The synthesis and structure-activity relationships of some long chain acyl carnitine esters on the coronary circulation of the rat isolated heart. J. Pharm. Pharmacol., 43, 636–639.
4. Leone, A.M., Francis, P.L., Rhodes, P., Furst, V. & Moncada, S. (1995). A rapid and simple method for the measurement of nitrite and nitrate in plasma by high performance capillary electrophoresis. Biochem. Biophys. Res. Commun., 200, 951–957.

C5a STIMULATES EOSINOPHIL ADHESION TO HUMAN LUNG MICROVASCULAR ENDOTHELIAL OR EPITHELIAL CELLS: PRIMING WITH TNFα OR IL-5

Anne C. Burke-Gaffney, David M. T. Macari, and Paul G. Hellewell

Applied Pharmacology
National Heart and Lung Institute
Imperial College
London, United Kingdom

Eosinophils are thought to play a key role in perpetuating inflammatory changes and airway epithelial damage seen in asthma. Binding of eosinophils to endothelial cell adhesion molecules facilitates the migration of these cells from blood vessels into the respiratory tract. Subsequent eosinophil adhesion to airway epithelium may contribute to epithelial damage. A knowledge of factors that may potentiate or "prime" eosinophil adhesion to endothelial or epithelial cells may have important implications for the development of new anti-asthma therapies. For example, it has been shown that transendothelial migration of eosinophils *in vitro* can be "primed" by cytokines such as interleukin-5 (IL-5), IL-3 or GM-CSF [1]. We have recently shown, *in vivo,* that tumor necrosis factor α (TNFα) causes priming of eosinophil accumulation in guinea pig skin [2]. In the present *in vitro* study we have investigated priming effects of TNFα or IL-3 on C5a-induced adhesion of human eosinophils to human lung microvascular endothelial cells (LMVEC) or to a lung epithelial cell line (A549), in culture.

Eosinophils, obtained from peripheral blood by plasma/Percoll gradient centrifugation and separated from neutrophils by negative selection with anti-CD16 antibody linked to magnetic beads (>97% purity) were labeled with a fluorescent dye, Calcein-AM (10 μM). Labeled cells (1.25×10^6) were incubated (30 min) with confluent monolayers of either LMVEC (Clonetics, California, USA; purchased from TCS Biologicals Ltd., Buckingham, UK) or A549 grown on 96-well plates. To measure priming, eosinophils were incubated with LMVEC or A549 monolayers for 30 min in the presence of TNFα (10 ng/ml), IL-5 (10^{-9}M) or Kreb's Ringer Phosphate Dextrose buffer. C5a (10^{-7}M) was then added for a further 30 min and adhesion measured at 60 min. Eosinophil adhesion was determined by measuring fluorescence with a Biolite Fl plate reader (at 485 ± 20nm excita-

Table 1. Effect of IL-5 or TNFα on C5a-stimulated eosinophil adhesion to LMVEC or A549

Eosinophil treatment		% adhesion			
		Lung endothelial cells (LMVEC)		Lung epithelial cells (A549)	
1st stimulus (0–30 min)	2nd stimulus (30–60 min)	Control[a]	TNFα[b]	Control[a]	TNFα[b]
Buffer	Buffer	9.7 ± 1.4	26.5 ± 4.4	25.7 ± 4.0	27.0 ± 5.6
Buffer	C5a (10^{-7} M)	31.2 ± 4.6**	43.7 ± 5.7*	47.0 ± 3.5**	59.7 ± 1.1**
Buffer	TNFα (10 ng/ml)	10.6 ± 1.9	30.7 ± 4.2	22.0 ± 3.0	27.9 ± 4.0
TNF	C5a	46.5 ± 3.6*§	54.1 ± 5.1**	49.7 ± 4.8**	64.9 ± 6.4**
Buffer	IL-5 (10^{-9} M)	13.4 ± 2.2	29.5 ± 1.0	21.0 ± 2.8	25.8 ± 4.6
IL-5	C5a	49.4 ± 5.0**§	50.9 ± 2.4**	54.6 ± 3.5**	56.4 ± 7.3*

*, ** denotes a significant difference (P<0.05, P<0.01 respectively) from basal adhesion and §P<0.05 from C5a-stimulated adhesion.[a, b]exposure of endothelial or epithelial cells for 6h to culture medium or TNFα (10ng/ml) respectively. n=5.

tion and 530 ± 25nm emission wavelengths) before and after washing plate with phosphate buffered saline containing 1% horse serum. Results (means ± S.E.M of n experiments) were expressed as percent adhesion (fluorescence after washing plate/fluorescence before plate washing × 100).

Adhesion, measured at 30 min, of unstimulated eosinophils to LMVEC treated with culture medium or TNFα (10ng/ml) for 6h was 7.8 ± 1.2% or 25.8 ± 3.4% (n=5), respectively. Eosinophil adhesion to TNF α-activated LMVEC was significantly increased (P<0.05) by C5a (10^{-7}M; 46.7 ± 6.1%). Adhesion of unstimulated eosinophils to A549 treated with culture medium or TNFα (10ng/ml) for 4h was 5.8 ± 0.2% or 4.7 ± 0.5% (n=3) respectively. Eosinophil adhesion to TNFα-activated A549 was significantly increased (P<0.05) by C5a (10^{-7}M; 46.7 ± 6.1%). Table 1 shows effects of pre-incubating eosinophils with TNFα or IL-5 (30 min) on C5a-stimulated eosinophil adhesion to LMVEC or A549.

This study demonstrates that C5a stimulates eosinophil adhesion to LMVEC and that this response may be increased by pre-exposure of eosinophils to IL-5 or TNFα. In contrast, C5a-stimulated eosinophil adhesion to lung epithelial cells was not significantly increased by TNFα or IL-5 in these experiments. Since basal adhesion of unstimulated eosinophils at 60 min was higher to A549 than to LMVEC a priming effect on eosinophil adhesion to the former may be masked. In future studies we will investigate priming effects of eosinophil adhesion to A549 at earlier times to reduce basal adhesion. These results may have important implications for the development of new anti-asthma therapies.

ABG and PGH are supported by the National Asthma Campaign, UK. DM holds a BBSRC/Yamanouchi studentship.

REFERENCES

1. Ebisawa, M., Liu, C., Yamada, T., Kato, M., Lichtenstein, L.M. *et al.*, (1994) *J. Immunol.*, 152, 4590.
2. Macari, D.M., Teixeira, M.T. & Hellewell, P.G. (1996) *Am. J. Respir. Crit. Care Med.* 153, A204.

SYNERGISTIC INTERACTIONS BETWEEN PRO-INFLAMMATORY CYTOKINES IN THE INDUCTION OF INOS IN HUMAN VASCULAR SMOOTH MUSCLE CELLS

A. H. Chester, G. S. Hoare, N. Marczin, and M. H. Yacoub

Department of Cardiothoracic Surgery
National Heart and Lung Institute
Harefield Hospital/Imperial College
London, United Kingdom

Inducible nitric oxide synthase (iNOS) is thought to play an important role in atherosclerosis. However, the precise role of LPS and individual cytokines in the induction of iNOS in humans remains undetermined. We have used human aortic vascular or internal mammary artery segments to assay the efficacy of LPS (1 g/ml), IFN-γ (400 units/ml), TNF-α (1000 units/ml) and IL-16 (100 units/ml), either alone or in combination to induce the expression of iNOS and subsequent accumulation of NO_2 detected by chemiluminescence. Combination of all 3 cytokines induced a significant increase in NO_2 accumulation after 36 (210%) and 48 (893%) hours incubation, which could be inhibited by L-NMMA (10^{-4}M) and cycloheximide (10^{-6}M). Medial smooth muscle of sections of vessel segments stained positive with an antibody raised against human iNOS. Incubation of IFN-γ and IL-16 alone both caused a small but significant increase in the level of NO_2 above control over a 48 hour period, by 50% and 75% respectively ($p<0.05$). TNF-α was unable to induce a significant rise in NO_2 over the same time period. Experiments with different combinations of the cytokines revealed a synergistic interaction between IFN-γ and IL-16, with the increase in NO_2 levels of 869%. A combination of IFN-γ and TNF-α, or IL-16 and TNF-α yielded no enhanced accumulation of NO_2 compared to the action of each cytokine alone. Incubation of LPS alone was unable to cause an increase in NO_2 accumulation after 48 hours. Combination of LPS with IFN-γ or IFN-γ and IL-16, gave no significant enhancement of NO_2 accumulation. However, LPS was capable of increasing the level of NO_2 accumulation in response to a combination of IFN-γ and TNF-α from 158% to 295%. These results highlight a central role for IFN-γ in the induction of iNOS and reveal a complex series of interactions between IFN-γ and other mediators.

CELL DEATH IN HUMAN ATHEROSCLEROTIC PLAQUES INVOLVES BOTH NECROSIS AND APOPTOSIS

Milita Crisby,[1,2] Bengt Kallin,[1] Johan Thyberg,[3] Boris Zhivotovsky,[4] Sten Orrenius,[4] Vasilios Kostulas,[2] and Jan Nilsson[1]

[1]King Gustaf V Research Institute
Department of Medicine
Karolinska Hospital
[2]Department of Neurology
Huddinge Hospital
[3]Department of Cell and Molecular Biology
[4]Institute of Environmental Medicine
Karolinska Institute
Stockholm, Sweden

The aim of the present study was to analyze the frequency and mechanism of cell death in atherosclerotic plaques with a recent history (<6 months) of rupture. Atherosclerotic plaques were obtained from patients with symptomatic ipsilateral carotid stenosis (>70% diameter reduction) undergoing carotid endarterectomy. *In situ* tailing nick translation of fragmented DNA, agarose gel electrophoresis of plaque DNA and electron microscopy were used to identify cell death by apoptosis (programmed cell death) and necrosis. The mean number of cell containing fragmented DNA in the plaques was 12.7 ± 3.5% (n=15). Focal accumulations of cells with DNA fragmentation occurred in the fibrous cap, at sites of rupture, close to lipid deposits and necrosis and was always accompanied by the presence of inflammatory cells. Electrophoretic separation of DNA isolated form part of plaques where the presence of DNA fragmentation had previously been demonstrated by *in situ* DNA nick translation resulted in multiple ladders of 180–200 base pairs characteristic of apoptosis. Electron microscopic analysis revealed presence of cells with morphological signs of degeneration in a frequency even higher than that found by *in situ* nick translation. Some of these cells had a characteristic apoptotic appearance with condensed chromatin and cytoplasm, but the large majority of the cells had an ultrastructure typical for cells undergoing cell death by necrosis with membrane disruption and swollen, disintegrating organelles. Thus, although apoptosis clearly takes place in atherosclerotic plaques necrosis appears to be a much more common mechanism for cell death. This may have negative consequences for plaque stability since cell death by necrosis induces a stronger inflammatory reaction than apoptotic cell death.

TUMOR NECROSIS FACTOR α INDUCES PAF-DEPENDENT ANGIOGENESIS

A. De Martino,[1] G. Montrucchio,[1] E. Lupia,[1] E. Battaglia,[1] F. Bussolino,[2] and G. Camussis[3]

[1]Department of Clinical Physiopathology
[2]Department of Genetics Biology and Medical Chemistry
University of Torino
[3]Division of Nephrology and Microbiology
University of Pavia

The *in vivo* angiogenic effect of tumor necrosis factor (TNF) α is considered due to the production of secondary mediators, as it is *in vitro* a potent inhibitor of endothelial cell growth (1). Since TNF-α stimulates the synthesis of platelet-activating factor (PAF), a phospholipid mediator of inflammation and septic shock, by monocytes and endothelial cells (EC) (2), we investigated the possible involvement of PAF in the angiogenetic effect of TNF-α.

In vivo angiogenesis was studied in a murine model in which Matrigel was used as a vehicle for the delivery of mediators and it was assayed as growth of blood vessels from subcutancous tissue into the solid gel (3).

In this model, the angiogenesis induced by TNF-α was inhibited by WEB 2170, a specific PAF receptor antagonist; PAF itself was detected within Matrigel, 6 and 24 h after TNF-α injection. The synthesis of PAF in the Matrigel was concomitant with the early migration of EC and infiltration of monocytes. Therefore, we investigated the angiogenic properties of this phospholipid.

In vitro, PAF induced a dose-dependent migration of the endothelial cells across the polycarbonate filters in Boyden's chambers; the migration was prevented by WEB 2170. D-PAF, the biologically inactive enantiomer, and Lyso-PAF on EC did not induce angiogenesis, PAF did not stimulate a proliferative response of EC.

In vivo, PAF induced a dose-dependent angiogenic response, which at pharmacologic concentration (1–5 μM) did not require heparin, but at physiologic concentration (5–50nM) required the presence of heparin at doses that were not angiogenic per se. The angiogenesis induced PAF was, indeed, inhibited both by protamine and by WEB 2170; the effects of D-PAF and Lyso-PAF were not significant. Neutralizing Abs to basic fibroblast growth factor induced a slight, but not statistically significant reduction of the angiogenesis induced by PAF.

In conclusion, in the present study it has been shown that PAF mediates, at least in part, the *in vivo* angiogenic effect of TNF-α and it may directly provide the signal for the migration of EC, but not the signal for their proliferation. Therefore, one can speculate that *in vivo* PAF may induce the recruitment of EC within Matrigel and may stimulate the production by infiltrating EC of heparin-binding growth factors, which can provide the signal for cell proliferation and final formation of canalized vessels.

REFERENCES

1. Frater-Schroder M., Risau W., Hallaman R., et al. (1987) *Proc. Natl. Acad. Sci. USA* 84:5227.
2. Camussi G., Bussolino F., Salvidio G., and Baglioni C. (1987) *J. Exp. Med.* 166:1390.
3. Passaniti A., Taylor R.M., Pili R., et al. (1992) *Lab Invest.* 67:519.

REDUCED PULMONARY REMOVAL OF CIRCULATING ENDOTHELIN-1: A NEW MARKER OF HUMAN PULMONARY HYPERTENSION

Jocelyn Dupuis, Gilbert Gosselin, Duncan J. Stewart, Jean-Claude Tardif, Peter Cernacek, Ihor Dyrda, Raoul Bonan, and Jacques Crépeau

Montreal Heart Institute
Royal Victoria Hospital
Montreal, Quebec, Canada
St. Micheal's Hospital
Toronta, Ontario, Canada

The pulmonary vascular endothelium is an important site for both clearance and production of endothelin-1 (ET-1). Pulmonary hypertension (PH) is associated with increased circulating ET-1 levels that may contribute to the disease process: it is presently unknown if a reduced clearance, and increased production or a combination of both are responsible. We studied the pulmonary metabolism of ET-1 in 13 controls (C), and 17 patients with PH (12 mitral stenoses, 4 CHF, I idiopatic) by combining the indicator dilution technique with measurements of immunoreactive ET-1 levels (IRET-1). We measured percent I^{125}-ET-1 extraction (Ext) as well as net ET-1 production and survival (amount of ET-1 surviving passage through the lungs). The permeability-surface area product (PS) for ET-1 removal was computed using the Renkin model. In controls, aortic IRET-1 levels (0.67 ± 0.10 pg/ml, mean ± SE) were not different from pulmonary artery (0.61 ± 0.08 pg/ml). Levels were higher in PH ($p<0.05$) with a tendency for higher aortic (1.23 ± 0.16 pg/ml) than pulmonary artery levels (1.07 ± 0.19 pg/ml, $p=0.07$).

	Ext (%)	PS (ml/s)	Production (ng/min)	Survival (ng/min)
C	47 ± 2	199 ± 10	1.10 ± 0.19	0.99 ± 0.15
PH	34 ± 3.6	112 ± 7	1.25 ± 0.22	1.86 ± 0.35
p value	<0.01	<0.001	0.60	0.03

Both ET-1, Ext, and the PS product were reduced in PH explaining an ET-1 survival while pulmonary ET-1 production was unchanged. Pulmonary ET-1 Ext was inversely related to the severity of PH (R=0.524, p=0.03). We conclude that the increase in circulating ET-1 is thus a marker rather than a mediator of PH and reflects pulmonary vascular endothelial dysfunction.

HIBERNATION AND STUNNING OF ARTERIAL MYOCYTES: CLINICAL REVERSAL BY EDTA CHELATION THERAPY

David A. Edwards and Corazon Ibarra Illarina

Bio Medical Health Center, Inc.
International Biomedical Research Institute, Inc.
Reno, Nevada

In recent years numerous reports have appeared demonstrating that even in the absence of previous myocardial infarction, patients with coronary artery disease often present with chronic regional wall motion abnormalities that are reversible spontaneously[1] after coronary revascularization[2], angioplasty[3,4], thrombosis[5], and chelation therapy.[6] This regional dysfunction, in the absence of infarction, is postulated to result from either prolonged postischemic dysfunction (stunning) or adaptation to chronic hypoperfusion (hibernation). Hibernation myocardium has been defined as a state of persistently impaired left ventricular function at rest that is due to reduced coronary blood flow.[7] In contrast, stunned myocardium is defined as postischemic left ventricular dysfunction.[8,9] Myocardial stunning and hibernation have been reported to last for months to years.[1,2,10,11,12] It has now been established that the time sequence for return of functional integrity to hibernating myocardial tissue can be independent of correcting the mechanical obstruction of flow.[2,3,13,14,15] Stunned myocardial cells are postulated to be a temporary phenomenon and the term simply refers to the unexplained delay in return of optimal function of cellular activity after flow has been restored. Myocyte hibernation appears to be much more difficult to explain. All reports to date have involved myocardial myocytes.

The cardiovascular system is the first system to function in the embryo. It is derived from specialized mesodermal mesenchyme called angioblastic tissue.[16] From an embryological perspective, the arterial myocyte and the myocardial myocyte are derived from identical mesenchymal cells.[17] There are a number of conventional theories of atherosclerosis. All currently held theories involve abnormalities in the pluripotential vascular smooth muscle cell or myocyte. The genesis of systemic atherosclerosis is generally considered to be identical to coronary atherosclerosis.[18] The symptoms of atherosclerosis have always been assumed to be due to an obstruction of flow to the involved tissue. The authors describe an unusual case of asymptomatic peripheral vascular sclerosis and uncontrollable hypercholesterolemia that demonstrated dramatic reversal following intravenous Ethylenediaminetetraacetic acid (EDTA) chelation therapy. The remarkable changes in lower limb noninvasive vascular testing and normalization of serum total and LDL cholesterol in this 58-year old female who underwent previous surgical repair for coarctation of the aorta years before suggest that the lower limb and mesenteric arterial vascular system was stunned/hibernating/embalmed since the coarctation repair.

The specific points in this case that suggest a metabolic rather than mechanical mechanism of dysfunction in the arterial tissue are: 1) the unusual configuration of the pre-treatment vascular studies showing adequate doppler flow at rest but a total absence of flow reserve by Post-Occlusive-Reactive Hyperemia (PORH), 2) the absence of clinical symptoms of ischemia prior to chelation despite extremely poor compliance of the arteries (markedly abnormal Pulse Volume Recording), 3) the absence of significant associated coronary, carotid and cerebral vascular disease as demonstrated by normal carotid/supraoptic doppler flow and normal Ultra-fast CT Scan of the heart, 4) dramatic response to a relatively low dose (44 grams) of EDTA Chelation therapy with complete normalization of PORH and PVR tracings, 5) normalization of the previously difficult to control total and oxidized (LDL) cholesterol following chelation therapy indicating that perhaps the oxidized (LDL) cholesterol was a marker of oxidant stress rather than a cause of disease as suggested by Cranton and Frackleton[19] and 6) the loss of bowel symptoms to cholesterol exchange resin therapy that had been so troublesome before EDTA therapy suggesting that this may have represented unrecognized ischemic (hibernating/stunned mesenteric arterial myocyte) bowel symptoms.

Conventional concepts of cellular pathology and physiology have not previously recognized the recently described clinical phenomena of myocyte hibernation or stunning. Despite this, their existence is now well established in the scientific literature. Both stunning and hibernation may represent different manifestations of the same phenomenon. The mechanism(s) by which they occur and the tissue(s) in which they occur remain to be established. The area of study of these phenomena has, here to fore, been confined to the myocardium. They may not be confined to cardiac myocytes, but may express themselves throughout the entire vascular myocyte syncytium. The authors believe this case may be an example of hibernation and stunning in peripheral arterial tissue.

REFERENCES

1. Galli M, Marcassa C, Bolli R, et al. Spontaneous delayed recovery of perfusion and contraction after the first 5 weeks after anterior infarction. Evidence for the presence of hibernating myocardium in the infarcted area. Circulation 1994; 90(3):1386–1397.
2. Takeishi Y, Tono-oka I, Kubota I, et al. Functional recovery of hibernating myocardium after coronary bypass surgery: Does it coincide with improvement in perfusion? Am Heart J 1991; 122(3 Pt 1):665–670.
3. Swift PC, Turner JH, Oxer HF, et al. Myocardial hibernation identified by hyperbaric oxygen treatment and echocardiography in post-infarction patients: Comparison with exercise thallium scintigraphy. Am Heart J 1992; 124(5):1151–1158.
4. de Feyter PJ, Surapranata H, Serruys PW, et al. Effects of successful percutaneous transluminal coronary angioplasty on global and regional left ventricular function in unstable angina pectoris. Am J Cardiol 1987; 60(13):993–997.
5. Conti CR. The stunned and hibernating myocardium: A brief review. Clin Cardiol 1991; 14(9):708–712.
6. Edwards DA, Crimm L, Davis PR. Noninvasive cardiology and chelation patients. AAMP Fall Convention, November 1985, San Francisco. Telebeam Audiotapes; Sherman Oaks, CA, 1986.
7. Rahimtoola SH. The hibernating myocardium. Am Heart J 1989; 117(1):211–221.
8. Kloner RA, Przyklenk K, and Patel B. Altered myocardial states. The stunned and hibernating myocardium. Am J Med 1989; 9(1A):14–22.

9. Kloner RA and Przyklenk K. Hibernation and stunning of the myocardium. New Engl J Med 1992; 325(26): 1877–1879.
10. Buxton DB. Dysfunction in collateral-dependent myocardium. Hibernation or repetitive stunning? Circulation 1993; 87(5):1756–1758.
11. Bashour TT and Mason DT. Myocardial hibernation and "embalmment." Am Heart J 1990; 119(3 pt 1): 706–708.
12. Ito H, Tomooka T, Sakai N, et al. Time course of functional improvement in stunned myocardium in risk area in patients with reperfused anterior infarction. Circulation 1993; 87(2):355–362.
13. Bolli R. Mechanism of myocardial "stunning." Circulation 1990; 82(3)723–738.
14. Chiariello M, Perrone-Filardi P, and Ambrosio G. Hibernating and stunned myocardium. Cardioscience 1993; 4(2):63–68.
15. Vanoverschelde JL, Wijns W, Depre C, et al. Mechanisms of chronic regional postischemic dysfunction in humans. New insights from the study of noninfarcted collateral-dependent myocardium. Circulation 1993;87(5):1513–1523.
16. Sadler TW. *Langman's Medical Embryology*, 6th Edition 1990, Williams and Wilkins, Baltimore, pp. 71–73.
17. Moore KL. The developing human. *Clinically Oriented Embryology*, 4th Edition 1988, WB Saunders, Philadelphia, pp. 286–294.
18. Robbins SL, Cotran RS, and Kumar V. *Pathologic Basis of Disease*, 3rd Edition 1984, WB Saunders, Philadelphia, pp. 506–567.
19. Cranton EM and Frackelton JRP. Free radical pathology in age-associated diseases: Treatment with EDTA chelation, nutrition, and antioxidants. J Advancement Med 1989, 1/2:17–54.

ANTITUMOR ACTIVITY OF IL-12 IN A MOUSE OPPORTUNISTIC VASCULAR TUMOR: ROLE OF IMMUNOCOMPETENT CELLS AND CYTOKINES

F. Fioretti, Q. G. Dong, M. Fusco, A. Vecchi, and A. Mantovani

Instituto di Ricerche Farmacologiche "Mario Negri"
Milan, Italy

IL-12 has been reported to have antitumoral activity in different mouse tumors. In this study IL-12 was tested on the *in vivo* growth of H5V cells, a tumor line obtained by polyoma middle T (PmT) infection of mouse endothelial cells (Garlanda et al., P.N.A.S., 91:7291, 1994). H5V injection in syngeneic hosts causes vascular tumors that may represent an experimental model for human opportunistic vascular tumors. T lymphocytes, mainly CD8, and Th1 cytokines have been shown to play a major role in the *in vivo* resistance to H5V to H5V tumor; IFNγ and TGFb1 display antitumoral activity, mainly in immunocompetent hosts.

IL-12 (1 μg/mouse, i.p.) given 5 times a week for 4 weeks, significantly delayed tumor growth and increased the survival of mice. About 40% of IL-12 treated animals were cured. In mice, IL-12 reduces the tumor growth, increasing the survival time, but never cured the animals. Treatment with specific monoclonal antibodies showed that LFA1+ and CD8+ but not NK cells play a major role in IL-12 activity. In anti IFNγ treated animals IL-12 effects are reduced, suggesting that IFNγ is an important mediator of IL-12 activity. The lack of effects of IFNγ treatment and the reduced effect of IL-12 in nude mice indicate that IFNγ can play a role in this model but other mechanisms are likely to contribute to the antitumor effects. Anti-LAF1 or anti CD8 antibodies completely inhibited IL-12 activity. The potent antitumor activity displayed by IL-12 in this vascular tumor suggests investigations on the potential of this cytokine in human opportunistic vascular tumors.

PULMONARY CAPILLARY RECRUITMENT AND EXTRACAPILLARY VASCULAR TONE

Lyle E. Fisher, Jr.,[1] Curt M. Steinhart,[1] and John D. Catravas[2]

[1]Department of Pediatrics
[2]Vascular Biology Center
Medical College of Georgia, Augusta, Georgia

INTRODUCTION

Elucidating the mechanisms controlling capillary recruitment could improve our understanding of the mechanisms underlying various pharmacological and pathophysiological processes.

Although earlier work has associated control of capillary recruitment with extrinsic factors, such as pressure in the larger blood vessels upstream and downstream from the capillary bed, more recent evidence has raised the possibility of the presence of regulatory factors intrinsic to the capillary bed.

HYPOTHESIS

Pulmonary capillary recruitment may *not* depend solely on extracapillary factors. In order to test this hypothesis, we examined the effects of changes in extracapillary vascular tone on dynamically perfused capillary surface area in a constant pulsatile flow, full cardiac bypass preparation.

METHODS

Male New Zealand White rabbits, (3.9 ± 0.1 kg.) were anesthetized with Urethane and diallylbarbital. They then underwent tracheostomy and were mechanically ventilated to maintain pH, P_aCO_2, and P_aO_2 and were maintained within physiologic ranges. After receiving indomethacin and heparin they were placed on cardiopulmonary bypass. Airway, systemic, pulmonary and left atrial pressures were continuously monitored and recorded.

MEASUREMENTS

Pulmonary blood flow, Q_p, was measured by electromagnetic flow meter, and held constant (mean 101 ml/kg/min) after an initial stabilization period.

1. Baseline — mean PAP.
2. Constricted — PA hypertension induced by serotonin infusion.
3. NO 1 — Nitric oxide inhalation adjusted until PAP had diminished approximately half-way back to baseline.
4. NO 2 — Nitric oxide inhalation adjusted until PAP approximately at baseline.

Segmental pressure determinations were made using a combination of arterial, venous and double occlusion maneuvers.

ACE Metabolism

The ACE substrate, [^{3}H]-Benzoyl–phenylalanyl–alanyl–proline was injected into the PA, and pulmonary venous effluent collected, 20 samples in 12 seconds. The % metabolism determined in each sample.

Segmental Resistances

1. R_{la}, large artery resistance, $(P_{AP} - P_{ao})/Q_p$
2. R_{sa}, small artery resistance, $(P_{ao} - P_{do})/Q_p$
3. R_{sv}, small vein resistance, $(P_{do} - P_{vo})/Q_p$
4. R_{lv}, large vein resistance, $(P_{vo} - P_{VP})/Q_p$

Dynamically Perfused Capillary Surface Area

Using the integrated form of the Henri-Michaelis-Menten equation, the modified first order rate constant A_{max}/K_m is determined. $A_{max}/K_m = (E * k_{cat})/K_m$ where E is angiotensin converting enzyme mass available to circulating substrate, k_{cat} is a catalysis constant and K_m is the Michaelis constant. Thus changes in Amax reflect changes in ACE mass. Since ACE is distributed homogeneously across the endothelial surface, changes in ACE mass reflect changes in capillary surface area.

Statistics

Results were analyzed using ANOVA, followed by Newman-Keuls Test. Statistical significance was accepted at a p value < 0.05.

SUMMARY

Serotonin caused a significant increase in large and small artery resistance without affecting capillary surface area. Nitric oxide then *reversed* the elevated upstream resistances without affecting capillary surface area.

CONCLUSION

Significant changes in upstream pressures may occur without affecting pulmonary capillary recruitment.

SPECULATION

It may be possible to manipulate pulmonary capillary recruitment independent of upstream segmental resistances.

EVIDENCE FOR ADVENTITIAL REMODELING IN HUMAN CORONARY ATHEROSCLEROSIS

P. Ghenev,[1] K. Valchanov,[1] A. Tonchev,[1] R. Pancheva,[1], and G. Chaldakov[2]

[1]Department of General and Clinical Pathology
[2]Laboratory of Electron Microscopy
Medical University of Varna
Bulgaria

Research in atherosclerosis for a long time has been focused mostly upon the intimal changes. The outer layer, the adventitia has been entirely neglected. Very recently, results confirming adventitial remodeling appeared. However, data concerning atherosclerotic involvement of arterial adventitial layer are few. The purpose of the present study is to look for adventitial remodeling in a sample of postmortem human coronary arteries. Irregular thickening, collagen overproduction, and hypercellularity, as well as mononuclear infiltration and abundance of newly-formed blood vessels were the most common findings in the adventitia of human coronary arteries. These results are similar to those found experimentally and strongly suggest that adventitia and adventitial remodeling are not innocent bystanders in atherogenesis.

BASIC FRIBROBLAST GROWTH FACTOR OVEREXPRESSION IN ENDOTHELIAL CELLS: AN AUTOCRINE MECHANISM FOR ANGIOGENESIS AND ANGIOPROLIFERATIVE DISEASES

A. Gualandris,[1] M. Rusnati,[1] M. Bastaki,[1] M. Ziche,[2] and M. Presta[3]

[1]Unit of General Pathology and Immunology
Department of Biomedical Sciences and Biotechnology
University of Brescia
[2]Department of Pharmacology
University of Florence
[3]Institute of Anatomy, Histology, and Embryology
University of Bari
Italy

Basic fibroblast growth factor (bFGF) is expressed in vascular endothelium during tumor neovascularization and angioproliferative diseases. The ultimate significance of this observation is poorly understood. We have investigated the biological consequences of endothelial cell activation by endogeneous bFGF in a mouse aortic endothelial cell line stably transfected with a retroviral expressionvector harboring a human bFGF cDNA. Selected clones expressing 24 kD, 22 kD, and/or 18 kD bFGF isoforms were characterized by a transformed morphology and an increased saturation density. bFGF transfectants showed invasive behavior and sprouting activity in three-dimensional fibrin gels and formed a complex network of branching cord-like structures connecting foci of infiltrating cells when seeded on laminin-rich basement membrane matrix (Matrigel). The invasive and morphogenetic behavior was prevented by anti-bFGF antibody, revealing the autocrine modality of the process. The biological consequences of this autocrine activation were investigated *in vivo*. bFGF-transfected cells gave rise to highly vascularized lesions resembling Kaposi's sarcoma when injected in nude mice and induced angiogenesis in avascular rabbit cornea. Injected into the allantoic sac of the chick embryo they caused an increase in vascular density and formation of mangiomas in the choriallantoic membrane. In conclusion, bFGF-overexpressing endothelial cells acquire an angiogeneic phenotype and reclute quiescent endothelium originating angioproliferative lesions *in vivo*. These findings demonstrated that bFGF overexpression exerts an autocrine role for endothelial cells and support the notion that tumor neovascularization and angioproliferative diseases can be triggered by stimuli that induce vascular endothelium to produce its own autocrine factors(s).

Supported by grants form AIRC (Special Project Angiogenesis), Istituto Superiore di Sanitá (AIDS Project), and European Communities (Human Capital Mobility Project "Mechanisms for the Regulation of Angiogenesis") to M.P.

ACTIVATION STATE OF A4β1 INTEGRIN CONTROLS CAPTURE, IMMOBILIZATION AND MIGRATION OF FLOWING LYMPHOCYTES ON PURIFIED VCAM-1

Patricia Lalor,[1] J. M. Clements,[2] R. Pigott,[2] M. J. Humphries,[3] J. H. Spragg,[4] and G. B. Nash[1]

[1]Department of Physiology, The Medical School
Birmingham, United Kingdom
[2]Neures Ltd.
Abingdon, Oxfordshire, United Kingdom
[3]Biological Sciences
Manchester University, Manchester, United Kingdom
[4]Yamanouchi Research Institute
Oxford, United Kingdom

Vascular cell adhesion molecule-1 (VCAM-1) is expressed on endothelial cells following cytokine stimulation, and can bind leukocyte α4β1 integrin. Adhesion of flowing human lymphocytes to VCAM-1 was analyzed through use of purified recombinant VCAM-1 immobilized onto plastic in a flow chamber. Integrin-mediated attachment supported adhesion of lymphocytes at physiological shear rates, but the nature of attachment was dependent on the concentration of VCAM-1 (10–100 μg/ml). At low concentration most adherent cells were rolling (~75%) but this decreased to <5% at 100 μg/ml VCAM-1. Activation of α4β1 integrin by Mn^{++} or monoclonal antibodies 12G10 and TS2/16 induced significant increases in lymphocyte adhesion only at low VCAM-1 concentration, and converted rolling adhesion to the stationary form. Treatment with phorbol ester had a similar effect, but also induced adherent lymphocytes to flatten and spread over the VCAM-1. These results suggest that the adhesive behavior of lymphocytes on VCAM-1, i.e. rolling adhesion vs. stationary adhesion and spreading, is dependent on the activation state of VLA-4, and that ligation of high concentrations of VCAM-1 is sufficient to induce activation of this integrin. Pretreatment of lymphocytes with the metabolic inhibitor sodium azide, or chelation of intracelluar Ca^{++} caused lymphocytes to remain rolling on 100 μg/ml VCAM-1, supporting the concept of ligand-induced, energy-dependent signaling. This implies that regulation of VCAM-1 concentration and integrin activation *in vivo* could control the nature of lymphocyte adhesion, with a co-expressed stimulus probably required for subsequent migration.

REGULATION OF NF-κB IN HUMAN VASCULAR TISSUE

Nandor Marczin, Ginette Hoare, Karen Morrison, Adrian Chester, and Magdi Yacoub

Department of Cardiothoracic Surgery
National Heart and Lung Institute
Imperial College of Science, Technology, and Medicine
Harefield Hospital
United Kingdom

Activation of NF-κB transcription factor may contribute to proliferative and inflammatory responses of vascular smooth muscle (SMC) during atherogenesis and systemic inflammatory response. This study was designed to elucidate the cellular mechanisms underlying cytokine-induced NF-κB activation in human aortic SMC. The inhibitor (I-κB) and the DNA binding p65 subunit of NF-κB were identified in the cytoplasm of unstimulated cells by immunocytochemistry and Western blotting. Exposure to increasing concentrations of IL-1β (1–100 U/ml), TNFα (10–1000 U/ml) or bacterial lipopolysaccharide (10–1000 ng/ml) caused increased phosphorylation of I-κB followed by complete depletion of I-κB within 30 minutes. Depletion of I-κB was associated with increased nuclear content of p65. To investigate the role of oxidant stress in cytokine-induced NF-κB activation, SMC were pretreated with antioxidants N-acetyl-L-cysteine (NAC) or metal chelator pyrrolidine-dithio-carbamate (PDTC). PDTC (20mM) but not NAC (20mM) caused a 10-fold rightward shift in the concentration response curve of cytokines. These results do not appear to support the ultimate role of oxidant stress in NF-κB activation in these cells. To elucidate the role of proteolysis in NF-κB activation, we tested the effects of distinct protease inhibitors with different activities towards the proteasome pathway. Calpain inhibitors I and II, but not leupeptin, temporarily stabilize phosphorylated I-κB and cause a strong positive immunoreactivity in the high molecular weight region of 250–100 kDa, consistent with accumulation of polyubiquitinylated I-κB. Moreover, both Calpain inhibitors, but not leupeptin attenuated p65 nuclear translocation. Calpain inhibitor I appears to be 10 fold more potent than Calpain inhibitor II, suggesting that the proteasome pathway rather than calpains are responsible for I-κB depletion. These findings identify a signaling pathway in human vascular SMC involving rapid phosphorylation and polyubiquitination of I-κB followed by degradation by the protease pathway which may underlay NF-κB activation and be targeted pharmacologically in important pathologies.

PATTERNS OF AN ANGIOTENSIN CONVERTING ENZYME (ACE) SUBSTRATE HYDROLYSIS BY PULMONARY CAPILLARY ENDOTHELIUM-BOUND ACE IN CRITICALLY ILL PATIENTS

S. E. Orfanos,[1] D. Langleben,[2] A. Armaganidis,[1] J. Khoury,[2] P. Sarafidou,[1] C. Glynos,[1] R. D. Schlesinger,[2] L. Dragatakis,[2] J. D. Catravas,[3] and Ch. Roussos[1]

[1]Critical Care Department
Evangelismos Hospital
University of Athens Medical School
Athens, Greece
[2]Lady Davis Research Institute and Division of Cardiology
Sir Mortimer B. Davis Jewish General Hospital
McGill University
Montreal, Canada
[3]Vascular Biology Center
Medical College of Georgia
Augusta, Georgia

The pulmonary vascular endothelium participates in various important physiologic processes, such as synthesis, release and degradation of hormones and vasoactive peptides, prostaglandin synthesis, release and uptake, lipid processing, xenobiotic metabolism, and antithrombogenic-thrombolytic activities (Orfanos and Catravas, 1993). Enzymes responsible for many of these functions are located on the luminal surface of the capillaries (ectoenzymes). Due to their location, they are directly accessible to blood-borne substrates and their activities may be measured *in vivo* by means of *indicator-dilution* type techniques. Among these enzymes, angiotensin-converting enzyme (ACE) a major regulator of systemic blood pressure has been extensively studied in animals: monitoring pulmonary capillary endothelium-bound (PCEB) ACE activity *in vivo* under normal conditions allows estimations of the dynamically perfused capillary surface area (DPCSA; Orfanos et al., 1994), whereas under toxic conditions provides an early index of lung injury. We have initiated similar studies in man, in an effort to validate the use of the aforementioned techniques as a tool of pulmonary endothelial function assessment in the normal and injured human lung.

PCEB ACE activity was assessed by measuring single pass transpulmonary hydrolysis of the specific substrate ^{3}H-benzoyl–Phe–Ala–Pro (BPAP, 20–30 μCi, 22.2 Ci/mmol), under first order reaction conditions, by means of *indicator-dilution* techniques. BPAP hydrolysis was expressed as % metabolism (% M), and v=–ln(1–M). The modified first order reaction parameter A_{max}/K_m (product of ACE mass and the constant of product formation) which under normal conditions is proportional to DPCSA, was also calculated.

PCEB ACE activity assessment was performed in three groups of patients. Group 1 (Jewish General Hospital): 12 adult subjects with no history of lung disease undergoing cardiac catheterization for coronary artery disease. Group II (Evangelismos Hospital): 5 mechanically ventilated adult subjects, previously declared brain-dead. Group III (Evangelismos Hospital): 9 mechanically ventilated adult subjects with various degrees of lung injury (ALI) and lung injury scores (LIS; Murray et al., 1988), ranging from 0 to 3 (i.e. from no injury to ARDS).

In Group I, BPAP was injected into a main pulmonary artery, and inside the right atrium in order to assess ACE activity in one and both lungs, respectively. Pulmonary blood flow (F_b) was estimated by thermodilution. F_b in one lung was estimated as $0.5 \times F_b$ (Lipscomb and Pride, 1977; Wernly et. al., 1980). Similar values of %M (65.2 ± 2.9 vs. 63.2 ± 2.8) and v (1.11 ± 0.10 vs. 1.04 ± 0.09) were observed in both and one lung respectively, denoting homogeneous pulmonary capillary ACE concentrations and capillary transit times in both human lungs. In contrast, an almost 50% reduction of A_{max}/K_m normalized per body surface area (BSA) was observed (2114 ± 140 ml/min/m^2 in both lungs to 1083 ± 81 ml/min/m^2 in one lung, $p<0.01$), suggesting that like in animals this procedure can quantify DPCSA in man.

In Group II, enzyme activity estimations were performed at different pulmonary blood flows per subject, ranging from 3.3 to 14.1 L/min. The variations in cardiac output and subsequently in pulmonary blood flow occurred as a result of administration of inotropes and/or fluid loading. As F_b increased from 3.3 to 14.1 L/min, BPAP %M fluctuated between 46 and 64% (Pearson r=0.15) and v fluctuated between 0.6 and 1.03 (r=0.12), but not in relationship to F_b. In contrast, as F_b increased, mean ^{3}H transit time, a reflection of mean lung transit time, decreased linearly from 27.8 to 13.8 sec (r=0.74, $p<0.01$). PVRI (pulmonary vascular resistance index) also decreased linearly from 4.34 to 2.5 mmHg/L/min/m^2 (r=0.84, $p<0.01$), while A_{max}/K_m/BSA increased proportionally in an inverse linear pattern from 1450 to 4662 ml/min/m^2 (r=0.85, $p<0.01$). These data suggest that like in the animals, substrate hydrolysis by PCEB ACE in the mechanically-ventilated human subject does not change over a wide range of pulmonary blood flows, implying unmodified capillary transit times, probably due to increases of the total fraction of concomitantly perfused capillaries (i.e. recruitment) that accommodate the increasing lung blood volume. The linear pattern of A_{max}/K_m/BSA increase, along with the inversely proportional decrease in PVRI, suggests recruitment of new capillaries rather than distention of already perfused ones, and reconfirms the hypothesis that A_{max}/K_m is a reliable DPCSA index in man.

In Group III, one estimation of PCEB ACE activity was performed per patient. As LIS increased from 0 to 3, BPAP %M decreased linearly from 67 to 15% (r=0.97, $p<0.01$) and v decreased linearly from 1.10 to 0.17 (r=0.96, $p<0.01$). In patients with clinically developed ALI/ARDS, as LIS increased from 1.5 to 3, the linear patterns of BPAP %M and v decrease were maintained (41 to 15%; r=0.84, $p<0.05$ and 0.38 to 0.17; r=0.84, $p<0.05$, respectively). In contrast, A_{max}/K_m varied from 562 to 1826 ml/min, but not in relationship to LIS (r=0.16). These date suggest that PCEB ACE activity is altered in patients with ALI and ARDS. The strong correlation observed between the severity of ALI/ARDS and the reduction of most ACE parameters, suggests that this technique provide sensitive indices that can quantify lung injury in man. The early profound decrease of A_{max}/K_m in ALI to the levels observed in ARDS, is probably due to the fact that this parameter depends on enzyme catalytic activity as well as on functional capillary surface area (i.e. enzyme molecules available for reaction), both of which are altered during ALI/ARDS.

In summary, our studies indicate that PCEB ACE activity may be assessed in humans by means of indicator-dilution techniques applied at the patient's bedside. Monitoring of human pulmonary capillary endothelium-bound ACE activity provides: 1) information on lung microvascular physiology, 2) quantification of the dynamically perfused capillary surface area, and 3) a sensitive index able to quantify endothelial dysfunction and thus lung injury early in the process of ALI/ARDS.

Supported by PENED 1997, the Thorax Foundation, Fonds de la Recherche en Sante du Quebec, and HL 31422).

REFERENCES

Lipscomb and Pride, Thorax 32:720–725; 1977.
Murray et al., Am Rev Respir Dis 138:720–723, 1988.
Orfanos and Catravas, In: *Annual of Cardiac Surgery*, Yacoub & Pepper (eds.), Current Science, London, 1993, pp. 52–59.
Orfanos et al., Toxicol Appl Pharmacol 124:99–111, 1994.
Wernly et al., J Thorac Cardiovasc Surg 80:535–543, 1980.

DETERMINATION OF THE FUNCTIONS MEDIATED BY EACH OF THE TWO VEGF RECEPTORS KDR/FLK-1 AND FLT-1

Nathalie Ortéga and Jean Plouët

Institut de Biologie Cellulaire et Génétique
Laboratoire de Biologie Moléculaire des Eucaryotes
Centre Nationale de la Recherche Scientifique
Toulouse Cédex, France

PURPOSE

Vascular endothelial growth factor (VEGF) binds to two different receptors flt-1 and KDR/flk-1. To determine the functions mediated by each of the VEGF receptors, we compared the effects of VEGF to that of selective flt-1 or flk-1 ligands.

METHODS

We expressed recombinant placenta growth factor (PIGF), which binds only to flt-1 and used the anti-idiotypic antibody strategy to construct specific flk-1 agonists. Two cell types were used, adrenal cortex capillary endothelial cells (ACE) and bovine corneal endothelial cells (BCE). Proliferation was measured by cell counts, migration was assessed by a wound assay and permeability by transendothelial resistance recording.

RESULTS

ACE cells proliferate upon VEGF or flk-1 agonist addition and this proliferation is inhibited when these cells are transfected by KDR/flk-1 antisense oligonucleotides. They do not proliferate upon PIGF which is a flt-1 specific agonist. BCE express flt-1 but not KDR/flk-1, they do not proliferate upon VEGF or flk-1 agonist but they respond to PIGF and VEGF by migration and permeability.

CONCLUSION

The functions mediated upon VEGF stimulation might be dissociated in flt-1 activation for migration and permeability and KDR/flk-1 activation for proliferation.

INDUCING AUTOREACTIVITY TO CONTROL CHOLESTEROL METABOLISM

Charles W. Rittershaus

T Cell Sciences, Inc.
Needham, Massachusetts

Cholesterol ester transfer protein (CETP) mediates the transfer of neutral lipids between HDL and the lower density lipoproteins VLDL and LDL. Elevated serum HDL, which is a negative risk factor for atherosclerosis is associated with lower levels of CETP activity. CETP-deficient humans have been identified and described as generally free of atherosclerosis and prone to longevity. The objective of this sutdy is to determine whether anti-CETP autoantibodies could be induced in experimental animals by vaccination as a novel approach to the reduction of CETP activity and thereby alter the cell helper epitope of tetanus toxoid, a B cell epitope derived from the C-terminus of CETP, and an N-terminal cysteine was synthesized and used to immunize transgenic mice expressing human CETP (CETP mice) and New Zealand white rabbits after primary vaccination and two booster injections. The anti CETP autoantibodies were directed against the neutral lipid binding site since they competed with monoclonal antibody TP2, known to inhibit CETP activity, for binding to recombinant human CETP. In a small study using a hypercholesteremia rabbit model, the vaccinated group has significantly reduced aortic atherosclerotic lesions compared to non-vaccinated controls ($p<.01$).

IL-4 AND IL-13 BIND TO A SHARED HETERODIMERIC COMPLEX ON ENDOTHELIAL CELLS MEDIATING INDUCTION OF VCAM-1 IN THE ABSENCE OF THE COMMON γ CHAIN (γc)

Bruno Schnyder, Serena Lugli, Ningping Feng, and Rene Moser

Institute of Toxicology of the Swiss Federal Institute of Technology and University of Zurich
Schwerzenbach, Switzerland

IL-4 and IL-13 exert similar, non-additive effects on endothelial cells, inducing VCAM-1 expression and subsequent transmigration of eosinophils. The receptor (R) for IL-4 and IL-13 was described as a shared heteromultimeric complex in which the γc subunit was essential for activity. Endothelial cells bound both cytokines with high affinity; by flow cytoflourometry and RT-PCR expressed IL-4Rα but not the γc of the IL-2R, which has previously been reported to participate in the IL-4R complex of different hematopoietic cells. Radioligand crosslinking experiments followed by immunoprecipitation with the mAb S697 to the IL-4Rα showed IL-4-specific binding at 130 kDa, the IL-4Rα, and to a minor extent to a double band co-immunoprecipitated at 65–75 kDA. [^{125}I]IL-13 bound predominantly to that 65–75 kDa and with a trace amount at 130 kDa. However, with [^{125}I]IL-13, no ligand crosslinked receptor was precipitated by the mAb S697 indicating an individual IL-13 binding subunit. The finding that excess unlabeled IL-4 completely displaced IL-13 binding suggests shared receptor subunits. Similarly, non-signaling IL-4 (Y124D)-mutant abolished IL-4 and IL-13 mediated signal transduction. The mAb TUGh4, specific for the γc, failed to precipitate ligant-crosslinked IL-4- and IL-13R's. In summary, the data show that on human endothelial cells, the receptor for IL-4 and IL-13 is heterodimeric complex of the IL-4Rα and 65–75 kD IL-13 binding protein, the putative IL-13α (1).

REFERENCES

1. Schnyder, B., S. Lugli, N.P. Feng, H. Etter, R.A. Lutz, B. Ryffel, K. Sugamura, H. Wunderli-Allenspach, and R. Moser. 1996, Interleukin-4(IL-4) and IL-13 bind to a shared heterodimeric complex on endothelial cells mediating vascular cell adhesion molecule-1 induction in the absence of the common gamma chain. *Blood*, 87(10):4286–4295.

THE RELATIONSHIP BETWEEN PLASMA ANTIOXIDANT STATUS, NITRATE LEVELS, AND LIPID PROFILE IN PATIENTS WITH HYPERTENSION AND CORONARY HEART DISEASE

Bülent Sözmen,[1] Eser Yildirim Sözmen,[2] Cahit Kazaz,[1] Dilek Taşkiran,[2] Leyla Aslan,[1] and Akan Akyol[1]

[1]Department of Internal Medicine
İzmir State Hospital
[2]Department of Biochemistry
Ege University, Faculty of Medicine
Bornova-İzmir, Turkey

In this study, we investigated first, the possible involvement of nitric oxide and antioxidant enzyme levels in the pathogenesis of essential hypertension and second, we determined serum nitrite–nitrate and antioxidant enzyme activities and lipid profiles (total, HDL and LDL cholesterol, tryglyceride) to investigate the relationship between these parameters in atherosclerosis. 13 patients with coronary heart disease, 18 patients with essential hypertension and 16 age-matched healthy subjects were taken into this study. Plasma nitrite, nitrate levels were measured by Griess reaction, erythrocyte superoxide dismutase and catalase activities were determined by the methods of Misra & Fridovich and Aebi, respectively. Serum lipid analysis were performed in the autoanalyzer. We found a decrease in superoxide dismutase activity ($p<0.01$) and nitrite and nitrate levels ($p<0.05$), an increase in total cholesterol, LDL-cholesterol and tryglyceride values of patients with hypertension ($p<0.05$). There was positive correlation between nitrite and HDL-cholesterol values in same patient group ($r=0.467$, $p<0.05$). We found a decrease in SOD activity and nitrate levels ($p<0.01$) and any significant difference in nitrite and LDL-cholesterol, total cholesterol, tryglyceride levels in patients with coronary heart disease. On the other hand, catalase activities in both groups remained without change. These findings focus the attention to interactions between nitric oxide, antioxidant status and atherogenesis and propose that nitric oxide may be important physiopathologically in essential hypertension and atherosclerotic heart failure.

VARIATIONS OF ENDOTHELIUM ANTIOXIDANT ENZYMES AND NITRITE/ NITRATE LEVELS IN COLLAR-INDUCED ATHEROSCLEROSIS OF RABBITS

Eser Yildirim Sözmen,[1] Zeliha Kerry,[2] Levent Üstünes,[2] Fevziye Uysal,[1] Asli Özer,[2] and Taner Onat[1]

[1]Faculty of Medicine, Department of Biochemistry
[2]Faculty of Pharmacy, Department of Pharmacology
Ege University, Bornova-İzmir, Turkey

Endothelial Derived Relaxing Factor (EDRF) has been identified as Nitric Oxide (NO) and this endothelial product relaxes vascular smooth muscle and prevents platelet adhesion to the endothelial surface through a cyclic 3⁻5⁻ guanosine monophosphate-mediated mechanism. There are some evidence linking excess vascular oxidative stress to the impairment of EDRF action associated with hypercholesterolemia. Recently, it's shown that arteries derived from cholesterol fed rabbits produce excess superoxide and the administration of superoxide dismutase (SOD) to atherosclerotic rabbits improves EDRF action. Although it has been shown that excess vascular oxidative stress contributes to impaired EDRF action in experimental hypercholesterolemia and atherosclerosis, there are no data in literature related to antioxidant enzymes in intima bearing vessels predisposing to atherosclerosis. Since intimal thickening is an essential prerequisite for the development of atherosclerosis, the aims of our preliminary study was to investigate the role of antioxidant activity and NO levels in collar-induced intimal thickening by measuring the antioxidant enzymes (SOD and catalase). The intimal thickening was induced by a silicone collar placed around the left carotid artery for 2 weeks and the sham-operated, right carotid artery served as control. After 14 days the rabbits were anesthetized, blood samples were taken and both carotid arteries were dissected. Together with nitrite/nitrate level, SOD and catalase activities in erythrocytes and dissected collared and control carotid artery tissues were measured. While the catalase activity and nitrate level in intima bearing collared carotid artery segments were higher than the control arteries (0.996 ± 0.25 vs. 2.104 ± 0.31 U/mg protein and 302 ± 8.17 vs. 431 ± 78.94 umol/mg tissue respectively), the SOD activity in collared arteries were significantly lower (14.06 ± 2.09 vs. 10.22 ± 0.88 U/mg protein). Since EDRF is readily inactivated by superoxide and the action of EDRF is dependent upon intact endothelium SOD activity. The data demonstrated that EDRF inactivation may occur by depending on oxidative stress and lowering SOD activity in collared-induced intimal thickening without hypercholesterolemia and LDL oxidation.

INORGANIC COMPOSITIONS AND PROPERTIES OF CARDIOVASCULAR CALCIFIC DEPOSITS

Branko B. Tomazic[1] and William D. Edwards[2]

[1]ADAHF Paffenbarger Research Center
National Institute of Standards and Technology
Gaithersburg, Maryland
[2]Division of Anatomic Pathology
Mayo Clinic
Rochester, Minnesota

Calcified deposits (CD) that form in the cardiovascular system are bioapatites with composition comparable to hydroxyapatite $Ca_5(PO_4)_3OH$. Previous studies have yielded information on the calcium, phosphate, sodium, and magnesium contents of CD samples (1). The present study examined the compositions of three distinctly different types of CD's that were isolated from: 1) bovine pericardium bioprostheses (bpb); 2) human heart valves (hhv), and 3) human aortas (ha). The present study additionally focused on the fluoride and carbonate contents, specific surface area and thermogravimetric and structural properties of deposits. The three types of CD's differed significantly in several aspects, notably Ca/P ratios, fluoride and carbonate contents. The Ca/P ratios were ha (1.61) > hhv (1.55) > bpb (1.45) ($p<0.01$); the fluoride contents (ppm) were ha (1390) > hv (890) > bpb (3.7) ($p<0.01$). these differences may be attributed to varying locations of formation and different CD resiendce times in the cardiovascular system. In a separate experiment, these CD materials were used as seeds for mineral growth under well defined super-saturation conditions. Preliminary data indicate that the surface properties of the CD have significant effects o the rate of mineral formation. Information on the composition and crystal growth properties of the various CD's and help elucidate the mechanism of the CD formation, which is not fully understood.

Supported in part by NIH grant HL30035.

REFERENCES

1. B.B. Tomazic. Characterization of mineral phases in cardiovascular calcification. In: *Hydroxyapatite and Related Materials*, P.W. Brown and B. Constantz, Eds., Boca Raton: CRC Press, 1994.

INVOLVEMENT OF POTASSIUM CHANNELS IN THE ACETYLCHOLINE-INDUCED DILATATIONS OF THE RAT RENAL VASCULATURE

M. Tuncer and I. Ay

Department of Pharmacology
Faculty of Medicine
Hacettepe University
Ankara, Turkey

We examined the mechanism of acetylcholine (ACh)-induced dilatations in rat isolated kidneys. Wistar rats of both sexes were anesthetized with ether and kidneys were isolated and perfused via renal artery. Optimal perfusion pressure was found to be 50–75 mmHg and optimal flow rate was 6–8 ml/min. In control experiments Ach caused concentration-dependent dilatations and the maximum dilatation was 65.38 ± 4.05% of maximum phenylephrine constriction. Atropine (10^{-6}M), hexametonium (10^{-4}M), indomethacin (10^{-5}M), and 4-aminopyridine (10^{-3}M) attenuated the ACh-induced dilatations significantly ($p<0.05$, n=6, ANOVA). Tetraethylammonium (10^{-3}), ouabain (10^{-6}), and quinacrine ($3*10^{-6}$M) did not change the dilatations. On the other hand, methylene blue (10^{-5}M), tetrodotoxin (10^{-6}M) and capsaicin (10^{-6}M) caused significant inhibition of ACh-induced dilatations ($p<0.01$, n=6, ANOVA). These results suggest that ACh-induced dilatations may occur vial the factors released from endothelium and presynaptic nerves and act through soluble guanylate cyclase and both ATP-sensitive and calcium-dependent potassium channels.

This study was supported by Hacettepe University Research Grant #95.3.011.007.

REFERENCES

Bauersachs, J., Hecker, M., and Busse, R. *Br. J. Pharmacol.*, 113:1548–1553, 1994.
DeMey, J.G. and Gray, S.D. *Prog. Appl. Microcirc.*, 8:181–187, 1985.
Tuncer, M. and Vanhoutte, P.M. *Blood Pressure*, 2:217–220, 1993.

ENDOTHELIAL CELL RESPONSES TO INFLAMMATORY AND THROMBOTIC MEDIATORS — EVIDENCE FOR LOCALISATION OF CELL–CELL SIGNALS

Nicholas Wickham

Department of Clinical Oncology
The Chinese University of Hong Kong
Shatin, N.T., Hong Kong

Human umbilical vein endothelial cells (EC) were isolated and cultured on glass cover slips. This permitted insertion of EC into a cuvette in a thermostatically controlled cuvette holder which enabled calcium responses to be assessed in a SPEX scanning spectrofluorimeter using the calcium-fluorescent probe, Fura-2.[1] In addition, EC phosphatidylinositol (PI) turnover and platelet activating factor (PAF) production were also measured using standard techniques based on incorporation of ^{3}H-myo-inositol and ^{3}H-acetate respectively. Thrombin stimulation of EC produced a dose-dependent rise in intracellular calcium between doses of 0.005 u/ml to 1 u/ml. Higher doses produced a more rapid and larger peak response and a correspondingly more rapid decline, reaching baseline much quicker than after lower doses, suggesting that following potentially damaging stimuli a scavenging mechanism may be being activated. A classical model of cellular activation involves a rise in intracellular calcium linked to an increase in PI turnover, however, these two signal transduction processes may be obligatory connected. Calcium transients may be detected in EC following stimulation with thrombin at doses as low as 0.005 u/ml, whereas an increase in PI turnover could not be detected until doses of thrombin of 0.1 u/m. or greater were used. Similarly, an increase in EC-PAF production was apparent which mirrored the dose-response to thrombin. When resting neutrophils (final concentration 1×10^7/ml) are suspended in the Hanks buffer surrounding the EC during an experiment, no calcium transient is detected. However, if the neutrophils are activated by the addition of the bacterial peptide, f-met–leu–phe (FMLP), then a low level transient calcium signal (25% above baseline) is generated.[2] Using neutrophil release products or supernatants from sonicates (the most potent method of releasing granule products), larger calcium signals could be obtained. With sonicate supernatants, both PI turnover and PAF production could also be stimulated, although neither of these responses could be detected with lower-level stimuli. This signal could be blocked by heparin, but not by either proteolytic inhibitors or free radical scavengers and further experiments suggested that one the release products chiefly responsible for this response was the cationic protein, Cathepsin G. To examine in more detail the effect of individual neutrophils on endothelial calcium signaling events, EC were grown on Lab-Tek slides and loaded with Indo-1 which permitted observation of calcium responses in individual cells using a Meridian scanning confocal laser microscope. Neutrophils in varying concentrations were layered on top of the EC and stimulated with FMLP (final concentration 10^{-7}M) and EC calcium responses imaged over the subsequent 5 minutes. In contrast to stimulation with thrombin when all the cells imaged showed an increase in calcium uniformly, activated neutrophils stimulated discrete areas of EC cytoplasm provoking a calcium response which then propagated through the cell and sometimes involved neighboring cells. This suggests that local release of inflammatory mediators by localized (and non-toxic) signals in EC which may be important in cell–cell communication relevant to normal physiological processes such as diapedesis, which other workers have shown can be inhibited if the calcium signal is blocked.[3] Higher concentrations of neutrophil release products provoke a stronger response with activation of PI turnover and PAF production which is known to play a positive feed-back role in enhancing the inflammatory response by priming neutrophil responses.[4]

REFERENCES

1. Wickham NWR, Vercellotti GM, Visser MR, Moldow CF, and Jacob HS. Measurement of intracellular calcium concentration in intact monolayers of human endothelial cells. J Lab Clin Med, 1988;72:29–33.
2. Wickham NWR, Moldow CF, Severson SP, Jacob HS, and Vercellotti GM. Exposure of human endothelial cells to stimulated neutrophils or their lysosomal products causes a rise in endothelial cell intracellular calcium. Clin Res, 1988;36(3):570A.
3. Huang AJ, Manning JE, Bandak TM, Ratau MC, Hanser KR, and Silverstein SC. Endothelial cell cytosolic free calcium regulates neutrophil migration across monolayers of endothelial cells. J Cell Biol, 120(6): 1371–1380.
4. Vercellotti GM, Wickham NWR, Gustafson KS, Yin HQ, Herbert M, and Jacob HS. Thrombin-treated endothelium primes neutrophil functions: inhibition by platelet activating factor antagonists. J Leuk Biol, 1989;45(6):483–490.

THE ROLE OF NITRIC OXIDE PRODUCTION DURING CARDIAC ALLOGRAFT REJECTION

David S. Winlaw, Phillip M. Sprati and Peter S. Macdonald

St. Vincent's Hospital
Sydney, Australia

The role of nitric oxide (NO) production in acute allograft rejection is not known. In normal human vasculature, nitric oxide synthase is constitutively expressed within endothelial cells (cNOS), and small quantities of NO produced by this pathway mediate vasorelaxation via a cyclic GMP-dependent mechanism. During cardiac allograft rejection, NO may also be produced by the inducible isoform of nitric oxide synthase (iNOS), present in endothelial cells, smooth muscle cells and cardiac myocytes. Unlike cNOS activity, which is an integral part of cardiovascular homeostasis, iNOS activity is a component of the inflammatory response and may have both beneficial and deleterious effects during rejection. Early reports suggested that acute cardiac allograft rejection was associated with an increase in nitric oxide production. The general aim of this study was to document the way in which nitric oxide production is affected by allograft rejection, and to determine the significance of these changes. Enhanced nitric oxide synthesis may have a graft protective effect by inhibiting clonal expansion of immune cells targeting alloantigen. However, immunosuppressive agents have been shown to block NO production, raising the possibility that NO may modulate graft function and/or survival. The following hypotheses were tested; that acute cardiac allograft rejection is associated with a rise in NO production and an increase in iNOS expression, that conventional immunosuppressive agents reduce NO production by limiting iNOS expression, that an increase in substrate availability for nitric oxide synthase would increase NO production and decrease graft survival, and that L-NMMA, a competitive inhibitor of both constitutive and inducible isoforms would prolong graft survival.

A heterotopic model of cardiac transplantation was established, with measurement of urinary nitrate excretion and graft survival. Nitrate is the stable end product of endogenous NO production, and an assay utilizing gas chromatography/mass spectrometry was used to measure nitrate excretion by rat cardiac transplant recipients. Expression of iNOS mRNA was measured within explanted rat myocardium and cultured neonatal rat cardiac myocytes, using a reverse transcriptase polymerase chain reaction technique. It was demonstrated that a short course of conventional immunosuppressive agents, at a dose which approximately doubled graft survival, delayed the onset- and reduced the magnitude of peak nitrate production in proportion to the length of graft survival. This was associated with a proportional reduction in iNOS expression. Conventional immunosuppressive agents also inhibited iNOS expression in cultured cardiomyocytes suggesting that inhibition of iNOS activity within cardiac myocytes may partly explain the effects of these agents. Inhibition of NO production with L-NMMA caused a small increase in graft survival, however substrate enhancement failed to increase NO production in this model of cardiac transplantation. L-NMMA did not alter iNOS expression, consistent with its post-translational mode of action. In separate experiments using isolated strips of rat myocardium in an organ bath, it was shown that the contractile performance of allografted tissue during electrical field stimulation was not improved by the addition of L-NMMA to the bath.

We concluded that in this animal model of cardiac allograft rejection, NO is not a key effector molecule determining graft survival and function. Recent work by other investigators has demonstrated that selective iNOS inhibition significantly prolongs survival and improves graft function. These results suggest that NO produced by cNOS and iNOS may have different and even opposing roles in the pathogenesis of allograft rejection and that non selective inhibition of both isoforms to some extent masks the benefit seen with inhibition of iNOS alone. Human cardiac allograft recipients were also studied using urinary and plasma nitrate assays taken at the time of endomyocardial biopsy. Whilst individual patients demonstrated a trend towards elevated nitrate production during rejection, there was no relationship between nitrate production and severity of rejection. The close relationship between nitrate production and allograft rejection observed in the animal model was not seen in human recipients.

INCREASED NITRIC OXIDE PRODUCTION IN HEART FAILURE AND CORRELATION WITH FUNCTIONAL SEVERITY

David S. Winlaw, Phillip M. Sprati, and Peter S. Macdonald

St. Vincent's Hospital
Sydney, Australia

The role of nitric oxide (NO) in heart failure is unknown. Studies of isolated blood vessels from patients with heart failure demonstrate impaired NO-dependent relaxation to acetylcholine. On the other hand, endomyocardial biopsy specimens from patients with dilated cardiomyopathy show increased activity of the inducible isoform of nitric oxide synthase (iNOS). Plasma levels of TNF-α, an activator of iNOS, are increased in heart failure, and TNF-α has been shown to have negative inotropic effects in hamster papillary muscle through an NO-dependent mechanism.

The aim of our study was to determine whether nitric oxide production is increased in patients with heart failure, and if this rise correlated with functional severity of heart failure. Measurement of plasma nitrate, the stable metabolite of nitric oxide, is an index of whole body nitric oxide production, although the site of production cannot be defined using this method. "Normal" plasma nitrate levels may be explained by degradation of NO produced by the constitutive isoform of NOS, involved in vasoregulation and neurotransmission.

Fifty patients (40 men, 10 women) attending a specialist heart failure clinic ($n = 5$, 8, 22 and 15 for NYHA Class I to IV respectively) were compared to normals ($n = 62$, 22 men 38 women). The aetiology of heart failure was IDCM in 28 patients and end stage coronary artery disease for the remainder. Administration of nitrate containing medication was noted. Random (non-fasting) plasma nitrate was analyzed by GC/MS. Plasma nitrate in patients with heart failure was 57.5 (95% CI 48.1–66.9) μmol/L, which was significantly greater than in normals 24.6 (19.1–30.0) μmol/L, ($p < 0.0001$). Exclusion of patients receiving nitrate containing medication did not alter the significance of this finding, although patients receiving such medication did have a significantly higher plasma nitrate level, 70.1 (54.1–86.1) μmol/L compared to 50.4 (38.9–61.9) μmol/L, $p = 0.04$. Plasma nitrate correlated with increased functional severity of heart failure ($r = 0.45$, $p = 0.001$).

There was no relationship between nitrate, age nor sex in either group. There was no difference in nitrate level between patients with IDCM and IHD, regardless of nitrate containing medication administration.

Patients with heart failure have increased plasma nitrate levels, and this increase correlates with severity of heart failure. These findings suggest increased endogenous production of NO, which may have a beneficial (vasodilator) effect by antagonizing neurohumoral adaptations to heart failure. On the other hand, increased NO production may exert a negative inotropic effect and have a detrimental effect on cardiac function.

INHIBITION OF ENDOTHELIAL CELL ACTIVATION BY -MEDIATED EXPRESSION OF IκBα, AN INHIBITOR OF THE TRANSCRIPTION FACTOR NF-κB

C. Wrighton,[1] R. Hofer-Warbinek,[2] T. Moll,[2] E. Hofer,[2] F. H. Bach,[1] and R. DeMartin[2]

[1]Sandoz Center for Immunobiology
Harvard Medical School
Boston, Massachusetts
[2]Department of Vascular Biology and
Thrombosis Research at VIRCC
University of Vienna
Vienna, Austria

Stimulation of endothelial cells (EC) by inflammatory cytokines leads to the transient expression of a set of genes encoding cell adhesion molecules, interleukins, and components of the coagulation cascade, e.g. ELAM-1, VCAM-1, ICAM-1, IL-1, IL-6, IL-8, iNOS, and tissue factor. Up-regulation of these genes is controlled predominantly on the transcriptional level.

A common transcriptional activator for gene expression in EC is the transcription factor NF-κB. Functional NF-κB binding sites are present in the promoter regions of all genes mentioned above. NF-κB activity is regulated by complex formation with an inhibitor, IκBα, that retains NF-κB in the cytoplasm. Upon stimulation, phosphorylation and proteolytic degradation of IκBα leads to nuclear translocation of NF-κB and to gene transcription. Vice versa, NF-κB regulates IκBα expression directly through distinct NF-κB binding sites in the IκBα promoter (1). This regulation of IκBα by NF-κB has led to the concept of an autoregulatory feedback mechanism that ensures the transient nature of NF-κB and NF-κB dependent gene expression.

We have now tested the concept of inhibition of EC activation via inhibition of NF-κB by overexpression of IκBα using a recombinant adenovirus (rAd.IκBα) expressing IκBα. rAd.IκBα was used to infect EC before challenging with inflammatory stimuli. Infected cells expressed high levels of IκBα as determined by Northern blotting and immunoperoxidase staining and showed complete lack of NF-κB binding activity in EMSA.

Concomitantly, LPS-induced expression of IL-1, IL-6, IL-8, VCAM-1 and tissue factor, as well as adhesion of HL-60 cells to activated EC, was reduced over 90%. We conclude that in EC, NF-κB is a common mediator of central importance for the expression of many of the relevant genes associated with the inflammatory phenotype.

REFERENCES

1. de Martin, R., Vanhove, B., Cheng, Q., Hofer, E., Csizmadia, V., Winkler, H., and Bach, F.H. 1993, Cytokine-inducible expression in endothelial cells of an IκBα like gene is regulated by NF-κB. EMBO J., 12: 2773–2779.
2. Wrighton, C., Hofer-Warbinek, R., Moll, T., Hofer, E., Bach, F.H., and deMartin, R. 1996, Inhibition of endothelial cell activation by adenovirus-mediated expression of IκBα, and inhibitor of the transcription factor NF-κB. J. Exp. Med., in press.

ENDOTHELIAL CELL PAI-1 PRODUCTION — MODULATION BY MALIGNANT CELLS *IN VITRO*

Lu Wei Zhen, Meng Chun Ling, and Nicholas Wickham

Department of Clinical Oncology
The Sir Yue Kong-Pao Cancer Centre
Prince of Wales Hospital
Chinese University of Hong Kong
Shatin, N.T.

INTRODUCTION

Plasminogen activator inhibitor-1(PAI-1) is a single chain glycoprotein of approximately 50,000 Daltons which is synthesized by endothelial cells and hepatocytes. It is one of the many factors that endothelial cells (EC) secrete which are involved intimately in the maintenance of the haemostatic balance. In malignant states, especially those with widespread metastasis, this balance is frequently upset with a resulting increase in thromboembolic events. There is not one single easily identifiable pattern of changes that accounts for this increased risk, but disruption of the fibrinolytic system, in particular an increase in circulating PAI-1 concentration, is a common finding in patients with cancer. We wanted to assess the degree to which this might be due to an effect of malignant cells upon EC PAI-1 production.

MATERIALS AND METHODS

HUVEC

Human umbilical vein endothelial cells were isolated and grown to confluence using standard methods in tissue culture medium M199 with 20% fetal calf serum in Costar 24 well plates. Confluence, confirmed microscopically, usually took 5 to 7 days to attain, with fresh tissue culture medium supplements on alternate days.

HEP G2 Cells

This human hepatocellular carcinoma cell line was obtained from the American Type Culture Collection (ATCC HB-8065) and stored frozen. Cells were thawed and cultured using RPMI and 10% fetal calf serum and subsequently mobilized by trypsinization. The suspended cells were washed twice and resuspended in medium prior to use.

Thrombin

Human thrombin was obtained from Sigma, reconstituted and aliquotted to give a concentration of 100 NIH units/ml in buffered saline and stored frozen. Working solutions were thawed, diluted with buffer as required and kept on ice until used.

PAI-1 Assays

These were performed using a commercial standard enzyme-linked immunoassay ASSERACHROM PAI-1 (Diagnostic Stago) according to the assay data sheet instructions. This assay detects total PAI-1, whether complexed or not. A calibrator curve was prepared from a reference standard supplied and normal pooled human plasma was used as an external control.

Experimental Samples

Endothelial cell supernatants were collected at the end of each experiment and subjected to high-speed centrifugation to remove any particulate matter and stored frozen in 2 aliquots at –20ºC for up to two weeks prior to assay. Samples required to be stored for longer periods were transferred to –85°C freezers.

EXPERIMENTAL DESIGN

1. EC PAI-1 Production

HUVEC were grown to confluence, washed twice with medium then left in fresh medium and incubated at 37°C in 5% CO_2 either for 24 hours or several days. Different wells were sampled in duplicate at different time points, stored and assayed as detailed above.

2. EC PAI-1 Production & Thrombin

Thrombin stimulation of EC, which is known to alter the rate of synthesis of PAI-1 as well as modulate many other EC-dependent functions, was used as a model for increased PAI-1 release and thus as a positive control. EC were washed × 2 with medium prior to each experiment then thrombin was added in final concentrations varying from 0.01 u/ml to 5 u/ml. Wells were sampled in duplicate at different time points up to 24 hours. In some experiments, thrombin was left *in situ* and in others it was removed after 30 minutes, the EC were washed twice, and fresh medium was added and incubation was continued for 24 hours before samples were taken for assay.

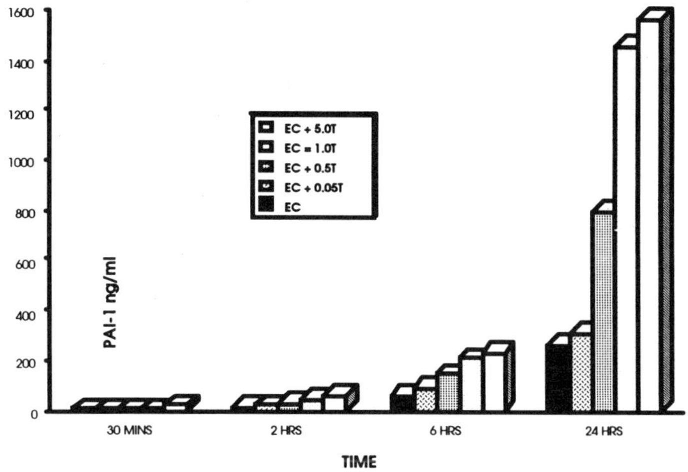

Figure 1. Data from mean of 4–8 samples.

3. EC PAI-1 & HEP G2 Cells

Hep G2 cells ($5x10^6$) were added to confluent cultures of EC in 24 well plates (after washing twice in medium). Control wells consisted of either no cells, EC alone or Hep G2 cells alone. In some experiments, thrombin stimulation was also employed with thrombin final concentrations of 0.1 to 1.0 NIH units/ml. Duplicate samples were taken for assay at 24 hours.

RESULTS

Human umbilical vein endothelial cells produced PAI-1 constitutively. Furthermore, PAI-1 secretion continued at a steady rate over at least 8 days with no apparent *in vitro* feed-back inhibition (not shown).

PAI-1 secretion increased in response to human thrombin in a dose-dependent fashion (0.01 to 1.0 NIH u/ml). Higher doses (2.0 to 5.0 NIH u/ml) did not produce any further significant increase in PAI-1 secretion (Figure 1). Removing the thrombin after 30 minutes reduced substantially the 24 hour increase in EC PAI-1 concentration, suggesting that EC continued to respond to the presence of thrombin over several hours. (Not shown) HEP G2 cells also secrete PAI-1 constitutively but this secretion did not increase with exposure to thrombin. PAI-1 concentrations in supernatant from EC co-cultured with HEP G2 cells showed a possible synergistic increase in PAI-1 production, the increase being approximately double that expected from a simple additive effect. This synergistic effect was not apparent following additional thrombin stimulation when results were additive only (Figure 2).

CONCLUSIONS

These preliminary findings suggest that there may be a modest synergistic interaction with respect to PAI-1 production between HUVEC and Hep G2 cells under the condi-

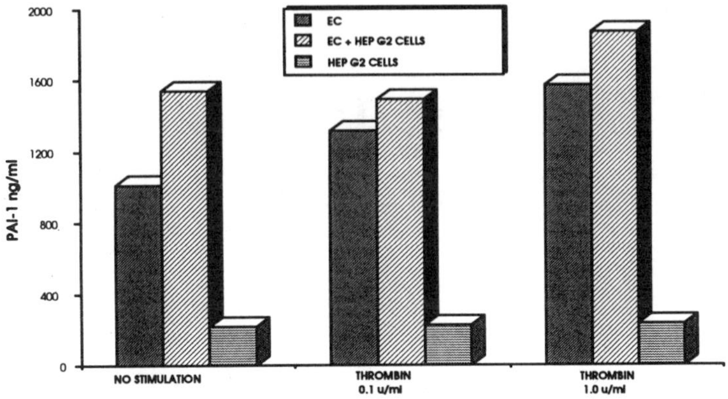

Figure 2. Data from means of duplicate samples.

tions described. This effect is small and was not detected against the larger increases when additional thrombin stimulation was used, presumably being swamped by the more powerful stimulus. However, the increased production of PAI-1, whether from EC or Hep G2 cells, is equivalent to the increase in EC PAI-1 production following a 0.1 NIH u/ml thrombin stimulus and may have considerable biological significance. Further experiments are planned to ascertain whether this apparent synergistic effect is reproducible, whether the excess PAI-1 comes from EC or Hep G2 cells, and whether it represents a true increased rate of synthesis of PAI-1.

REFERENCES

1. Declerck PJ, et al. Measurement of plasminogen activator inhibitor 1 in biologic fluids with a murine monoclonal antibody-based enzyme-linked immunosorbent assay. Blood 71(1):220, 1988.
2. Nand S, et al. Hemostatic abnormalities in untreated cancer: incidence and correlation with thrombotic and hemorrhagic complications. J Clin Oncol 5(12):1998, 1987.
3. Zurbon KH, et al. Investigations of coagulation of system and fibrinolysis in patients with disseminated adenocarcinomas and non-Hodgkin's lymphomas. Oncology 47(5): 376, 1990.
4. Mannucci PM, et al. Changes in fibrinolysis in patients with localized tumors. Eur J Cancer 26(2): 83, 1990.
5. ATCC Cell Lines & Hybridomas, 8th Edition 227, 1994, ATCC HB-8065, (Reference: Nature (Lond.)282:615, 1979.

PARTICIPANTS

Jorg Adamczewski, Ph.D.
Biochemische Endokrinologie
Schering AG
Berlin, Germany D-13342

Kenneth B. Adler, Ph.D.
Department of Cell Biology
North Carolina State
4700 Hillsborough Street
Raleigh, North Carolina 27606

Futuan Al-Mohanna
Biological and Medical Research Centre
King Faisal Specialist Hospital
MBC 03, Riyadh, 1211 Saudi Arabia

Sean Allen, Ph.D.
Cell Biology Group, Heart Science Center
Harefield, Middlesex UB9 6JH

Vivian Andriopoulou, Ph.D.
Instituto Di Ricerche
Mario Negri, Milano, Italy 20157

Eleni Bardouniotou, M.D.
Critical Care Department
University of Athens Medical School
G. Gennimata Avenue
Elefsina, Greece

Panayotis K. Behrakis, M.D., Ph.D.
Intensive Care Unit
Red Cross Hospital of Athens
Erithros Stavros No 1
Athens, Greece 11526

Kare Berg, M.D., Ph.D.
University of Oslo
PO Box 1036, Blindern
Oslo, Norway N-0315

Roy Bicknell, Ph.D.
Molecular Angiogenesis Group
Imperial Cancer Research
University of Oxford
Oxford OX3 9DU
United Kingdom

David Bishop-Bailey
Department of Applied Pharmacology
National Heart and Lung Institute
Dovehouse Street, London SW3 6LY
United Kingdom

Kate Blease
Department of Applied Pharmacology
National Heart and Lung Institute
Dovehouse Street, London SW3 6LY
United Kingdom

Richard Bogle, Ph.D.
Department of Pharmacology
St. George's Hospital Medical
Cranmer Terrace
London SW17 ORE

Jacques-Paul Borel, M.D.
Laboratorie de Biochimie
51 Rue Cognacq Jay
Reims Cedex, France 51095

Tommy Brock, Ph.D.
Department of Pharmacology
Texas Biotechnical
7000 Fannin Drive, Suite 1920
Houston, Texas 77030

Arnold R. Brody, Ph.D.
Department of Pathology
Tulane University School of Medicine
1430 Tulane Avenue
New Orleans, Louisiana 70112-2699

Anne Burke-Gaffney, Ph.D.
Department of Applied Pharmacology
Imperial College of Science
Dovehouse Street
London SW3 6LY

Allan D. Callow, M.D., Ph.D.
Department of Surgery
Boston University School of Medicine
80 E. Concord Street, NEB 606
Boston, Massachusetts 02118-2394

John D. Catravas, Ph.D.
Vascular Biology Center
Medical College of Georgia
Augusta, Georgia 30912-2500

Romeo Cecchelli, Ph.D.
Institut Pasteur De Lille
Serlia et U. 325 Inserm
Lille Cedex, France 59019

Adrian Chester, Ph.D.
Imperial College of Science
Harefield Hospital
Harefield, Middlesex UB9 6JH

Milita Crisby, M.D.
Department of Neurology
Hiddinge University Hospital
Huddinge, Sweden 14186

Luis M. Cunha-Ribeiro, M.D.
Department of Physiology
Porto Medical School, Hospital S. Joao
Porto, Portugal 4200

Pedro M. daSilva, M.D.
Bristol-Myers Squibb
Ruo soeiro Pereira Gomes
Lisbon, Portugal 1600

Antonella De Martino
Instituto di Nefro-Urologia
Cso A.M. Dogliotti, 14
Torino, Italy 10126

M. P. Dehouck, Ph.D.
Institut Pasteur De Lille
Serlia et U. 325 Inserm, Lille Cedex
France 59019

Durisala Desaiah, Ph.D.
Department of Neurology
University of Mississippi
2500 North State Street
Jackson, Mississippi 39216-4505

Christiana Dimitropoulou
UNI Mainz
Joh. Becher Weg 9, Mainz, Germany 55099

Jocelyn Dupuis, M.D., Ph.D.
Institut de Cardiologie de Montreal
Centre de Recherche
5000 est rue Belanger
Montreal, Quebec H1T 1C8

David Edwards, M.D.
Biomedical Health Center, Inc
6490 S. McCarran Blvd.
Reno, Nevada 89509-6118

Gerasimos Filippatos, M.D.
28 Doukissis Plakentias Str
Ambelokipi, Athens, Greece 11523

Francesca Fioretti
Instituto di Ricerche
Via Eritrea 62
Milano, Italy 20157

Gene Fisher, M.D.
Department of Pediatric Critical Care
Medical College of Georgia
1120 15th Street
Augusta, Georgia 30912

Arlene Frater, Ph.D.
Vascular Biology Research
Campden Hill Road
London W8 7AH, United Kingdom

Leslie C. Fuchs, Ph.D.
Vascular Biology Center
Medical College of Georgia
1120 15th Street
Augusta, Georgia 30912-2500

Petra Ghele, Ph.D.
Department of Cardiothoracic Surgery
Harefield Hospital
Harefield, Middlesex UB9 6JH
United Kingdom

Peter I. Ghenev, M.D.
Department of Pathology
Medical University of Varna
Marin Drinov str. 55, Varna, Bulgaria 9002

Gehard Hannig, Ph.D.
T Cell Sciences, Inc, 115 4th Avenue
Needham, Massachusetts 02194-2725

Ginette S. Hoare
Imperial College of Science
Heart Science Center
Harefield, Middlesex UB9 6JH
United Kingdom

Erhard Hofer, M.D.
Department of Vascular Biology and
Thrombosis Research
University of Vienna
Brunnerstrasse 59
Vienna, Austria A-1235

Tibor Hortobobagyi, M.D.
National Stroke Center
Huvosvolgyi ut 116
Budapest
Hungary H-1021

Ivan Horvath, M.D.
Vascular Biology Center
Medical College of Georgia
1120 15th Street
Augusta, Georgia 30912-2500

Corazon Ibarra-Ilarina, M.D.
Bio Medical Health Center
6490 S. McCarran Blvd
Reno, Nevada 89509-6118

Alain F. Junod, M.D.
Department of Medicine
University Hospital
Geneva
Switzerland 1211

Panagiotis Kaltsas, M.D.
Critical Care Medicine
University of Athens Medical School
Ipsilantou Str 45-47
Athens
Greece 10676

Nicholas Kefalides. M.D., Ph.D.
Department of Medicine
University of Pennsylvania
3624 Market Street
Philadelphia, Pennsylvania 19104-2614

Miklos Kellermayer, M.D., Ph.D.
Department of Clinical Chemistry
University Medical School of Pecs
Ifjusag u. 13, Pecs
Hungary 7624

Joseph Khoury
Division of Cardiology
The Sir Mortimer B. Davis Jewish General
Hospital
3755 Cote Ste-Catherine
Montreal, Quebec H3T 1E2
Canada

Patricia Lalor, Ph.D.
Department of Physiology
University of Birmingham
Edgbaston, Birmingham B15 2TT

Maria G. Lampugnani, Ph.D.
Laboratorio di Biologia
Instituto Di Recerche
Mario Negri
Milano, Italy 20157

David Langleben, M.D.
Division of Cardiology
The Sir Mortimer B. Davis Jewish
General Hospital
3755 Cote Ste-Catherine
Montreal, Quebec H3T 1E2

Joseph Loscalzo, M.D., Ph.D.
Department of Medicine and Biochemistry
Whitaker Cardiovascular Center
700 Albany Street
Boston, Massachusetts 02118

Aldons Lusis, M.D.
Department of Medicine
UCLA School of Medicine
Los Angeles, California 90024-1679

Margareta Mahrabinn, Ph.D.
Department of Medicine
UCLA School of Medicine
Los Angeles, California 90024-1679

Savvas Makrides Ph.D.
Praceis Pharmaceuticals, Inc.
1Hampshire Street
Cambridge, Massachusetts 02139

Alberto Mantovani, M.D.
Laboratory of Immunology
Instituto di Ricerche
Via Eritrea 62
Milano, Italy 20157

Michael E. Maragoudakis, Ph.D.
Department of Pharmacology
University of Patras Medical School
Patras, Greece 26500

Sadis Matalon, Ph.D.
Department of Anesthesiology
University of Alabama at Birmingham
619 South 19th Street
Birmingham, Alabama 35233

Michael E. Maragoudakis, Jr.
Department of Pharmacology
University of Patras Medical School
Rio Patras, Greece 26110

Nandor Marczin, M.D.
Imperial College of Science
Harefield Hospital
Harefield, Middlesex UB9 6JH

Henry Marsh, Ph.D.
T Cell Sciences, Inc.
115 4th Avenue
Needham, Massachusetts 02194-2725

Jeffrey Matthieu
Boston University School of Medicine
Department of Surgery
80 E. Concord St.
Boston, Massachusetts 02118-2394

Rene Moser, M.D.
Department of Pharmacy
Winterhurerstrasse 190
Zurich, Switzerland CH-8057

Zoltan Nagy, M.D.
National Stroke Centre
Huvosvolgyi ut 116
Budapest, Hungary H-1021

Stylianos Orfanos, M.D.,Ph.D.
Critical Care Department
Evangelismos Hospital
University of Athens Medical School
45-47 Ipsilandou St
Athens, Greece 10676

Nathalie Ortega, M.D.
Institut de Biologie Cellulaire
118 route de Narbonne
Toulouse
Cedex, France 31062

Stamatis E. Papaioannou
University of Patras
Rio Patras, Greece 26110

Marco Presta, M.D.
General Pathology, Science
University of Brescia
via Valsabbina 19, Brescia, Italy 25123

Chuck Rittershaus, Ph.D.
T Cell Sciences, Inc
115 4th Avenue
Needham, Massachusetts 02194-2725

Una Ryan, Ph.D.
T Cell Sciences, Inc.
115 4th Avenue
Needham, Massachusetts 02194-2725

Eleni Sakkoula
Department of Pharmacology
University of Patras Medical
Patras, Greece 26110

Inan Soydan, M.D.
Department of Cardiology
Ege University
Bornova, Izmir 35100, Turkey

Saliha Soydan, M.D.
Department of Pathology
Ege University
Bornova, Izmir 35100, Turkey

Bulent Sözmen, Ph.D.
Department of Biochemistry
Ege University Faculty of Medicine
Bornova, Izmir 35100, Turkey

Eser Y. Sözmen, Ph.D.
Department of Biochemistry
Ege University Faculty of Medicine
Bornova, Izmir 35100, Turkey

Donald J. Sutherland, M.D., Ph.D.
Ontario Cancer Treatment and Research Foundation
2075 Bayview Avenue
Toronto, Ontario M4N 3M5, Canada

Stergio Theodoropoulos, M.D.
A' Cardiac Surgery Unit
Athens Medical Center
Marousi, Athens 15125, Greece

Branko Tomazic, Ph.D.
Paffenbarger Research Center
Gaithersburg, Maryland 20899

Meral Tuncer, M.D.,Ph.D.
Department of Pharmacology
Hacettepe University, Faculty of Medicine
Ankara 06100, Turkey

B. S. Uydes-Dogan, Ph.D.
Department of Pharmacology
Faculty of Pharmacy
Hipodrom, Ankara, 06330, Turkey

Victor W.M. van Hinsbergh, Ph.D.
Gaubius Laboratory TNO-PG
PO Box 2215, Leiden 2333
The Netherlands

Monika Vastag, Ph.D.
National Stroke Centre
Huvosvolgyi ut 116
Budapest, Hungary H-1021

Tom Vink, Ph.D.
Department of Hematology
University Hospital Utrecht
Heidelberglaan 100
CX Utrecht 3584, The Netherlands

Nicholas Wickham, Ph.D.
Department of Clinical Oncology
The Chinese University of Hong Kong
Shatin, Hong Kong

David Winlaw, M.D.
Royal North Shore Hospital
St. Leonards NSW, Australia 2065

Magdi Yacoub, FRCS
Department of Cardiothoracic Surgery
Harefield Hospital
Harefield, Middlesex UB9 6JH
United Kingdom

Eleni E. Zanni, Ph.D.
Section of Molecular Genetics
Boston University Medical School
700 Albany Street, W-509
Boston, Massachusetts 02118-2394

Vassilis Zannis, Ph.D.
University of Crete
School of Medicine
Heraklion, Greece

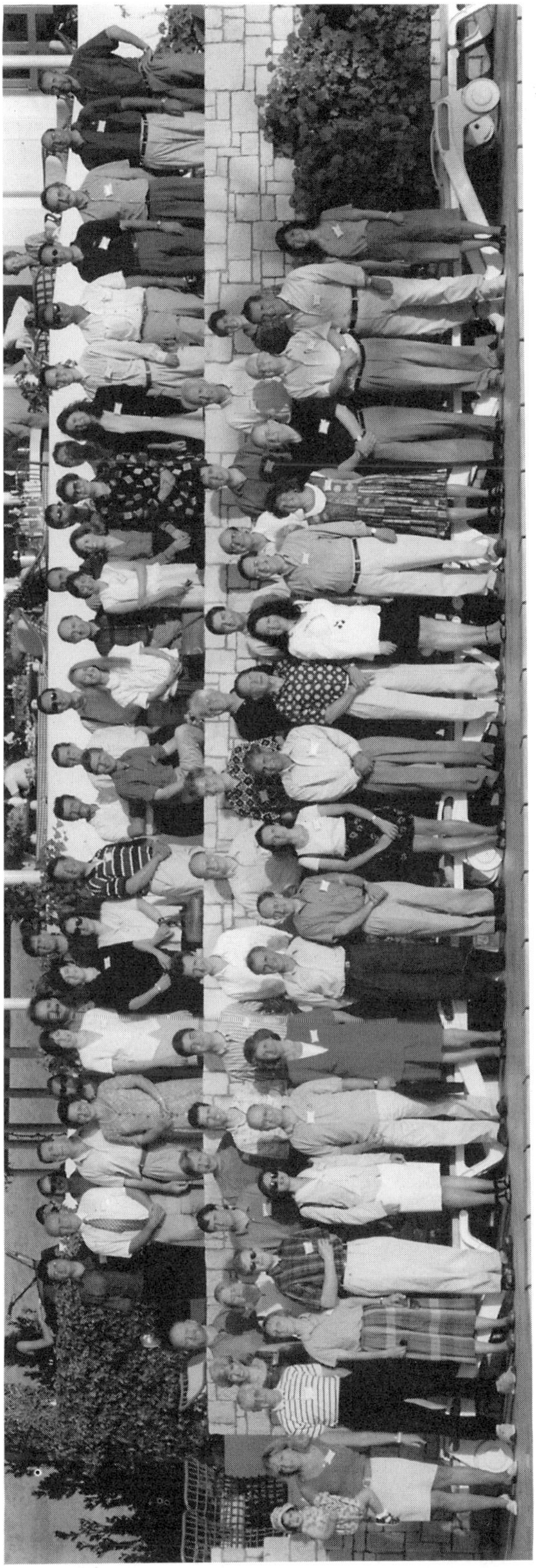

Participant photo

INDEX